SIX-
LEGGED
SOLDIERS

SIX-
LEGGED
SOLDIERS

*Using Insects as
Weapons of War*

JEFFREY A. LOCKWOOD

OXFORD
UNIVERSITY PRESS

OXFORD
UNIVERSITY PRESS

Oxford University Press, Inc., publishes works that further
Oxford University's objective of excellence
in research, scholarship, and education.

Oxford   New York
Auckland   Cape Town   Dar es Salaam   Hong Kong   Karachi
Kuala Lumpur   Madrid   Melbourne   Mexico City   Nairobi
New Delhi   Shanghai   Taipei   Toronto

With offices in
Argentina   Austria   Brazil   Chile   Czech Republic   France   Greece
Guatemala   Hungary   Italy   Japan   Poland   Portugal   Singapore
South Korea   Switzerland   Thailand   Turkey   Ukraine   Vietnam

Copyright © 2009 by Oxford University Press, Inc.

First issued as an Oxford University Press paperback, 2010

Published by Oxford University Press, Inc.
198 Madison Avenue, New York, New York 10016

www.oup.com

Oxford is a registered trademark of Oxford University Press

Library of Congress Cataloging-in-Publication Data
Lockwood, Jeffrey Alan, 1960–
Six-legged soldiers : using insects as weapons
of war / Jeffrey A. Lockwood.
p.   cm.
Includes bibliographical references and index.
ISBN 978-0-19-973353-8
1. Biological warfare. 2. Insects as carriers of disease.
3. Agroterrorism—Prevention. I. Title.
UG447.8.L63 2009
358'.3882—dc22   2008006935

1 3 5 7 9 8 6 4 2

Printed in the United States of America
on acid-free paper

To
John (Jack) E. Lloyd
Genuine Friend, Patient Mentor, and Exemplary Scientist

CONTENTS

FOUR
COLD-BLOODED FIGHTERS
OF THE COLD WAR

FIVE
THE FUTURE OF ENTOMOLOGICAL WARFARE

PREFACE

Although the historic and prospective use of insects as weapons is not the sort of topic that tends to lull one into a sense of well-being, I would like to put the reader's mind at ease with regard to a few important considerations.

This book is, in large part, about history and science. And I am of the studied opinion that neither venture is particularly objective. As such, I cannot claim neutrality without abject hypocrisy. So in the spirit of honest disclosure, the reader should know the following. Despite claims to the contrary by early readers and reviewers, I am neither antireligious nor un-American. In fact, I attend church (Unitarian Universalist) almost every Sunday, and I vote in every election (being a registered Independent with Democratic leanings, although I increasingly struggle to discern the difference between the parties). I am, however, a skeptic with a sense of humor, a quality that might seem irreverent when I doubt the veracity of a particular reader's favored institution.

It is my sense that human organizations—including universities, religious associations, corporate enterprises, government laboratories, federal agencies, and international bodies—have as their primary goal the acquisition and maintenance of power, not the search for and reporting of the truth. That said, I am not equally dubious of all sources. For example, I would believe an account provided by the U.S. government over one provided by the North Korean government, all other things being equal. But, of course, things are not often equal, and during times of hot and Cold Wars the honesty of both sides must be questioned. Historical and political accounts most often provide a complex set of partial truths from which one must attempt to assemble a best guess of what actually transpired.

In this light, my interpretation of historical events in which insects have been used as weapons—with or without the knowledge of the combatants—may not accord with the cultural, religious, or political sensitivities of all readers. It is not my intention to be insulting, but neither is it my goal to be politically correct. Rather than stepping on nobody's toes, I suppose that I've probably managed to offend almost every reader in some way. After all, in thousands of years of human history across the face of the earth, it seems all but certain that some group with which we feel an affinity was up to something nefarious.

And so I am a patriotic (but not a jingoistic) and faithful (but not dogmatic) skeptic of human institutions that are, by and large, our primary sources of information about the world (individuals can be more reliable but they often represent institutional, or at least social and cultural, views). I think that smart people—like those who read books—can differ in their interpretations of events. Those readers who have confidence in western governments will find that I've put too much stock in the veracity of some alternative accounts (e.g., the communists' claim that the U.S. military used entomological weapons during the Korean War) while those with greater distrust of American politicians will find that I've not taken other reports seriously enough (e.g., the Cuban accusations of the United States' dropping insect vectors of disease and crop pests on the island nation).

By way of further disclosure, I am not a military historian. For that matter I'm not any kind of a professional historian, although I might fall among those who consider themselves impassioned amateurs. I am an entomologist and a writer, and it is from this background that I undertook the research for this book. As such, I relied heavily on secondary sources—books and articles produced by historians. Where possible, I sought to acquire primary sources, although for many of the events recounted in this book accessing these materials was not possible. To facilitate the reader's further engagement with this topic, I've provided a list at the end of this preface of the top 10 books that one should explore for further understanding of the people, times, places, and events of entomological (and biological) warfare.

Given my limitations—and I suppose that a historian writing such a book would have had to acknowledge his or her reliance on others' entomological expertise—the reader with an affinity for a particular historical period may find some of my descriptions overly simplified. Keep in mind that even with a focus on the last few centuries, the book covers 100,000 years of human history and there's only so much detail that can be included. Moreover, warfare, especially when covert, is a complicated, confused, and often controversial human endeavor. My interest, however, is in the role that insects and their relatives played in these conflicts. As such, the perspective of the book is one in which war, terror, and torture are viewed from the perspective of entomology. This is, admittedly, an odd point of view, but therein lies both the uniqueness and (I hope) your fascination with the subject of the book. This particular take on biological warfare has not been systematically undertaken by previous writers. That said, I've attempted to provide enough social, political, and cultural context for the entomological events to be both meaningfully related to one another and to the grand sweep of human history. Although the book is not about epidemiology, political theory, or sociology, to make sense of how we've conscripted insects in our efforts to harm one another it is important to have some understanding of disease, agriculture, foreign policy, and cultural values.

The reader may be rightfully dubious of various accounts in this book. I know that I am. In this regard, I should hasten to note that I've consciously chosen to be inclusive in my research, allowing all plausible—even if hard to believe—claims their place in the story. At least sometimes truth really is stranger than fiction. In many instances, particularly with regard to early human history, I've included events in which the role of insects is not established with certainty or the antagonists may not have been aware that these creatures were the agents of suffering and death. I intentionally cast a wide net because these incidents often played an important role in the larger narrative of warfare, serving as examples that prompted military tacticians to pursue various lines of development with respect to entomological and biological weapons.

Given my approach to this topic, I've tried to phrase various accounts and explanations with appropriate caution, and the reader may find my careful wording (e.g., the qualification of claims with *probably, possibly, presumably,* and *perhaps*) makes the narrative less convincing or authoritative. Had I been an expert—or wished to appear as one—on all of the subjects and events addressed in this book, I might have written in confident tones whether or not my knowledge justified such academic aplomb. But people who claim to know what happened with regard to most of these historical events are making themselves into false authorities—the information simply precludes anyone from speaking with certainty (except those who were there, and for the most part, they aren't talking, and when they do their veracity is often questionable).

Scientists can be frustratingly circumspect in their writing such that we sound like we're speculating. Often we are, although we prefer to call it "reasoned inference from available evidence to the best explanation." Perhaps this has infused my narrative, but this is not the full explanation of my approach. Rather, I don't take it as my role to convince the reader of how or if particular events transpired. My goal is to clearly present and critically evaluate the various incidents in entomological and historical terms. I presume that the reader is an intelligent and informed individual who will bring to bear his or her knowledge and experiences to my accounts in determining the believability of the stories.

In all cases, I've attempted to provide as even-handed and objective an analysis as possible. However, there are two problems in writing about the use of insects as weapons. First, biological warfare is, for the most part, poorly documented. In early human history there were not many detailed accounts of the roles played by insects per se, so one must draw conclusions based on circumstantial evidence from sometimes disparate sources. And as we approach modern times, humans' use of living organisms to kill other humans becomes increasingly proscribed. In light of these moral reservations, political and military leaders are less than forthcoming about their development and use of insects as agents of war—let alone terror.

Second, given the unsavory nature of using insects as weapons, entomological warfare is a tempting topic for propaganda. As such, nations are prone to make claims regarding their enemies' use of such tactics—and their enemies are motivated to strenuously deny such accusations. All of this makes it difficult for a historian or scientist to sort out exactly what happened. The use of propaganda and the back-and-forth charges of governments are part and parcel of the history of biological/entomological warfare. The point is that in many instances, indeed most cases, we simply do not know who is telling the truth. So I have attempted to recount the events, through the lens of my entomological background and skeptical proclivities, in an engaging manner and allow the reader to decide what happened.

Finally, various people who have seen early drafts of this book have expressed concern that it could be a "how to" guide for terrorists or others who intend to do us harm. Although I shared this apprehension early in my research, I would like to assuage the concern with a few observations.

First, I don't believe that terrorists and rogue nations are as uninformed and inept as we might think or hope. At least, I don't think that the obvious and simple methods that I describe have not occurred to them. The evidence in support of this position is that we've already seen them use chemical and biological weapons (e.g., the Tokyo subway attacks with sarin in 1995, the Oregon attacks on the public with salmonella in 1984, the California attacks on agriculture with Medflies in 1989).

Second, virtually all of the information in this book was extracted from publicly available resources. I don't reveal anything that a reasonably intelligent and educated individual would not be able to find from a library and the Internet. Indeed, I would be disappointed if any student graduating with a master's degree in entomology could not conceive of and execute any of the entomological attacks that I've described.

Third, one might reasonably contend that various scenarios and threats should be communicated to government agencies, rather than the public (and potentially terrorists). However, the government has been informed of the risks (via a thorough study by the National Research Council) and their response was to shift agricultural border inspection from the U.S. Department of Agriculture's Animal and Plant Health Inspection Service to the U.S. Department of Homeland Security, where it is evident from testimony and a General Accounting Office report that we are more vulnerable than ever to pest introductions. It seems that for anything to change in our federal priorities, we need a reasonably informed and appropriately concerned citizenry.

Fourth, the experts on bioterrorism with whom I've spoken have not expressed any concern with regard to the nature of the material that I describe. Indeed, some of these individuals have provided even more extensive and detailed accounts of attack scenarios in widely available formats. Their concern is clearly focused on calling the attention of the government and the public to these risks in a proportional (not alarmist) manner—and I share this objective. While insects arriving through natural and accidental routes are far more likely to harm people and their economic interests than are organisms released by terrorists, this does not mean that the latter should be dismissed as a concern (more people die from accidental poisoning, falling, and drowning than were killed in the 9/11 attack, but surely this is no reason for ignoring the risks of terrorism). The experts seem to harbor little doubt that a terrorist organization would be able to mount such an attack, lacking neither the technical information nor the logistical capacity.

Recommended Reading on Entomological Warfare, Terror, and Torture

Barenblatt, Daniel. *A Plague Upon Humanity: The Secret Genocide of Axis Japan's Germ Warfare Operation.* New York: HarperCollins, 2004.

Croddy, Eric. *Chemical and Biological Warfare: A Comprehensive Survey for the Concerned Citizen.* New York: Springer-Verlag, 2002.

Endicott, Stephen, and Edward Hagerman. *The United States and Biological Warfare: Secrets from the Early Cold War and Korea.* Bloomington, Ind.: Indiana University, 1998.

Engelberg, Stephen, Judith Miller, and William Broad. *Germs: Biological Weapons and America's Secret War*. New York: Simon & Schuster, 2002.

Gold, Hal. *Unit 731 Testimony*. Singapore: Yen Books, 1996.

Harris, Robert, and Jeremy Paxman. *A Higher Form of Killing: The Secret History of Chemical and Biological Warfare*. New York: Random House, 2002.

Harris, Sheldon. *Factories of Death: Japanese Biological Warfare, 1932–1945, and the American Cover-Up*. New York: Routledge, 2002.

Mayor, Adrienne. *Greek Fire, Poison Arrows, and Scorpion Bombs: Biological and Chemical Warfare in the Ancient World*. New York: Overlook Duckworth, 2003.

National Research Council. *Countering Agricultural Bioterrorism*. Washington, D.C.: National Academies, 2003.

Regis, Ed. *The Biology of Doom: The History of America's Secret Germ Warfare Project*. New York: Henry Holt, 2000.

ACKNOWLEDGMENTS

This book is the result of four years of reading, research, conversations, interviews, and writing. Although I have many people to thank for their assistance, none of the acknowledgments should be interpreted as meaning that these individuals, their agencies, or their institutions agree with any of the claims made in this book.

I should begin by thanking those at my own institution, the University of Wyoming, who lent their expertise to the project, including colleagues in the entomology section of the Department of Renewable Resources: Alex Latchininsky, Jack Lloyd, Scott Shaw, and Scott Schell. I am also indebted to Mark Byra (Division of Kinesiology and Health), Paul Flesher (Religious Studies Program), and Tim Kearley (College of Law) for their willingness to share their time and knowledge. The University of Wyoming's Legal Office (Rod Lang) and the university's reference librarians were instrumental in my efforts. And finally I am truly compelled to thank a cadre of individuals who proved that, contrary to a near-consensus among faculty, effective administrators can authentically facilitate and truly support scholarship: Harvey Hix (MFA program in Creative Writing), Ed Sherline (Department of Philosophy), and Tom Thurow (Department of Renewable Resources).

I would like to thank an array of academic colleagues at other institutions for taking the time to respond to my queries with information, images, and sometimes polite declinations. These first-rate scholars include: John Abbott (University of Texas), David Block, Petrina Jackson, and William Maddison (Cornell University), Chuck Bomar (University of Wisconsin, Stout), Jerry Bromenshenk (University of Montana), Michael Burgett (Oregon State University), John Capinera (University of Florida), Jim Carey and John Skarstad (University of California, Davis), Thomas Daniel (University of Washington), Ron Fearing (University of California, Berkeley), Alan Ferg and Roger Myers (University of Arizona), Phillip Greenfeld (South Dakota State University), Garnet Hertz (University of California, Irvine), Barrett Klein (University of Texas), Gene Kritksy (College of Mount St. Joseph), Joseph LaForest (University of Georgia), Bill Lewinger and Roger Quinn (Case Western Reserve University), Harley Moon (Iowa State University), Bob Peterson and Joe Shaw (Montana State University), Alejandro Ramirez-Serrano (University of Calgary), Cliff Sadof (Purdue University), Panos Seranis (University of Cambridge), Victoria Smith (University of Nebraska), Robert Wood (Harvard University), and Raymond Zilinskas (James Martin Center for Nonproliferation Studies).

Although it is de rigueur to denigrate the efficiency and competency of government employees, my experience was that our agencies include some remarkably intelligent and effective people. The research for this project was greatly facilitated by wide-ranging discussions and e-mail exchanges with extraordinarily competent, helpful, and perceptive individuals in the U.S. Department of Agriculture's Agricultural Research Service (Dave Carlson, Sharon Drumm, Tim Gottwald, Eric Jang, Edward Knipling, Peter J. Landolt, Richard Nunamaker, Norma Ross, Alan Showler, John Sivinski, Nancy Vanatta, and Robert Vander Meer), the USDA's Animal and Plant Health Inspection Service (Melissa O'Dell, Jim Reynolds, Matt Royer, and Bruce Shambaugh), Office of Inspector General (Tim Danaher), and the FOIA Staff.

Individuals in other state and federal agencies also provided assistance in a variety of forms. These people included: Jeffrey Grode (U.S. Customs and Border Protection), James Hinchman (National Academy of Sciences), Dennis A. LaPointe (U.S. Geological Survey), Wilbert Mahoney (National Archives and Records Administration), Robin Schoen (National Research Council), Sarah Maxwell and David West (U.S. Army), Janie Santos (U.S. Air Force), Janine Anderson and Mary Ann Showalter (U.S. Department of Energy), Erlinda Byrd (Department of Homeland Security), and Pat Minyard, Cliff Ramos, and Jim Wiseman (California Department of Food and Agriculture).

A number of private individuals (Eric Bredesen, Galen Frysinger, Tim Hale, Ailsa Hopkin, Ethan Lockwood, Walter Müller, Tom Murray, Matthias Ziegler, and Ari Zivotofsky) and people from various organizations were kind enough to further my efforts with everything from intriguing images to valuable leads to critical information, including Susan Bennett and John Moffett (Needham Research Institute), Naima Boumaiza (UNESCO), Jocelyne Bruyère and Lisa Schwarb (World Health Organization), Ginger Cisewski (USGenNet.org),

Nancy Farrell (WGBH, Boston), Mike Frosch (civil-war.net), Wilfried Funk (insektenbox.de), Matthias Gabriel (Zuschauerredaktion), Dan Gower (DUSTOFF Association), Arvel "Jim" Hall and Stephanie Simon (Disabled American Veterans), Emily Hall (Emerald Group Publishing Limited), Edward Hammond and Jan van Aken (Sunshine Project), Peter Hicks (Napoleon Foundation), Kate Jackson (MiniMonsters, UK), Jeff Leys (Voices for Creative Nonviolence), Grant Lockwood (Sandia National Laboratories), Nazeem Lowe (Iziko Museums of Cape Town), Malick Kane (Getty Images), Sharlissa Moore (Student Pugwash USA), Nicole Neitzey (Mine Action Information Center), Ed Rouse (psywarrior.com), Leigh Russo (PARS International), Vincent Smith (Natural History Museum, London), Jane Stevenson (Bristol Friends of the Earth), Joan Taylor (*New York Times*), Lynn Waterman (Legacy Preservation Society), Jo Wilding (Garden Court Chambers, London), Robert Windrem (NBC Universal), Martin Winters (BBC), and Suren Varma (International Center for Agricultural Research in the Dry Areas).

I am particularly indebted to those experts who were willing to spend hours allowing me to probe the considerable depths of their professional experience through personal interviews. Their patient and knowledgeable individuals include: Charles Bailey (director of the National Center for Biodefense at George Mason University), Robert Kadlec (staff director for the U.S. Senate Subcommittee on Bioterrorism and Public Health), Geoff Letchworth (former director of the USDA's Arthropod-Borne Animal Diseases Research Laboratory), Michael Oraze (Biological/Agricultural Terrorism Director, Office of Field Operations, Customs and Border Protection, U.S. Department of Homeland Security), and William Patrick III (retired chief of Fort Detrick's Product Development Division).

I was fortunate to have three hard-working and intellectually tenacious research assistants over the course of this project: Kyran Ellison, Seth Hansen, and Erin Lockwood. Their technical skills and ability to ferret out the most obscure information from the strangest sources were critical to uncovering the history of how insects have been used in warfare, terror, and torture. Funding for these research assistants and much of my travel was provided through Chemtura, a company that understands the value of studies conducted at the interface of the sciences and humanities.

This book would never have come into being without an editor who believed in the project and had faith in the author. Peter Prescott of Oxford University Press provided the balance of constructive critique and enthusiastic encouragement that kept the writing not only moving forward but headed toward the best possible outcome. Tisse Takagi at OUP provided timely and clear answers to my questions, no matter how many times I asked the same thing because of having misplaced her e-mails. I am also grateful to Oxford's extremely capable production editor, Christine Dahlin, and my copyeditor, Carole Berglie, who dealt patiently with my constitutional inability to correctly use "that" and "which" along with other important details. The index was developed and refined with the generous and able assistance of Margery Niblock. And I should be sure to thank the "close readers" of earlier drafts of this book for their insights, criticisms, and suggestions, which proved vital to telling the story of entomological warfare: Dr. Gene Kritksy, Professor of Biology at College of Mount St. Joseph and Dr. Daniel Strickman, Colonel (retired) of the U.S. Army Medical Service Corps.

I reserve my deepest expression of gratitude for my family. Their love and support are the foundation for my work. I am sustained by my wife, Nan, my daughter, Erin, and my son, Ethan.

Finally, I apologize to anyone whose contribution to this project I may have overlooked. And I feel compelled to thank—in advance, however presumptuous this may be—general readers and disciplinary experts for their willingness to pardon my occasional oversights and overreaches. Any project that hopes to integrate science with history, politics, and sociology is destined to simplify certain aspects of the story and perhaps even introduce blatant, if forgivable, errors.

ILLUSTRATIONS

SIX-
LEGGED
SOLDIERS

INTRODUCTION

Silent, insidious, devastating. This is how an entomological attack is likely to unfold today. But to imagine modern uses of insects as weapons, we must look to the past. History reveals an unholy trinity of strategies—transmission of pathogenic microbes, destruction of livestock and crops, and direct attacks on humans—through which six-legged soldiers have wreaked havoc on human society.

The woman nervously checks her watch. In an hour, the human tide of the New York City rush hour will begin to pour underground. But, for the moment, there are few people who can see her reach into a shopping bag and take out a soda can. The woman peels off a strip of tape covering the opening and rolls the can beneath a bench on the subway platform. She heads to the escalator a bit more quickly than she and her fellow terrorists had been trained to move, but she's anxious to complete the other deposits. The woman and the five hundred hungry fleas make their respective escapes. She can only guess how many commuters will find red lumps on their legs in the morning, but whoever is bitten will be wracked by fever within days. Swelling lymph nodes might tip off a perceptive physician, but most of the victims will succumb to the ravages of bubonic plague. She knows that only a few hundred Americans will die, but millions will panic when they realize their vulnerability.

Bacteria-laden fleas spread throughout a subway system would echo the most terrible military use of insects in human history—disease vectors. The most devastating entomological attack took place in 1343, when Janibeg, the last Mongol khan, unwittingly allied with insect-borne disease in attempting to take the city of Kaffa. The Asian leader never suspected the role that fleas played in the ensuing pandemic that killed 25 million people. But neither did Europe's foremost military leader understand that his greatest defeats were caused by insects.

In 1799, Napoleon Bonaparte's campaign against the Ottoman Empire was defeated by flea-borne plague, and three years later his bid to establish a stepping stone into North America was crushed by yellow fever mosquitoes in Haiti. Napoleon's worst defeat by insects came in 1812. Rather than taking Russia, his Grande Armée lost 200,000 men to louse-borne typhus—a disaster that was replayed a hundred years later.

If the Austrians had established a western front against Russia, the course of World War I might have changed dramatically. But typhus kept the Central Powers from invading Serbia. In the Second World War insects were weaponized by the Japanese. General Ishii Shiro's Unit 731 produced hundreds of millions of infected insects and dispersed them across China (and attempted to infiltrate the United States). By the end of the war, Ishii's fleas and flies were responsible for more deaths than the atomic bombs dropped on Japan.

Not to be outdone, scientists at Fort Detrick, Maryland, developed entomological weapons and conducted secret, open-air trials with (uninfected) vectors over U.S. populations in the 1950s. During the Korean War, the North Koreans and Chinese assembled a massive dossier to support the accusation that the U.S. military released an entomological potpourri infected with a microbial menagerie. The Americans passionately denied the charges, as they did when accused of using insects to spread disease in Cuba and Vietnam. But governments often disavow politically problematic knowledge.

In recent years, a troubling theory was quashed by federal agencies: the possibility that bioterrorists were responsible for the outbreak of West Nile virus. Saddam Hussein's minions had the motive, means, and opportunity, but the evidence is too circumstantial to accept—and too intriguing to ignore. We were virtually unable to check this mosquito-borne disease, and West Nile virus was a case of the sniffles compared to what would happen should Rift Valley fever be introduced. But an entomological attack need not target humans to inflict a terrible toll.

A white cardboard tent hangs in a tree like a Lilliputian bivouac for an elfin tree-climber. But the structure has no fanciful function, as the printed warnings make clear. Striding officiously beneath the luxuriant canopy of the orange grove, a man stops and removes the tent. The badge affixed to his breast pocket affirms that he is empowered to disturb federal property. He scowls upon seeing a series of cabalistic symbols scrawled over the warning label. The man peers inside, then rips open the tent and utters a mixed curse and prayer. The sticky interior of the tent has trapped dozens of flies. Their wings look like tiny stained-glass windows crafted *in amber tones. These are the only Mediterranean fruit flies he's ever seen outside of the tedious training sessions in Sacramento. By the end of summer, $100 million in fruit will fall from the trees and rot in the California sun.*

The second major use of insects as weapons is as assailants of agriculture. Starving one's enemy or crippling his economy by unleashing insects to destroy crops or livestock is a relatively recent innovation. Although farmers have battled pests for millennia, not until we mastered the industrial-scale mass production of insects could inducing hunger and poverty through entomological warfare become a military strategy.

In 1938, a British scientist warned: "It would be very surprising, for example, if insect pests . . . were not [dispersed] by hostile aeroplanes in the course of a future war." And by the summer of 1944, Germany had stockpiled 30 million secret weapons: the Colorado potato beetle. Whether an insectan Blitzkrieg landed on Britain's farms is a matter of debate, but when the Second World War devolved into the Cold War, using insects as covert weapons against an enemy's agriculture became very tempting.

As Cuba was poised to start a nuclear Armageddon in 1962, the U.S. Army was prepared to release planthoppers to destroy the Cuban sugarcane crop and cripple the nation's export economy. Neither missiles nor insects were launched that October, but for years Castro accused the Americans of infesting Cuban farms with aphids, beetles, moths, and mites. And in 1997 Cuba formally charged the U.S. State Department with releasing thrips to decimate the island nation's agriculture. The United Nations concluded that the pest outbreak "most likely" arose from an accidental introduction.

Encouraged by their ally's success in drawing world attention to the U.S. entomological warfare program, the North Vietnamese reported that Americans had loosed "killer insects" on the countryside (the dead crops probably succumbed to Agent Orange). Developing countries might be vulnerable, but most military strategists believe that modern pest-management practices make it impossible to starve an industrial nation via entomological warfare. Economic losses are another matter.

The Asian longhorned beetle that was accidentally introduced to the United States in 1995 will, if not stopped, inflict an economic toll exceeding the cost of the attack on the World Trade Center—an entomological scenario not lost on bioterrorists. In 1989, a covert group of environmental radicals threatened to release Medflies—voracious pests of valuable fruit crops—unless insecticide spraying was halted in California. Had they succeeded in establishing this pest, losses could have reached $13.4 billion.

Today's international terrorists are keenly aware of the potential of entomological weapons to inflict staggering economic losses and social turmoil. And the United States is far more vulnerable than the government would like to admit. Perhaps a demented terrorist might even consider adapting the oldest use of insects as weapons, a strategy that makes the suffering rather more immediate.

> An 88-year-old woman was checking on some vacant property she owned when she noticed a door to a shed that was normally locked was open. She entered the shed to check inside and was immediately attacked by a large swarm of bees. After getting a neighbor's attention, she was able to walk about 45 m (150 ft) from where the attack began and then collapsed in the yard. . . . Upon arrival in the emergency department, the patient [was] moaning and complaining of pain but alert and cooperative. Her lungs were clear to auscultation. Swelling of the tongue was noted. The patient was given morphine sulfate 4 mg intravenously for pain. . . . The patient was transferred to the ICU in stable but guarded condition. . . . Blood pressure at the time of transfer was 190/110 mm Hg, heart rate was 105/min with sinus tachycardia and occasional premature atrial contractions, and respirations were 22/min. The emergency department staff estimated that the patient had sustained approximately 1000 bee stings.[1]

This case study of an Arizona woman attacked by Africanized bees does not have a happy ending. She died less than 96 hours after the attack. Killer bees were not introduced into the United States as an act of war, but there are plenty of insects capable of inflicting debilitating pain, even death, that could be weaponized. Indeed, humans have used stinging insects as weapons for thousands of years.

The oldest tactic in biological warfare was the heaving of beehives and wasp nests at an entrenched enemy. So effective were stinging creatures that, when a people couldn't find insects, they conscripted the next best thing. In the second century, the Middle Eastern stronghold of Hatra forced Rome's finest legions to hightail it back home—an apropos image for an army that had been stung into submission by scorpions dropped from the city walls.

Stinging insects can be unreliable combatants, being unable to follow orders. So the ancients developed poisoned-tipped projectiles (legend has it that Hercules invented poisoned arrows by emulating wasps). The Romans yearned to obtain a mysterious poison from India. They never found the source, but scientists believe that a beetle excreted the poison—a chemical 15 times more potent than cobra venom. But there is another way to overcome the problem of one's proximity to untrustworthy conscripts.

Bees, wasps, and hornets were catapulted for centuries in battles stretching across Europe. But few history books reveal that King Richard was both Lion-Hearted and Bee-Armed, that stinging projectiles ensured the defeat of Henry I by the Duke of Lorraine, or that the entomological predecessor of the Gatling gun was a windmill-like device that propelled straw beehives from the ends of its rapidly rotating arms. And there are even fewer accounts of how insects were used as instruments of torture.

Although the "Great Game" in which England and Russia vied for control of Central Asia in the 1800s is an important chapter in history, the chilling tale of the Bug Pit and its victims—including a pair of unfortunate British officers—is rarely told. Created by the emir of Bukhara, the torture chamber was stocked with assassin bugs that slowly ate their victims alive. But modern militaries don't use insects to inflict agony, right? Wrong. The Vietcong wired boxes of scorpions to trip wires as booby traps in the subterranean tunnels of Cu Chi. And for their part, the Americans figured out how to issue chemical commands to order one of the world's biggest and meanest species of bees to attack the enemy.

The three major forms of entomological warfare (disease vectors, agricultural pests, and direct attacks) capture much of human ingenuity in conscripting these creatures for military use. However, insects are far too diverse and humans far too clever to stay within these neat categories. Today, scientists are designing insect-machine hybrids—tiny cyborgs to infiltrate enemy positions, gather military intelligence, and assassinate key individuals. The poisons devised to control insect pests have been transformed into deadly chemical weapons, and arthropod toxins serve as molecular models for the next generation of poisons. And as we delve deeper into genetics and become able to create new life forms, the possibilities may soon be limited only by our desire to relieve—or inflict—suffering. Imagine the power that would come to a terrorist organization whose genetic engineers altered the biochemistry of a common species of mosquito in the United States so that the blood-feeding insects were capable of transmitting AIDS.

Such modern possibilities are rooted in a remarkable story of human inquiry, ingenuity, and brutality. This is the tale of one of history's most potent military alliances: the intelligence of humans and the power of insects.

ONE

STINGING DEFEATS AND VENOMOUS VICTORIES

The insects fell into the Romans' eyes
and the exposed parts of their bodies . . .
digging in before they were noticed,
they bit and stung the soldiers,
causing severe injuries.

—Herodian, a historian of ancient Antioch,
describing the defense of Hatra in 199 CE

1

BEE BOMBS AND WASP WARHEADS

The first era of entomological warfare saw insects drafted into battle to directly afflict the enemy. More sophisticated tactics of transmitting diseases and destroying crops would have to wait for breakthroughs in human knowledge. However, we shouldn't disparage the cleverness of ancient peoples. After all, they laid the foundations for modern weaponry—and insects were the first organisms used to wage biological warfare.

The military historian John T. Ambrose contends that insects have long "served as models for man to emulate in . . . the art of warfare."[1] The social insects (those living in complex groups: ants, bees, wasps, and termites) were praised by early military strategists, as these creatures were thought to be unwaveringly loyal to their kin, extremely courageous, and exquisitely lethal. Early evidence of insects being held in military esteem is seen in Egyptian hieroglyphics dating to the first dynasty, more than 5,000 years ago. Some scholars contend that King Menes—the "Scorpion King" (so named for his family totem)—chose a hornet (subfamily Vespinae) as a symbol of his rule, reminding both countrymen and enemies of the pain that he could inflict. Others suggest that the ruler-insect connection was hardly an alliance, as the death of this Egyptian ruler is reportedly the first written record of human mortality caused by a wasp or hornet sting. To the military strategist of old, a swarm of bees (family Apidae, most likely the honey bee, *Apis mellifera*) driving away a pilfering bear evoked the image of a disciplined and ferocious phalanx of soldiers forcing a larger army into retreat. As anyone who has experienced the wrath of bees, wasps, or ants well knows, these creatures induce pain and fear disproportionate to their size. But the Hymenoptera (named for their tissue-like wings: *hymen* = thin membrane and *ptera* = wing) were not merely models of ideal armies.[2] More to the point, these insects became the earliest zoological conscripts of warring peoples. The challenge, of course, was how to compel these battle-crazed warriors—insects being unable to discern one side from the other in human warfare—to take out their aggression on the other side.

Brett Favre, Roger Clemens, and Michael Jordan made millions of dollars by virtue of their skill at one of the most primitive acts—heaving stuff. We've been variously called the thinking ape, the tool-using ape, and the naked ape, but we might best be named *Homo ejectus*—the throwing ape. A key to our success was the capacity to kill at a distance by flinging variously modified rocks and sticks. While the vestiges of our past are most evident in today's shot put and javelin throw, the real money is in propelling balls with speed and accuracy. To a primitive warrior with a "good arm," a nest of stinging insects (about the size and heft of a basketball) might have seemed the ideal natural bomb—easy to find, simple to store, convenient to transport, and full of fury.

The earliest hypothesized use of insects as weapons of war was around 100,000 years ago during the Upper Paleolithic period. By this time, humans were well practiced in making and throwing objects at one another and animals. And so the extrapolation from inanimate to animate projectiles was inevitable. According to Edward Neufeld, a scholar of Mesopotamian history, "It may be assumed with reasonable confidence that man has perceived the value of certain insects as an instrument of warfare long before recorded history. . . . It is almost a logical certainty that insect weaponry belonged to early man's 'natural' objects like those made from wood, bone or stone."[3] Humans would not domesticate animals for another 90,000 years, which rapidly led us to discover that dogs, horses, camels, elephants, and dolphins (a recent military conscript) could be used in warfare. But these creatures are prone to desertion in the midst of combat, given that they have an acute sense of self-preservation. Social insects, on the other hand, have evolved a remarkable tendency for aggressive self-destruction in the face of an enemy assault.

The worker bees in a hive are sterile sisters. The workers' only hope of passing on their genes, albeit indirectly, is through their common mother. This unusual biological situation predisposes a bee to attack an intruder that threatens her only source of genetic survival—the fertile queen. A bee's assault amounts to suicidal evisceration as her sting, modified from a defunct egg-laying structure, is torn from her body along with its poison sac. Humans, who are fertile and share less genetic similarity with one another,[4] usually avoid self-sacrificial aggression. Instead, we developed an impressive array of weapons that put some distance between ourselves and our enemies.

During the Paleolithic period, humans lived in caves and rock shelters—prime targets for a hurled nest of bees or hornets and related wasps (family Vespidae). With a concentrated enemy further defended by barriers of thorny shrubs or logs, a frontal assault was not a wise move. Although an inanimate object thrown over the stockade was unlikely to find its mark, a hive of bees was another matter altogether. The problem with heaving a bee hive in the midst of hand-to-hand combat was that the insects would fail to comprehend their military mission, stinging friend and foe alike. But when the enemy was holed up, they would bear the brunt of the attack. An angry swarm might even break the siege and drive a frantic enemy into the open. But there remained a serious problem: collecting a nest of stinging insects to heave at an entrenched opponent was fraught with risks.

To avoid becoming a victim of one's own weapon, early humans presumably gathered the insects at night when they are slowed by cooler temperatures and unable to see their abductor's approach. By the early Neolithic some 10,000 years ago, humans probably had discovered that smoke could be used to pacify bees. Transporting a nest of upset hymenopterans would have been a dicey proposition, so the people surely plugged the nest openings with mud or grass. Even more sensibly, the nests may have been carried within sacks or baskets, which also would have made the projectile easier to heave into a cave or over a windrow of brambles.[5] While all of this anthropological speculation seems entirely reasonable in light of what we know about the technology of ancient humans, the physical evidence is limited. Once writing was invented about 5,000 years ago, people began to leave records indicating that entomological warfare was part of military strategy. And the Bible is a treasure trove of data.

The god of Exodus in the Old Testament was a vengeful deity who would sooner smite his enemies than negotiate a settlement, an approach that works fine if you're omnipotent. But Yahweh was also a just god, and he gave his opponents fair warning of their imminent demise and a chance to surrender. Furthermore, and central to the history of entomological warfare, this god was a savvy strategist. Presumably an all-powerful deity could make the enemy disappear, but this would not evoke the darkest memories of human history. What God needed was nature's arsenal—blights that aroused a deep sense of mystery and fear. Winning a war by "shock and awe" would render a conquered foe psychologically beaten and culturally disheartened. If the natural world was loosed on the enemy, they would not soon forget.

Although secular and religious scholars differ on the ultimate causative agent—natural versus supernatural—for the plagues that struck Egypt,[6] the sequence of disasters was so compelling from the perspective of the Israelites that it became embedded in western culture's most abiding historical text: the Book of Exodus. Some scholarly suppositions and reasoned speculations are needed to make sense of the experiences of ancient people based on their stories, but those who recounted the plagues were certainly speaking of events within their scope of experience. Whether the scale and timing of the plagues are historically accurate is arguable, but ancient people were familiar with these kinds of disasters and could at least imagine what would happen if they were to unfold across the land. So if we take the Old Testament to be an account of human experience interpreted through a particular theological lens (a common approach among biblical scholars), Yahweh was perceived as an entomologically astute deity. The Creator's tenacious Egyptian opponent might have guessed that the god of Moses would call upon the forces of nature, but could the Pharaoh have imagined that six of the ten battles would be waged with insects as the warriors?

"Thus saith the Lord, in this thou shalt know that I am the Lord: behold, I will smite with the rod that is in mine hand upon the waters which are in the river, and they shall be turned to blood. And the fish that is in the river shall die, and the river shall stink; and the Egyptians shall loathe to drink of the water of the river" (Exodus 7:17). And so it was that God kicked off the plagues of Egypt with what some scientists surmise to have been a microbial bloom. The water of the Nile turned blood-red with dinoflagellates, snuffing out nearly all other aquatic life.[7]

The plague of frogs that followed might have been the result of these amphibians beating a hasty exit from their toxic habitat. Having Yahweh threaten that "if thou refuse to let them go, behold, I will smite all thy borders with frogs" (Exodus 8:2) might seem a bit comical. After all, being smitten with frogs doesn't stack up with being struck by thunderbolts. But an onslaught of living creatures—even something as generally innocuous as frogs (or, as Hitchcock understood, birds)—can be horrifying.

Without fish and frogs to consume insects, the ecosystem began to unravel. The third plague was probably an irruption of biting midges (family Ceratopogonidae) from the rich, moist soil along the riverbanks: "Then the Lord spoke to Moses, say to Aaron, stretch out your staff and strike the dust of the land that it may become gnat-swarms throughout all the land of Egypt"

(Exodus 8:16).[8] Apparently pleased with capacity of the gnats to mercilessly harass the Egyptians and their animals, God sent a second wave of six-legged mercenaries.

In the fourth plague, Yahweh warned, "Else, if thou wilt not let my people go, behold, I will send swarms of flies upon thee, and upon thy servants, and upon thy people, and into thy houses: and the houses of the Egyptians shall be full of swarms of flies, and also the ground whereon they are" (Exodus 8:21). The best hypothesis offered by entomologists is that these flies (order Diptera, meaning "two-winged") arose from the rotting vegetation laced with decomposing fish that had accumulated along the riverbanks. There are various swarming flies, but the most likely candidate is the stable fly (*Stomoxys calcitrans*), with horse and deer flies (family Tabanidae) being plausible contenders. Stable flies deliver wickedly painful bites, and their populations increase rapidly when the larvae have an abundance of filth to eat.

God targeted the Egyptian's livestock with the fifth plague, warning, "For if thou refuse to let them go, and wilt hold them still, behold, the hand of the Lord is upon thy cattle which is in the field, upon the horses, upon the asses, upon the camels, upon the oxen, and upon the sheep: there shall be a very grievous murrain. And the Lord shall sever between the cattle of Israel and the cattle of Egypt: and there shall nothing die of all that is the children's of Israel" (Exodus 9:2–3). Roger Breeze of the U.S. Department of Agriculture has formulated a viable hypothesis for this godly assault: The cattle, sheep, and camels could have been suffering from bluetongue, while the mules, horses, and asses suffered from African horse sickness.[9] Both viral diseases occur in northern Africa. And, most critically, both are transmitted during the blood feeding by insects—such as gnats. These biting flies are well-known vectors of livestock diseases.[10] Moreover, gnats are not particularly adept fliers, traveling only a few miles from their larval habitat, and this limited dispersal might explain why the Israelites' livestock were spared. Egyptian agriculture bordered the Nile, where rich soils were worked with draft animals and cattle and sheep fed on lush pastures. At the time of the plague, the Israelites were building a storage city in the desert east of the delta in Pithom. Hence, their animals would have been considerably farther from the marshy lands where the gnats abounded.

The sixth plague is a bit of a puzzle: "And the Lord said unto Moses and unto Aaron, take to you handfuls of ashes of the furnace, and let Moses sprinkle it toward the heaven in the sight of Pharaoh. And it shall become small dust in all the land of Egypt, and shall be a boil breaking forth with blains upon man, and upon beast, throughout all the land of Egypt" (Exodus 9:8–9). Repeated biting by midges can cause large, itchy spots, but these are not typically "blains" or open sores. There are, however, two diseases that cause open sores in both humans and livestock: glanders and anthrax. Symptoms of the former disease depend on the route of infection, but when a wound—such as that caused by a biting fly—is the point of origin, a pus-filled, oozing ulceration forms. When anthrax enters a wound, it causes localized lumps, which fill with bloody fluid until they rupture and become necrotic ulcers. For these diseases to irrupt over a large area would presumably require that the microbes were spread efficiently, not merely by accidental, physical contact among livestock and humans. Remember the stable flies? These bloodthirsty insects are capable of spreading the bacteria responsible for glanders and anthrax, although this is not the most common means of transmission.

The seventh plague of Egypt was a horrendous hailstorm, which foreshadowed a change in the weather that would bring the most infamous pestilence. And so it was that "Moses and Aaron came in unto Pharaoh, and said unto him, thus saith the Lord God of the Hebrews, how long wilt thou refuse to humble thyself before me? Let my people go, that they may serve me. Else, if thou refuse to let my people go, behold, tomorrow will I bring the locusts into thy coast. And they shall cover the face of the earth, that one cannot be able to see the earth: and they shall eat the residue of that which is escaped, which remaineth unto you from the hail, and shall eat every tree which groweth for you out of the field" (Exodus 10:3–5). According to the account, an east wind carried what we now know to be the desert locust (*Schistocerca gregaria*) from its breeding grounds along the Red Sea coast and the Arabian Peninsula. When the Pharaoh capitulated, the wind shifted to the west, sweeping the eighth plague into the Red Sea. But alas, the Egyptian ruler's heart hardened again and God sent the ninth plague—darkness—which fell over the land. With enormous areas of cropland stripped bare by the locusts, howling winds carried the loose soil into the air, and the skies were blackened for days on end.

The tenth and final plague was the most terrible: the death of the first born in all of the Egyptian families. There are two competing explanations relying on the interaction of sociology and biology, with insects playing a supporting role.[11] Some scholars contend that the killer was a flea-borne disease, bubonic plague. Rats are the disease reservoir, but fleas (order Siphonaptera, meaning "wingless siphons," which is terribly apropos) transmit the pathogen. Every spring, the Jewish people fastidiously removed all grain from their homes, cognizant of the relationships among grain, rats, and disease. The Egyptians had no such tradition, so an outbreak of plague would have afflicted them far more severely. Those who advance this hypothesis take the selective death of the first born as a metaphor to express the depth of suffering. But other scholars believe that this demographic clue is vital to explaining the event.

Rather than a bacterium, the lethal agent could have been a relative of the black mold that grows following water damage in modern homes. Fungal toxins cause burning sensations, nausea, vomiting, diarrhea, abdominal pain, and even acute hemorrhaging of the lungs, particularly in infants. The Egyptians stored their grain, a wonderful substrate for mold, in pits that previous plagues had presumably moistened (thanks to hail), fertilized (courtesy of locust droppings), and covered with sand (via dust storms), creating a perfect environment for the fungus to flourish. But why would the toxic grain kill only the first born? The mold would have been abundant at the top of the grain stores, and the most important person in an Egyptian family—the first born—was served first. Moreover, in times of famine (the locusts had devastated the crops) these valued children were given a double serving. In a wicked irony, the Egyptians' efforts to honor and save their prized offspring may have ensured that these children received the most lethal meals. The Israelites used different methods of food storage, preparation, and allocation, which may have prevented the toxins from being concentrated into deadly doses.

Exodus provides the best-known account of entomological warfare with Yahweh using these weapons to defeat the Pharaoh—at least insects were central to the plagues that beset ancient Egypt, whatever their cause. But this is not the only book of the Bible that scholars have scrutinized for evidence of insects having been weaponized in the ancient world.

Biblical accounts of entomological warfare reveal tempting tidbits from which history can be tentatively reconstructed. Various books in the Old Testament allude to the use of insects in battle. Edward Neufeld contends that stinging insects were used for ambushes and guerrilla raids: "it may be confidently assumed that these texts give a strong impression of illustrating an authentic tradition of the use of insects as warfare agents."[12] And the most common tactic was probably dislodging an entrenched opponent. Joshua 24:12 recounts that "I sent the hornet before you, which dr[o]ve them out from before you, even the two kings of the Amorites; but not with thy sword, nor with thy bow." From this passage, historians surmise that insects were used like modern-day shock troops as a means of routing an enemy from a stronghold. The use of hornets in this manner is consistent with the passage in Exodus 23:28: "And I will send hornets before thee, which shall drive out the Hivite, the Canaanite, and the Hittite, from before thee." Presumably, the enemies of the Israelites were driven out of caves or other fortifications and forced into open combat. The use of hornets to flush an adversary from hiding seemed to be a sufficiently common practice that this tactic was attributed to the Almighty, as well: "Moreover the Lord thy God will send the hornet among them, until they that are left, and hide themselves from thee, be destroyed" (Deuteronomy 7:20). Making sense of such accounts means stepping into the murky realm of interpreting ancient scripture, and there are linguistic reasons for exercising caution in scriptural interpretations.

Scholars are divided as to the meaning of the Hebrew word צרעה. Some translators take this to literally mean "hornet," thereby supporting the case for entomological warfare. Other linguists interpret it as "panic," suspecting that the original word might have been a metaphor referring to any calamity. A tenth-century Arabic translator apparently took a middle position, using the term "plague" or "pest" to suggest an unspecified biological disaster. However, the interpretation becomes problematical because Arabic uses the same word to refer to a hornet and to a mass of people in panicked flight.

While one might question the extent to which humans or deities made effective use of insects as weapons, the analysis of biblical passages leaves little doubt as to the importance of entomological warfare in the military culture of ancient times. In particular, as the people thought of insects in terms of their potential as weapons, they appreciated that insects could inflict considerable damage to an enemy. They understood that these organisms were not only able to cause physical suffering but also had the capacity to instill fear and panic. Then, as now, winning a conflict is about defeating both the bodies and the minds of one's opponent.

Although ancient textual evidence is limited, we might reasonably surmise that for a long period of history entomological warfare primarily was a means of assaulting a concentrated, stationary foe. Gaining the upper hand during a siege by raining insects on your enemy was a rather simple tactic. Real human ingenuity would be needed to take these surly creatures onto the battlefield and make an opponent suffer their wrath in open combat.

A nest of stinging insects is much like a grenade inside a hatbox: you can't throw it very far, and once it explodes, projectiles whiz in every direction. What you need is a means of directing the six-legged shrapnel toward your target. To address this challenge, various cultures have developed stunningly astute methods.

The Tiv people of Nigeria developed the "bee cannon," an elegant battlefield weapon designed to ensure that the projectiles—angry bees—were directed at one's opponent.[13] The bees were loaded into large, specially crafted horns. In the heat of combat, the Tiv would aim the mouth of the horn toward the opposing tribe, the shape and length of the cannon effectively directing the bees toward the enemy. The accounts of this remarkable contrivance are recent, but the bee cannon is likely a very old device, perhaps even as venerable as the entomological weaponry of Mesoamerica.[14]

By 2600 BCE, the Mayans had weaponized bees or wasps.[15] The Mayan language does not distinguish these insects, although we might presume that wild bees were used, given that the Mayans had domesticated a stingless species that would not have been of much use in battle. The sacred text, Popul Vuh, tells of the people building dummy warriors topped with headdresses to conceal both the inanimate condition and the real purpose of the manikins. The heads were hollow gourds filled with stinging insects. When attackers smashed the gourds, the insects retaliated, precipitating a chaotic retreat that allowed the Mayans to annihilate the would-be invaders.

In addition to bee-based booby traps, ancient texts suggest that the Mayans used bees in open battle. The details of this armament are not clear, but given their pottery skills and what we know of other cultures, we can infer that they devised the "bee grenade." At about this time, some people in the Middle East were molding special containers from clay that were heavy enough to throw and fragile enough to burst upon landing. The shell casings were set outside to be colonized by bees or hornets. When a fight was in the offing, warriors plugged the openings with wads of grass to contain the insects while allowing air into the colony.[16] Such "pots of pain" could be readily carried into combat and conceivably thrown quite effectively at an enemy cornered in a draw or otherwise conveniently grouped on the battlefield.

With the integration of pottery into entomological warfare, the assailants were not limited to using bees or wasps. A nest of ants (family Formicidae; some species deliver a double-whammy of a burning bite and a searing sting) would make a fine weapon when encased in a clay pot. More than one ten-year-old kid has contemplated the possibility of using an ant farm to create mayhem—multiply this fantasy by a few orders of magnitude. Thousands of frenzied ants with a score to settle would have been just the ticket for flushing an enemy from battlefield cover or a fortified stronghold. But just as the integration of pottery and entomology was heading toward a technological apogee, the balance of power in the art of siegecraft was undergoing a reversal of fortunes.

The military advantage of allying with the Hymenoptera shifted from those who could mercilessly harass an entrenched force to those who could construct fortifications from which devastating counterattacks could be launched. Architectural engineering provided building methods that allowed the construction of formidable walls around cities and military installations. Try tossing a clay pot over a 30-foot wall while a bunch of guys on top of the rampart are shooting arrows and dropping beehives, and you get the picture.

In one of the earliest "how to" books, Aeneias the Tactician wrote *How to Survive under Siege* in the fourth century BCE. The 200-page manual is chock-full of helpful hints, including the author's advice for "besieged people to release wasps and bees into the tunnels being dug under their walls, in order to plague the attackers."[17] The tactic of burrowing was developed in response to assailants' being greeted with a hail of nasty items—stinging insects, along with boiling oil and molten tar—from the ramparts. And no conflict better exemplifies the capacity of insects and their kin to repulse a potentially overwhelming attack than does the battle for Hatra.[18]

At the end of the second century, the Roman emperor Septimus Severus set his sights on wresting control of Mesopotamia from the local monarchs. The Romans waged several ineffective campaigns, but Severus's Waterloo was the desert stronghold of Hatra—a city that wallowed in wealth by virtue of controlling the caravan routes connecting Mesopotamia with Syria and Asia Minor. Today, the remnants of Hatra lay 50 miles southwest of Mosul, Iraq's second largest city. The crumbling ruins reveal a once-formidable fortress. Hatra boasted a defensive perimeter of nearly five miles, formed by a moat sandwiched between 40-foot-high walls. King Barsamia and his citizens holed up inside this redoubt as the Roman legions advanced. But a proud desert people don't simply cower behind the city walls.

Thanks to Herodian, a historian from Antioch (Syria), we know that the Hatrians prepared for the onslaught by crafting earthenware bombs loaded with "insects" to hurl onto their attackers. The mystery, however, is precisely what creatures seethed within the clay pots. As with other ancient writers, entomological nuances were not Herodian's concern, and modern historians are left trying to reconstruct what he meant by "insects." The most obvious interpretation is that the Hatrians used beehives made of clay. But the problem is that this desert region supported only solitary bees, which would have been far too difficult to collect in sufficient numbers for arming the residents.

The favored hypothesis is that the bombs were not loaded with insects at all, but that they were filled with scorpions.[19] Scorpions were so prevalent and dangerous that Persian kings regularly ordered scorpion hunts and offered bounties to ensure safe passage for the caravans. According to the best-known scientist of Roman antiquity, Pliny the Elder, "[scorpions] are a horrible plague, poisonous like snakes, except that they inflict a worse torture by dispatching their victim with a lingering death lasting three days." Although he exaggerated a bit, the sting of a scorpion is intensely painful, and the venom can induce convulsions, slowed pulse, irregular breathing—and occasionally death (see Figure 1.1). Being lethal, dreaded, and plentiful are strong qualifications for an agent of war, but practical concerns had to be addressed to convert scorpions into weapons.

The people of Hatra would have had no difficulty finding plenty of scorpions in the surrounding desert, but avoiding being stung in the course of harvesting these creatures was surely a challenge. Fortunately, the Hatrians

Figure 1.1. Desert scorpions commonly found on the U.S. airbase in Tallil, southeast of Baghdad, including *Scorpio maurus* (left), *Mesobuthus eupeus* (top), and *Odontobuthus doria* (right). At least two species from Iraq can deliver deadly stings, but even nonlethal species, such as those pictured, could have inflicted more than enough pain to convince the Roman legionnaires to abandon their assault on Hatra in the second century BCE. (Photo by Senior Airman Matthew Hulke, courtesy of the U.S. Air Force)

didn't depend on the advice of Roman experts. Pliny advised that the venom was most deadly in the morning, "before the insects [*sic*] have wasted any of their poison through accidental strikes"[20]—an odd claim given that scorpions are nocturnal hunters. Another Roman natural historian, Claudius Aelianus, who enjoyed the patronage of Severus, suggested that one carefully spit on the tip of the sting to temporarily block the tiny opening through which venom was injected. Crawling around the desert spitting on short-tempered scorpions seems wholly ill-advised.

Aelianus's alternative recommendation was to sprinkle the creatures with powdered monkshood, a poisonous plant that would temporarily stun the scorpions, allowing them to be safely captured. Given that monkshood was used to kill body lice in the Middle Ages, it seems plausible that a low dose might have sedated an irascible scorpion. Such a method would have had a further advantage given the propensity of scorpions for cannibalism. Without sedation, a jarful of these creatures gathered in the morning might have yielded only a few fat survivors by the end of the day.

The "scorpion theory" of the city's defense is a tidy reconstruction of the events preceding the arrival of Severus's legions. But there is one baffling detail provided in Herodian's account. The Syrian historian further specified that the creatures used in the defense of Hatra were "poisonous flying insects." And, alas, there is no such thing as a flying scorpion. The ancient naturalists, however, were of a different opinion.

Aelianus reported 11 types of scorpion: white, red, smoky, black, green, pot-bellied, crablike, fiery red-orange, seven-segmented, double-sting, and those with wings.[21] We might dismiss the last two types as being fantastical, except that there are reliable, scientific reports of a rare developmental anomaly that results in two-tailed scorpions. And while no malformation produces winged scorpions, there are such things as scorpionflies (order Mecoptera). In these insects, the male's genitalia curl over his back to resemble a scorpion's tail. Although scorpionflies can fly, they are typically associated with damp habitats, rarely exceed an inch in length, and are quite harmless. Aelianus might have seen these insects at some time and mistaken them for scorpions, but they would not have been the allies of the Hatrians.

Another explanation of flying scorpions comes from Pliny, who maintained that these creatures became airborne in high winds and then extended their legs to function like wings. No modern text makes mention of such a phenomenon, and the winds necessary to lift a scorpion into the air would probably have blown Severus's army back to Rome. Other scholars speculate that the clay vessels were filled with assassin bugs (family Reduviidae). These insects don't particularly resemble scorpions, but they can fly and deliver an extremely painful bite. Perhaps the containers were loaded with a potpourri of biting and stinging creatures, which included some flying insects along with a liberal measure of scorpions.

We may never know what beasts rained down on Severus's men as they tried to scale the walls of Hatra, but the historical account makes clear the results. Herodian describes creatures of some sort inflicting severe punishment on the Romans wherever there was exposed skin—lower legs and arms, as well as their faces, and, worst of all, their eyes. Although the immediate effect of the arthropod arsenal was impressive, its ultimate role in the outcome of the siege remains a matter of dispute. Severus was held at bay for 20 days, but his troops were finally able to breach the walls. However, just as victory was at hand, the Roman emperor broke off the battle. Military historians can offer no convincing explanation for this turn of events. Some posit that a secret treaty was reached, or that the troops had become mutinous, or perhaps the brutality of the defenses had demoralized the legionnaires. In any case, the Romans slunk back home, and Hatra remained autonomous for another half century until the Sassanids arrived from Iran with armor-clad troops and reduced the city to ruins (see Figure 1.2). By that time, the use of insects to defend walled cities had become common throughout Europe.

For centuries, besieged Europeans conscripted bees in an effort to repulse invaders. In 908, the residents of Chester, England, were assailed by an army of Danes and Norwegians.[22] Having found the city's fortifications impenetrable, the Scandinavians tried undermining the walls. Like a horde of obsessed gophers, the tunnelers could not be dissuaded from their subterranean escapades. Projectiles launched at the mouth of the tunnel outside the walls did not impress the burrowers. Finally, the English collected all of the city's beehives and hurled them into the tunnel, summarily ending the military mining operation.

More than 700 years later, the lesson was repeated with another Scandinavian army. During the Thirty Years War, a Swedish general led his troops in an assault on the walled city of Kissingen.[23] One of the inhabitants, Peter Heil (for whom a street is now named), outmaneuvered General Reichwald. Following Heil's advice, the people threw their beehives at the invaders. Unlike the Danes and Norwegians centuries earlier, the Swedes were well protected by heavy clothing and armor. But their horses were not. The animals panicked as the

Figure 1.2. The ruins of Hatra, 50 miles southwest of Mosul, Iraq, still provide a sense of the defensive perimeter that the Romans faced during their siege of this formidable stronghold. From the top of the 40-foot walls (background), the citizens rained scorpions onto the invading soldiers until they broke off the attack. The city remained autonomous for another 500 years, although how often the Hatrians had to conscript the stinging denizens of the desert to defend the city is not known. (Photo courtesy of UNESCO)

bees stung them mercilessly and the assault collapsed into a melee of frenzied cavalry mounts.

Between these Scandinavian defeats, bees saw action throughout Europe. In 1289, these insects were used along with hot water and fire by the inhabitants of Gussing, Hungary, to drive Albert, the Duke of Austria, and his army into retreat.[24] Around the same time, the castle-dwellers on the Aegean island of Astypalaia fended off pirate attacks by dropping beehives from the parapets.[25] And the Turks were on the cusp of certain victory, having broken through the walls of Stuhlweissenburg, when the Hungarians plugged the breach with beehives and repelled the invasion.[26]

These methods of defense required hauling hives from bee yards to the scene of battle, a difficult task if the insects were needed urgently, given that a typical hive weighs over 100 pounds. So, reasoned some savvy planner, why not put the hives where they could serve dual purposes? One might reasonably infer that this is why some noblemen maintained bees on the parapets, allowing the insects to be ready for producing honey or havoc, as the situation demanded. The walls of a few medieval castles in Scotland, England, and Wales were equipped with recesses, termed "bee boles," as permanent homes for the bees. These structures were generally on the south-facing perimeter walls, which provided a warm setting for cold-blooded insects trying to make a living in the northern climes.[27]

After the Middle Ages, defenses were rather less creative. Low-tech tactics were revived during the reign of Emmanuel the Fortunate, when the Portuguese king's troops were repulsed by bee-flinging Moors.[28] Perhaps the simplest and most famous use of bees to defend a town came in the 1600s.[29] Many years earlier, the Count of Berg had given away part of the town of Wuppertal to the Brothers of the Cross. In time, a convent was built and nuns took up cultivating bees, along with virtue. When the community came under attack, the sisters refused to admit the marauding troops. The soldiers, not dissuaded by a bunch of women in habits, attempted to enter the town by force. The fast-thinking nuns toppled the beehives in their apiary, dashed inside, and allowed the bees to vent their anger. The soldiers were driven from the gates and the town changed its name to Beyenburg (or "bee-town") in honor of the insects.

Bees were part of city defenses into the 18th century, at which time Belgrade had become the object of a long series of bloody clashes between the Austrians and Turks. During one of the battles, the Turks succeeded in breaching the walls and were poised to celebrate their entry into the city when they encountered an impenetrable barricade of beehives. The victory was short-lived, however, as the Turks claimed Belgrade in 1739. Although simple tactics of entomological warfare were demonstrably effective for centuries, the full integration of human ingenuity and insectan ferocity required the imposition of machinery between man and beast.

While tossing bees and tipping hives are fine defensive strategies for repelling unwanted visitors, a major breakthrough in entomological warfare came with the development of machinery capable of launching insectan payloads into the enemy's ranks. What the slingshot did for the humble rock, instruments for heaving hives did for bees—and shifted the balance of entomological power in favor of the attacking forces.

The earliest machines for launching projectiles emerged from the workshops of Philip II, father of Alexander the Great.[30] These inventions worked well for firing arrows, but were limited in terms of their payloads. Once the Greek engineers substituted torsion springs and counterweights for composite bows, the "siege engines" became capable of heaving almost anything, including insects. The Greeks also ensured that the etymology and entomology of war become intriguingly entwined in the word *bombard*, which comes from *bombos*, meaning "bee"—an allusion to the threatening hum associated with both an angry swarm and an incoming projectile.

The Greeks might have been the inventors of the first mechanized weapons, but the Romans were history's most fervent launchers of bees and wasps. Perhaps the Roman generals were encouraged by Pliny's authoritative and exaggerated claim that precisely 27 hornet stings would be lethal to a human. Nonetheless, the Romans made extensive use of bees, whose hives were far easier to acquire as armaments. So widespread were beehives as catapult payloads that the well-documented decline in the number of hives during the late Roman Empire was probably a consequence of having heaved too many of these nests into enemy fortifications.[31]

European military history is replete with accounts of beehives having been used as missiles. Between 1000 and 1300 CE, bees were catapulted in battles stretching across the continent in settings ranging from sieges to battlefields. In the 11th century, the forces of Henry I of England were backed into a corner by the Duke of Lorraine's marauders. The battle turned when the English general ordered his men to launch "nest bombs" into the midst of the Duke's men, who abandoned their assault rather than suffer the wrath of the enraged bees. In the 12th century, King Richard catapulted hives into Moslem ranks and strongholds during the Third Crusade. In the 13th century, bees were used for both waging attacks and mounting defenses in the kingdom of Aragon in modern-day Spain. Perhaps the technological high point in hive-heaving machinery emerged in the 14th century with the development of the Gatling gun's entomological predecessor—a windmill-like device that propelled straw hives from the ends of the rapidly rotating arms.

The virtue of catapulting bees—the safe distance that this provided between those launching the hives and the point at which the unhappy insects landed—was also exploited in naval warfare.[32] As early as 332 BCE, earthen hives were thrown onto the enemy decks, and with the advent of the catapult, bees and hornets became standard projectiles on the high seas. Across the Greco-Roman and Syro-Palestinian worlds and into the Middle Ages, warships carried beehives as part of their arsenals. Although cannon balls largely replaced hives among the world's navies, bees were sporadically used for centuries. A well-documented case from the 1600s involved the crew of a privateer fighting off a 500-man galley by heaving earthen hives onto the larger vessel.

The most amusing, and probably apocryphal, maritime use of stinging insects took place during the War of 1812.[33] As the story goes, while in an American port, a group of British seamen was intrigued by a papery nest hanging from the branch of a tree. When they asked a passing boy about the object, the scamp realized that he had the raw material for a fantastic prank. In dead earnestness, the lad declared that it was the nest of a hummingbird. With natural history fast becoming the rage back in England, the sailors were delighted to acquire such a treasure to display upon their return home. We might presume that it was a crisp fall evening when the men cut the nest free from its moorings, for the residents remained calm. The scallywag implored the sailors to "go easy, lest they disturb the bird or break her eggs," but there was, of course, no hummingbird within the structure—only a colony of wasps sedated by the cool temperatures. The men proudly took their prize aboard the ship, only to have the insects later aroused by the warmth and jostling of their nest. Most of the crew was driven from the ship as the wasps took their revenge.

For all of their virtues as weapons, bees, wasps, and ants had one major limitation. Relying on insects to deliver the painful venom that drove the enemy from the battlefields and fortresses meant accepting a great deal of uncertainty. The little conscripts might disperse from the scene before creating havoc, they might be lethargic if it was too cold, they might escape from the munitions before the payload was launched, or they might turn traitor if used in close quarters. What military commanders needed was a way of exploiting the insects' capacity to deliver debilitating pain, and even death, without the fickleness of the six-legged soldiers mucking up a brilliant battle plan.

2

TOXIC TACTICS AND TERRORS

Stinging insects proved decisive in ending many sieges and battles, but few military historians know that a war was started by a bee.[1] In 637 CE, a rancorous fellow named Congal, heir to the throne of Ulster, was paying a state visit to the king of Ireland and his family. Domnall, the Irish king, was a gracious host, except for one small oversight: he failed to put adequate distance between his beehives and his guests. As fate would have it, Congal was stung in the eye by an errant bee.

If it had just been painful perhaps the offense could have been forgiven, but Congal was blinded. To add insult to injury, he was given the nickname "Caech," meaning "one-eye," after the incident. Enraged by their patriarch's being blinded and called a cyclops, Congal's kin demanded retribution. In keeping with the notion of "an eye for an eye," they demanded that one of Domnall's sons be blinded. Domnall was not keen on this solution and tried a less drastic punishment in hope of appeasing Congal. The king ordered the destruction of the entire colony of bees to ensure that the guilty insect would be killed.

Domnall apparently didn't know (or at least was hoping that Congal was entomologically unwitting) that the offending insect had eviscerated herself in the course of leaving her stinger in his guest's eye. Not placated by the legalistic ruse of executing bees, the Ulstermen decided to settle their grievance on the battlefield. Unfortunately for them, their capacity for righteous indignation surpassed their military aptitude and they were summarily defeated.

We might suppose that such a minuscule puncture as that caused by a single bee ought not to cause such egregious medical—or political—damage. After all, the stinger of a bee is a phenomenally fine needle, about 2 millimeters long and the diameter of a human hair. Of course, the trauma of a bee sting comes not from being pierced but from the chemicals released into the wound. When the stinger is torn from the bee's body, the attached tissues include a venom sac that pumps out a witch's brew of chemicals that digest cell membranes, induce inflammation, affect heartbeat and respiration, and inflict pain. There's even a dastardly component with an odor reminiscent of bananas that functions as a chemical alarm, signaling other bees to attack. The bee is effectively a missile that delivers a chemical warhead. And the warhead—in this case positioned at the back, rather than the front, of the missile—comprises a poison-filled reservoir and a lancet to deliver the payload to the target. Once insects are viewed as sources of toxins, the opportunities to exploit them in warfare expand dramatically.[2]

The earliest use of insect toxins was their integration with spears, arrows, and darts—manmade "stings" combined with the insect-made poisons. The San bushmen of Africa rely on gathering native plants for food, but hunting is an integral part of their culture and provides an important source of protein. The challenge is that they have lightweight arrows and heavyweight prey, such as antelopes, wildebeests, and giraffes. So various tribes use different poisons to enhance the efficacy of their arrows. The venoms of snakes and spiders, along with the toxins of plants, are often used, but in the northern Kalahari the most common source of toxin is a beetle.[3]

Digging as deep as three feet among the roots of the shrubby corkwood tree, San seek creatures that they call *Ngwa* or *Kaa*—the mature larvae of a leaf beetle (family Chrysomelidae). The young larvae, which look like miniature burnt marshmallows, fatten themselves on corkwood leaves, then drop to the ground, burrow into the soil, and construct a sandy mantle in which they can survive for years. This living encrustation is the prize that the bushmen seek. When conditions are favorable, the larva pupates and matures into a mundane brown beetle that crawls out of the soil and into the corkwood tree to mate, lay eggs, and begin the next generation.

The San prepare their arrow poison in a variety of ways. Some bushmen squeeze the freshly extracted larvae and pupae like tiny tubes of toothpaste while others opt to dry the insects in the sun and pulverize them into a lethal dust. In either case, the gooey or desiccated entrails are applied to the arrow shaft to ensure that if the hunter is accidentally pricked, he does not poison himself (see Figure 2.1). And this seems highly advisable, given the lethality of the weapon. Small game, such as porcupines or birds, die of cardiac arrest in minutes. Antelope or giraffe succumb in a matter of hours or days, depending on the number and location of arrows in the animal. Known as exceptional trackers, the bushmen can follow their weakening quarry over miles of desert scrub.

Figure 2.1. A San hunter squeezing the innards of a leaf beetle larva onto an arrow. The insect, *Diamphidia vittatipennis*, is excavated from among the roots of the corkwood tree. The poison within the larvae—which look like miniature burnt marshmallows (see foreground)—is potent enough to kill a small mammal in minutes and larger animals, such as antelope, within hours. (Photo courtesy of H. Robertson, Iziko Museums of Cape Town)

For the most part, the San were remarkably peaceful until the arrival of the Dutch and French in the 17th century. Various accounts tell of native people using poisoned projectiles in warfare, and the beetle-tainted arrows of the San probably played a role in these battles. But Europeans had little basis for crying foul, given their own history of using insect-derived toxins in warfare.

Although Claudius Aelianus probably did not witness the defeat of Severus's army by Hatrian scorpions, he was curious as to the nature of venom and its potential as a military weapon. The Roman naturalist thought that the flow of poisons through the living world was a process originating from the realm of the gods, and he imagined that clever humans could tap into the flow.[4] Aelianus believed that Hercules cadged the idea of poisoned arrows from wasps, who were understood to acquire their venom in a most interesting way. As anyone with a backyard grill can attest, wasps are attracted to meat, and the Romans observed these insects buzzing around dead animals, including the corpses of vipers. Aelianus believed that the wasps drew venom from the dead snakes, which had consumed poisonous plants, which had tapped into the original source of poisons—the noxious vapors at the entrance to the Underworld.

Although venomous snakes aren't herbivorous, the ancient notion of poisons being relayed through a food chain isn't entirely absurd. Some plants acquire toxic elements from the soil, such as the poisonvetches and goldenweeds that concentrate selenium in their tissues. Likewise, insects often acquire their toxins from feeding on poisonous plants—the monarch butterfly larva obtains its toxins from milkweed, for example. And gods and humans can apply the same principle.

Aelianus contended that "Hercules dipped his arrow in the venom of Hydra, just as wasps dip and sharpen their sting." So compelling was this account that Greek mythology provides the term that is still applied those who devoted to archery: toxophilus (*toxo* refers to both "poison" and "bow," while *philus* refers to "love"). But applying wasp venom to arrows was not practical for Roman warriors. Various plant poisons were tried, but they lacked potency. There was, however, a mysterious substance that captured the military's imagination.

In the fifth century BCE, a Greek physician named Ctesias described a remarkable poison from the mountains of India.[5] He had no idea whether a mineral, plant, or animal was the origin of the poison. But so potent was this substance—the smallest droplet could kill a man—that 700 years later Aelianus still marveled at the enigmatic extract. The intervening centuries had provided time for some further details to emerge. The poison—one of the most expensive gifts given by the king of India—was extremely rare. It was a treasured constituent of the royal pharmacy, ideal for suicide or assassination. If only the Romans could acquire their own supply of the toxin, their archers would be nearly invincible. If only they could find the dikairon, a tiny orange bird whose droppings were said to be the source of the legendary poison.

The Romans never found the bird, but neither did anyone else. Well into modern times naturalists speculated as to the identity of the dikairon, which the ancients described as being about the size of a large grape. Some argued that such a tiny creature could not have been a bird at all; rather, dung beetles were about the right size and some had orange markings. Moreover, "droppings" might have been a mistranslation of the word for secretions. The only problem was that dung beetles don't secrete toxins.

The solution to the poison puzzle was finally revealed in the 20th century, when scientists realized that *Paederus*, a genus of rove beetles (family Staphylinidae), matched the description of the dikairon (see Figure 2.2). These

Figure 2.2. A representative specimen of the genus *Paederus*, a type of rove beetle that harbors symbiotic bacteria producing a chemical called pederin. This was likely the legendary poison of the dikairon, a mysterious creature from India that the Romans believed to be a tiny bird. Although they were wrong about the source of the poison, pederin is more powerful than the venom of the black widow spider, which is itself 15 times more potent than cobra venom. (Photo by Tom Murray)

black-and-orange beetles are found in northern India, where most species live in damp forest litter. Some species of *Paederus* are capable of flying, although at ½ inch in length they could hardly be mistaken for birds and they are rather smaller than grapes. However, rove beetles have been associated with bird nests. Misunderstandings were common in ancient communications, and perhaps the nature of this creature changed as stories moved westward from India. Heavy rains can trigger mass emergences of the beetles, and in 1966 some 4,000 people in Okinawa were given painful lessons by *Paederus fuscipes* (most likely the dikairon).[6]

The Okinawans rediscovered what Asian cultures knew in earlier times. In the eighth century, Chinese physicians used the skin-blistering secretions of *Paederus* to remove tattoos and kill ringworm. Today, the people of India refer to a condition called "spider lick," a string of suppurating sores caused when someone brushes away a beetle and inadvertently smears the insect—and its toxin—across the skin. Less than a hundred-thousandth of a gram of this chemical, pederin, can cause festering lesions. Although painful and potentially disfiguring, these encounters are not lethal.

However, ingestion of the beetle leads to severe and deadly internal damage, and pederin is lethal if injected into the bloodstream. Indeed, this chemical is more powerful than the venom of the black widow spider, which is itself 15 times more potent than cobra venom. Pederin has recently been discovered to be a product not of the beetle itself but of symbiotic bacteria living within the insect[7]—Aelianus's theory of insect poisons being acquired from ever "lower" organisms was not so far off. But even Aelianus could not have guessed that the poison might turn out to have beneficial effects; as little as one billionth of a gram of pederin inhibits the growth of malignant cells. But not all modern interest in *Paederus* has been benevolent.

In 2002, the Indian Defense Ministry funded a study of ancient texts to identify natural substances as chemical weapons.[8] The toxin of *Paederus* beetles was among the poisons of interest, as were other insect-derived ingredients, including a potion of fireflies (family Lampyridae) and wild boar's eyes purported to bestow night vision. Although none of these chemicals or concoctions has become standard issue, the Indian military evidently values the lessons of history. Had previous generals been attentive to historical tactics—including the ability of a cunning enemy to exploit the poisonous potential of insects—thousands of soldiers might have lived to fight another day.

The use of toxins in waging war extends back to at least 600 BCE, when the Athenians uprooted cartloads of poisonous hellebore from the countryside around the besieged city of Kirrha and dumped the plants into the river that supplied the city.[9] The contaminated water induced violent diarrhea, providing the Athenians with the opportunity to overrun the city and put a lethal, but perhaps merciful, end to their enemies' intestinal agony.

The Greeks and Romans had a botanical arsenal at their disposal, including extracts of belladonna, hemlock, monkshood, and yew berries. In addition, they were well acquainted with rhododendron, a shrubby tree possessing gorgeous pink and white flowers—along with neurotoxic sap and nectar. This plant flourished throughout the Mediterranean, around the Black Sea, and into Asia, where its poisonous properties were widely known. Although the sap was used as an arrow poison, killing one enemy at a time is a laborious way to secure victory. For the crafty military mind, an intriguing property of rhododendron gave it the potential to become a weapon of mass destruction.

Although it was poisonous to humans, the nectar was harmless to bees. And this attribute made for a honey of a weapon.

At the turn of the fifth century BCE, a Greek army returning from Persia camped in the territory of Colchis, a region of modern Turkey.[10] Their commander, Xenophon, noted in his journal that "there is nothing remarkable about the place, except for the extraordinary numbers of swarming bees." He was blissfully unaware that his men had intruded on the homeland of Medea—a mythical sorceress known for her powerful poisons. In short order, the Greeks raided the wild bees and satiated themselves on the golden booty, upon which the troops "succumbed to a strange affliction . . . as though under a spell." The men staggered about as if drunk, and according to Greek historians they "lost their senses and were seized with vomiting and purging." Within hours, thousands of men were sprawled on the ground, completely debilitated.

Most survived and began to recover the next day, but many could not stand unaided for three or four days. Xenophon was worried that his army was vulnerable to attack. Upset with Greek pillaging and armed with farm implements, even disorganized villagers could easily slaughter near-comatose soldiers. So he ordered his pitifully weak troops into formation and began a shuffling march back to Greece. Xenophon never suspected that six-legged sorceresses had conjured the potion that debilitated his army.

By the first century CE, the Romans had figured out the nature of what they called "mad honey." Pliny the Elder surmised that the bees acquired toxins from poisonous plants, such as rhododendron. He even proposed that the bees did so to protect their larder, and modern entomologists have confirmed that the toxic honey deters vertebrate raiders. Of course, the people of the Black Sea region were well aware of this syrupy scourge, but they also had discovered that nonlethal intoxication could be a recreational experience.

Alcohol is deadly at high doses, but humans find sublethal inebriation to be quite enjoyable. In modest doses, rhododendron honey induces a pleasant buzz.[11] Called *deli bal* in Turkey and the Caucuses, "mad honey" was exported to Europe in the 18th century, with up to 25 tons being shipped each year. Reputable folks found that a small spoonful in a glass of milk made an effective pick-me-up, and tavern keepers discovered that their patrons enjoyed the extra kick that a dollop provided to a pint of ale. In America, the early settlers discovered that honey made from mountain laurel, another toxic plant, added a new twist to their evenings. Nor is our association with toxic honey entirely a thing of the past. In 1992 a Virginia man was hospitalized after consuming several spoonfuls of honey made from the nectar of mountain laurel.

As both ancient and modern lessons reveal, honey can be dangerous stuff. However, the near-disasters of Xenophon's men and the Virginian fellow were the result of insect ambushes—nobody tricked these people into consuming the honey. But between these incidents, the story of mad honey took a nefarious turn. Having trounced the Greeks, the Romans marched confidently and arrogantly into power. They had little use for their Grecian predecessors, but the Roman Empire would regret not having learned from Xenophon's experience.

A *mithridate* is a universal antidote. Such a master key to toxicology remains in the realm of legend, but the etymology of the term is grounded in history.[12] Mithridates Eupator VI was born in 132 BCE, in the kingdom of Pontus on the southeastern coast of the Black Sea. The poor lad had a spectacularly dysfunctional family. When Mithridates was 12, his mother ascended the throne in his name after killing his father. Shortly thereafter, his supposed guardians tried to kill him, but the boy escaped and hid in the countryside until he was strong enough to retaliate. At 21, Mithridates returned to the city of his birth, imprisoned his mother, claimed the throne, poisoned his younger brother, and married his sister—not the makings of a mentally healthy future.

Although Pontus was a fertile and thriving land, the young king was not satisfied. He ventured into the surrounding districts to scout for expansion opportunities. Meanwhile his mother and sister-wife conspired to take back the throne, forcing Mithridates to poison the women upon his return. For the next decade Mithridates busied himself with conquering neighboring kingdoms, until his growing power caught the attention of the Romans. They maneuvered to replace one of Mithridates' puppet rulers in neighboring Bithynia. With the backing of Rome, the new king attacked Pontus. When Mithridates sent a messenger asking the empire to call off the Bithynians, the Romans—who were itching for an excuse to attack Mithridates—interpreted the appeal as a declaration of war. And so the first Mithridatic War was on. The Romans, however, soon discovered that their opponent was not some half-baked tyrant.

Few enemies struck fear into the heart of the Roman Empire, but Mithridates did. After having a Roman legate arrested for bribe-taking, Mithridates executed him by pouring molten gold down his throat. And Mithridates was capable of scaling up his savagery, slaughtering 80,000 Roman citizens in an outlying province in single day. As the Pontic forces advanced toward a fearful Italy, the Roman army received a firsthand lesson in Mithridates' brutality when chariots with rotating scythes attached to the wheels sliced through the lines. The Roman legions had enough survivors to continue the fight, but they fled the battlefield in horror upon seeing the grisly spectacle of their comrades "chopped in halves but still breathing, and others mangled and cut to pieces."

The legionnaires, however, had their pride and did not retreat all the way to Rome. Once they recovered from their "shock and awe," the Romans turned against their depraved enemy. From 74 to 66 BCE, Lucinius Lucullus and his army pursued Mithridates and his troops, driving them back toward the east. Having cornered his enemy near the Black Sea, Lucullus had only to strangle Mithridates' army by laying siege to Eupatoria. Anxious for victory, Roman sappers bored beneath the walls of the city. Mithridates' men drilled holes into the tunnels and released swarms of bees and—somewhat less plausibly, despite the account provided by the historian Appian of Alexandria—rampaging wild bears. With the tide of battle turned by bees, along with poisoned arrows and burning tar for good measure, the Pontic forces routed the Romans. Had Lucullus seen the bees as a premonition of Mithridates' potential for exploiting insects as implements of warfare, he might have warned his successor.

The next campaign against Mithridates was led by Cnaeus Pompeius Magnus, the rival of Julius Caesar. More generally known as Pompey the Great, this battle-tested Roman general finally vanquished the Pontic army in 65 BCE. But the victory was tainted. In the chaos of battle, Mithridates slipped away, escaping over the Caucasus. In Colchis, the fallen monarch began plotting his revenge and return to power. Meanwhile, Pompey was not satisfied with having crushed his counterpart's army. True victory meant finding and killing the mastermind. Having tracked Mithridates into Colchis, Pompey was oblivious to the potential for entomological skullduggery. He had not learned of the experiences of Xenophon or Lucullus. And what you don't know can kill you.

As Pompey's army advanced, a sweet snare was set. As a student of toxicology, Mithridates was keenly aware of the poisonous potential of *deli bal*—a specialty of the Heptakometes, a barbarous people that he'd befriended. After his allies gathered masses of honey from the local rhododendron forests, the cunning Mithridates directed them to place jars of the insidious syrup either along a narrow mountain path on Pompey's route or more surreptitiously in what appeared to be a hastily abandoned cache. Historians differ on this tactical detail, but what came next is clear from all accounts.

The Roman legions were not especially well compensated by the empire. Allowing them to profit from raiding and looting provided incentive for soldiers to engage the enemy and kept the military budget within bounds. The jars of honey were found by Pompey's men, who eagerly took the bait. His troops were soon reeling, babbling, and vomiting. As they collapsed, the Heptakometes rushed in and put them out of their misery. Mithridates had secured a sweet revenge for his humiliating defeat at the hands of Pompey's army.[13]

Having poisoned a thousand Roman soldiers (along with his mother, brother, four sons, and various political opponents), Mithridates retired to a castle in Crimea. Obsessed with the possibility of being poisoned himself, he'd concocted what he thought was a universal antidote: a *mithridate*. The concoction was composed of 54 ingredients, in a base of—what else?—honey (most likely *deli bal*). In the end, he didn't need the antidote. The one son whom he hadn't gotten around to killing led a revolt against Mithridates, who, rather than facing death at the hands of the rebels, ordered his bodyguard to run him through with a sword.

As for the Romans, they eventually subjugated the region and imposed a war tax to pay the costs of having subdued the people. But the Romans were not to be fooled again. Having learned the hard way what Xenophon and Pompey might have taught them, the Romans prohibited the payment of tribute to the empire in the form of honey.

There is something particularly dishonorable about enticing an enemy with sweets, debilitating him with toxins, and then hacking him to death. In this regard, Mithridates might be crowned the champion of entomological depravity, if it were not for a ruler who conceived of insects not merely as weapons of war but also as instruments of torture.

3

INSECTS AS TOOLS OF TORTURE

The ancient Persians were perhaps the earliest people to use insects as torture devices. The gruesome practice of subjecting a condemned man to "the boats" was given the technical term *scaphism* (based on the Greek *skaphe*, from which we get the word "skiff," meaning a small, flat-bottomed boat).[1] The victim was initially force-fed milk and honey to induce severe diarrhea. Then the poor soul was stripped, lashed to a skiff (or hollowed-out tree trunk) so that his head, hands, and feet protruded over the sides, smeared with honey, and set adrift on a stagnant pond or simply left in the sun. Wasps attracted to the honey delivered excruciating stings, but the coup de grâce came with the insects drawn to the feces accumulating in the boat. Flies would breed in the filth and then begin laying eggs in the victim's anus and increasingly gangrenous flesh. Although the misery could be prolonged by providing the victim with continuing allotments of milk and honey, the condemned would eventually succumb to septic shock associated with being infested with maggots.

Other Asiatic cultures employed insects for torture without such elaborate preparation. Centuries ago, Siberian tribes simply tied a condemned prisoner to a tree and let nature take its course. Forests in that part of the world support phenomenal densities of biting flies, and so mosquitoes (family Culicidae), black flies (family Simuliidae), biting midges, deer flies, and their kin ensured an excruciating ordeal until shock or dehydration provided a merciful ending. Recent studies from the Canadian arctic suggest that an unprotected person can receive as many as 9,000 bites per minute—a rate sufficient to drain half of the blood from a large man in about two hours.[2] In modern China, biting insects are still used as a form of punishment, although apparently not as a means of execution.[3]

Most recently, followers of the banned religious sect Falun Gong have been arrested and reportedly subjected to torture with mosquitoes:

In the summer time, especially around dusk or dawn, the police strip practitioners of every piece of their clothing and handcuff them to a pole in a
place where mosquitoes and other insects swarm. Practitioners are bitten
over and over again . . . sometimes they cannot even open their eyes because
their eyelids are so swollen.[4]

But powerful regimes have long exploited insects as instruments of torture.
According to Aleksandr Solzhenitsyn, Soviet gulag prisoners were tormented
by being locked into closets swarming with hungry bed bugs.[5]

The use of arthropods to inflict agony was not unique to the Old World.
The Apache Indians of North America apparently used ants to ensure lingering, painful death.[6] Beyond the Hollywood westerns, there are several credible
reports from the late 1800s of Apaches staking captives over anthills. The victims either had honey smeared on their eyes and lips or had their mouths held
open with sharpened skewers. Typically, the tales came from white settlers who
found dead bodies and surmised the details of their hellish final hours. However,
in at least one case an Apache reported: "Old Eskimi[n]zin says he buried an
American alive in the ground once and let the ants eat his head off." The extent
to which Apaches used ants for torture is not clear, but Indian tactics probably
became increasingly vicious in response to continuing abuse at the hands of the
Spanish and Mexicans who practiced torture, albeit without insects.

In light of cultural bias, we might doubt some reports of Indian torture,
but an anthropologist working for the U.S. Bureau of Ethnology provided a
particularly compelling account of how one tribe used ants to inflict pain.[7]
While living with these Indians, Frank Hamilton Cushing earned his acceptance into an order of Zuñi priests through a series of arduous trials, including
the following: "Still fasting, bareheaded, and stripped nearly to the skin, I was
set at sunrise on a large ant-hill of the red fire ants of the Southwest, so named
because of their bites, and there all day long I had to sit, motionless, speech-
less, save to priests in reply to instructions."

If gruesome martyrdom is sufficient qualification for canonization, then Charles
Stoddart and Arthur Conolly are surely the patron saints of insectan torture.[8]
These two Englishmen became pawns in the Great Game, a political chess
match played out between two empires. The contest opened in 1837, when
Queen Victoria ascended to the throne and Britain began to establish a stra-
tegic presence in India. The opening moves from the Russians were made by
Tsar Nicholas Pavlovich I, who sought to expand his empire to the south. The

British feared that the tsar had designs on India or would at least impede their own colonial plans by controlling Central Asia. The Great Game was on.

Both sides realized that the region that we now refer to as the "Stans" (Kazakhstan, Kyrgyzstan, Tajikistan, Turkmenistan, and Uzbekistan) was vital to their imperial schemes. Each monarch dispatched agents to convince the local rulers that benevolent occupation was necessary to keep out the British brutes or Russian reprobates, depending on who was doing the talking. The Russians were based in Orenburg, just north of the present-day border with Kazakhstan, and the British operated from northern India. If one drew a straight line from Orenburg to Delhi, the midpoint was the walled city of Bukhara—the place where Stoddart and Conolly would meet their fate.

Bukhara was viewed as a sinister place well before its entomological abominations came to light. No European had set foot in this forbidding and strategic stronghold for a hundred years. In a desolate landscape bloodied by marauding tribes, Bukhara was a cultural oasis endowed with palaces, mosques, and bazaars (see Figure 3.1). Such an important political and economic center could not long remain a mystery to the empires who sought control of Asia.

Russia sent a diplomatic mission to Bukhara in 1820, to recover their imprisoned countrymen and foster diplomatic ties. Britain countered by

Figure 3.1. At the entrance to Bukhara, little has changed since the arrival of Lieutenant Colonel Charles Stoddart in December 1838. The Englishman was charged with ingratiating himself—and his nation's geopolitical interests—to the despotic emir in control of the strategic Central Asian city. Within hours of passing through the gates, Stoddart had offended the ruler and begun his terrifying journey into the infamous Bug Pit. (Photo courtesy of Galen Frysinger)

sending delegations in 1824 and 1832. British concerns were deepened after the Russians sent a second mission in 1835 and also began forging ties with Persian and Afghani rulers. In response, Sir John McNeill, the British ambassador to Persia, decided to send his own emissary to Bukhara to secure the release of Russian slaves and prisoners. There was no altruism involved; the British wanted to deny the Russians an excuse for invasion. In addition, the envoy was to offer his country's assistance in case of Russian invasion and assure the emir that he had nothing to fear from a British presence in his city.

Lieutenant Colonel Charles Stoddart was chosen for the job. McNeill figured that Stoddart's service in Persia and Afghanistan provided the necessary experience for dealing with an emir. The ambassador was tragically mistaken. As one of Stoddart's fellow officers remarked, "To attack or defend a fortress, no better man than Stoddart could be found; but for a diplomatic mission, requiring coolness and self-command, a man less adapted to the purpose could not readily have been met with." Nor did McNeill fully comprehend the nature of the emir, Nasrullah Bahadur-Khan. His official title—the Shadow of God Upon Earth—should have been a tip-off. Presumably, the moniker used by the emir's subjects was not known to the British ambassador. Behind closed doors, the citizens of Bukhara referred to their ruler as "the Butcher."

Nasrullah ascended the throne in 1826 through the expedient of murdering those who had prior claims, thus leapfrogging the corpses of his father and two older brothers. Concerned that turnabout might be fair play, he extended his homicidal streak to include his three younger brothers and several other relatives. An earlier British emissary described the extent of the emir's domination:

> In order to exemplify in the best manner the tyranny of the Ameer of Bokhara, I need only mention the following facts: That every letter sent from Bokhara, and every letter arriving for their merchants and dignitaries, and every private note which the wife writes to her husband, or the husband to the wife, must first be opened and perused by the King of Bokhara. . . . Another act of tyranny committed by the Ameer is that boys are [required] to report to him every word which other boys talk in the streets even brother to brother at home, and servants in families are also obliged to write down for the King any conversation they hear between husband and wife, even in bed; and the people set over me were ordered to report to him what I might happen to speak in a dream.[9]

Stoddart himself, in a letter smuggled back to England in 1839, concisely described Nasrullah: "The Ameer is <u>mad</u>." The path that Stoddart took to arrive at this conclusion began on December 17, 1838, with his arrival in Bukhara. Upon riding into the walled city, Stoddart proceeded to the main square in front of the palace to present himself. He was not aware that riding there was forbidden, nor did he know that, when the emir rode up, a visiting horseman was expected to dismount. Instead, in accordance with British military tradition, he remained in the saddle and saluted the emir. Whereupon, according to one source, the emir "looked at him fixedly for some time, and then passed on without saying a word."[10] An inauspicious start.

Stoddart gained an audience with the emir and handed the aggrieved ruler a letter of introduction. Upon seeing that the missive had not been signed by Queen Victoria herself, Nasrullah was deeply insulted—again. The diplomatic disaster was fully consummated when the emir was slipped a message by one of Stoddard's native servants. The poor Englishman was doubly betrayed; the letter was from the Emir of Herat, one of Stoddart's earlier acquaintances, who denounced the Brit as a dangerous spy. Nasrullah had a special place for such odious enemies. And that's how Charles Stoddart became the first westerner to experience the terror of the Bug Pit (see Figure 3.2).

Figure 3.2. The prison of Bukhara, where the emir maintained his entomological chamber of horrors, a 21-foot-deep pit covered with an iron grill and accessible only by a rope. The "Black Well," as the locals called it, would have been an awful place in its own right, but the brutal ruler stocked the pit with assassin bugs, the toxic saliva of which generated festering sores. Victims, according to the jailer, had "masses of their flesh . . . gnawed off their bones." (Photo courtesy of Galen Frysinger)

The local people referred to the infamous pit as "Si(y)ah Cha," which meant "Black Well" or "Black Hole." It was located in the prison compound behind the emir's palace, so that he had ready access to the chamber of horrors. Twenty-one feet deep, covered with an iron grill and accessible only by a rope, the pit would have been an awful place even without the creatures lurking in the depths. Nasrullah seeded the pit with rats and reptiles, rather standard fare for dungeons. The emir's pièce de résistance—the innovation for which he attained infamy—was the insects that he used to ensure a constant, torturous experience for his victims. The foulest of the emir's six-legged minions were the assassin bugs, although their eight-legged cousins, the sheep ticks (probably *Dermacentor marginatus*), added to the torment. When there were no unfortunate souls to feed to the cold-blooded menagerie, chunks of raw meat were dropped into the pit. The arthropods would not have found such fare to be particularly appealing, so we might suppose that the emir managed to find live victims on a regular basis.

Assassin bugs belong to the Reduviidae, a family of carnivorous insects (see Figure 3.3).[11] The proclivity of some species for cannibalism accounts for the common name of the group. These creatures range in size from 1/10 to nearly 2 inches and are endowed with a stout, curved beak for piercing their prey. Assassin bugs inject toxic saliva that paralyzes and kills other insects, along with enzymes that liquefy the innards of the prey, allowing the predator to suck it dry. A few

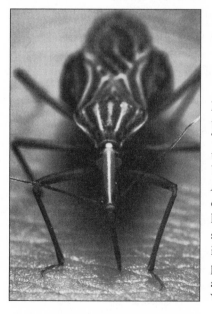

Figure 3.3. An assassin bug preparing to take a blood meal through human skin. These creatures were used to torture victims in the Bug Pit, developed by the Uzbek emir Nasrullah Bahadur-Khan. Although his official title was "the Shadow of God Upon Earth," behind closed doors his subjects called him "the Butcher." The species used by the emir to inflict agonizing pain on his enemies is not known; the pictured species is native to the Americas and transmits Chagas disease. (Photo by WHO/TDR/Stammers)

assassin bugs feed on mammals, but the bite of these insects—also known as kissing bugs—is not usually painful. Stealth makes sense when securing a meal from a creature thousands of times larger than you. Most likely, the emir used species that do not normally bite humans, but when starved will feed on any animal tissue. The bite of these insects has been compared to being pierced with a hot needle, and the digestive enzymes that they inject cause suppurating sores.

Covered in oozing ulcers, Stoddart was eventually released from Nasrullah's pit. The subsequent treatment of the British officer varied with the political climate. When the emir perceived that the British forces were weak, Stoddart was either imprisoned or, if Nasrullah was feeling particularly peeved, dumped back into the Bug Pit. During periods in which the British appeared to be a powerful presence in the region, the emir released Stoddart into the city to ponder his fate.

Stoddart was deeply devoted to serving God and country, but martyrdom was another matter. After being led to the edge of his own freshly dug grave and given two options, Stoddart chose to become Muslim rather than die. However, neither Stoddart nor Nasrullah could have guessed that this coerced conversion would soon bring another British officer to Bukhara.

In December 1839, the Russians sent 5,000 men, 22 cannon, and 10,000 camels on an expedition to Khiva, the capital of a key region to the northwest of Bukhara. The Russians, after losing half their men and thousands of camels to the cold, were forced to turn back. However, the British figured that such an audacious gambit by the Russians had to be countered.

Enter Captain Arthur Conolly of the Bengal Light Brigade in India, a man with a flair for adventure. Although the phrase "Great Game" was popularized in Rudyard Kipling's *Kim*, it was Conolly who coined this term to describe the British-Russian contest for control of the region. The captain had made a name for himself by ousting an Afghani monarch and installing a king friendlier to the British. So the British government sent Conolly into present-day Uzbekistan with orders to "establish . . . a correct impression of British policy and strength; reach amicable agreements with the rulers of Khiva and Kokand [before the Russians succeeded in this aim] . . . and, if circumstances permit, to return to Afghanistan by way of Bokhara."[12] It was this last element that gave Conolly a personal reason for the mission. For the captain was not only courageous but also very religious. He had received word that Stoddart had converted to Islam, and Conolly felt duty-bound to save Stoddart's soul. But first, he'd have to rescue his countryman's body.

Conolly had cordial, if not particularly successful, meetings with Khans of Khiva and Kokand. Both rulers warned the British officer not to go to

Bukhara, as they were on hostile terms with Nasrullah. But Conolly was unde-terred. Moreover, while in Kokand, he received a letter from Stoddart, whose ability to interpret a diplomatic situation had not improved much despite his experiences. "The favor of the Ameer," wrote Stoddart, "is increased in these days towards me. I believe you will be treated well here."[13] Stoddart had an incomprehensible lack of political judgment, but he can't be faulted for a shortage of optimism.

Conolly entered Bukhara on November 10, 1841. At first, Stoddart and Conolly were not treated badly, though Nasrullah doubted the latter's inten-tions. Things took a decided turn for the worse when the British suffered a ter-rible loss in Afghanistan and the monarch that Conolly had brought to power was overthrown. The emir surmised that the British were no longer a military threat. To add fuel to the fire, war broke out with Bukhara's neighbors, and Nasrullah blamed Conolly, who had just visited the emir's enemies, for instigat-ing the conflict. The final straw for the emir was learning that the governor-general of India, rather than Queen Victoria herself, had answered his recent diplomatic communiqué. Nasrullah directed his rage at Stoddart and Conolly.

The emir would have consigned the two officers to the Bug Pit, but another political enemy had prior reservations. Undeterred, Nasrullah found a way to use insects to torment his British prisoners. In a letter dated April 6, 1842, Connolly wrote, "This is the hundred and seventh day of our confinement without change of clothes, but the weather having become warmer, we can do without the garments that most harbored the vermin we found so distressing." A message from Stoddart in late May was the last direct communication that the outside world received from the lice-ridden prisoners.

In the fall, members of a Russian diplomatic mission returned to St. Petersburg from Bukhara. They had tried to get the emir to release the British officers, but Nasrullah said he would not free the Brits until the Queen sent an answer to his letters. On October 1, 1842, Nasrullah was a sent a letter plead-ing for the release of the prisoners, but it was not from the Queen and went unanswered. Then in November, a former Persian servant of Conolly reported that both men had been publicly beheaded.

While Stoddart and Conolly were officially listed as dead, their fate was not accepted with only a servant's word as evidence. A group of hopeful friends put up £500 to support an expedition. Joseph Wolff, an Anglican priest, was commissioned to ascertain definitively the fate of the British officers. Wolff fully understood the emir's temperament, so he was well prepared to meet "the Butcher." Having been told that a visitor should bow three times in the presence of the emir while saying in Arabic, "Peace be to the King," Wolff bowed and recited the phrase repeatedly until Nasrullah finally laughed and told him to stop.

The priest was allowed to meet with the emir's commander of the artillery, who was also Stoddart and Conolly's jailer and privy to the complete story of the men's fate. Wolff learned that the Russian delegation's explicit concern for the emir's prisoners inadvertently doomed the two men. Nasrullah concluded that the Brits had become a serious, political liability. So, as soon as the Russians left, Stoddard and Conolly were thrown into the Bug Pit. For two horrific months they were slowly eaten alive, until, in the jailer's own words, "masses of their flesh had been gnawed off their bones."[14] Finally, Nasrullah had the men taken from the pit and beheaded, righteously proclaiming that "strangling gives more pain, and the rascally Khan of Khiva strangles his people; and therefore, out of mercy, I command the heads of the evil-doers to be cut off with a common knife."

Victims of Nasrullah's dungeon were afflicted with physical pain and psychological suffering, as the insects both pierced a captive's flesh and crawled across his body in search of a meal. And as we recently learned, a deep-seated fear of insects can, in the hands of a skilled torturer, become a singular source of agony.

Documents released by President Barack Obama in April 2009 revealed that the CIA intended to use insects as part of an Orwellian strategy to extract information from Abu Zubaydah, an alleged high-ranking member of al-Qaeda.[15] A 2002 U.S. Justice Department memo condoned the plan, ruling that exploiting the captive's entomophobia (an irrational fear of insects) did not constitute torture based on the contention that a "reasonable" person would not fear bodily harm. This legal rationalization was vacuous, given that the point of the CIA tactic was to tap into an unreasonable, but profoundly powerful, psychological vulnerability of the prisoner.[16]

For thousands of years, insects have been co-opted into human torture. These were extrapolations of our most memorable entomological encounters—being bitten or stung. While the insect conscripts effectively delivered pain and fear to enemy targets, this capacity only scratched the surface of insects' potential as agents of war. The realization of their darkest powers would dawn more slowly but persisted into the modern era. For once we discovered insects' deadliest payloads, these creatures would utterly transform the history of warfare—and humanity.

TWO

VECTORS OF DEATH

This land is covered with a rank, dense, tangled growth of trees, reeds, grasses and water plants. . . . Its stagnant water is poisonous, [water] moccasins and malaria abound; flies and mosquitoes swarm. . . . Here was to be the home of our New Hampshire men, who had never sniffed malaria nor breathed miasma. Here for two months they were to dwell in the midst of alarms in "this horrible place," during the very hottest of the Southern year.

—Journal entry of a Union surgeon recounting the conditions that the Confederate forces defending Richmond forced the Army of the Potomac to endure in the summer of 1862

4

HORSESHOES AND HAND GRENADES

Humanity is fortunate that solving problems often does not require understanding how or why our solutions work. Being close is good enough. We used aspirin for centuries before medical science revealed its mode of action; the steam engine preceded the physicists' understanding of vacuum pressure by years; and bronze was invented centuries before chemists and metallurgists could explain its properties. Likewise, for thousands of years, weapons of war were lethal enigmas. The ballistic equations describing the trajectory of a spear or arrow, the mechanical principles underlying a catapult, and the chemistry of gunpowder were mysteries, but scientific ignorance did not preclude using and refining these weapons. And no form of killing more clearly exemplifies our capacity to exploit natural processes beyond the limits of our knowledge than does entomological warfare.

Humans have long understood that disease could be passed among individuals.[1] A clay tablet from the 17th century BCE suggests awareness that sick people were contagious and that quarantine could prevent the spread of disease. The ability of insects to transmit pathogens was vaguely suspected by the ancients. Various records allude to flies as having a magical role in illness. And for millennia, the cultures of the Mediterranean region drew connections among mosquitoes, marshes, and malaria. But they erred in concluding that the illness was caused by breathing the swamp vapors; hence, the term *mal* ("bad") *aria* ("air") for the disease.

Despite the biological blunder as to the origin of malaria, ancient civilizations made impressive progress in controlling the disease, and the work of Empedocles is a prime example.[2] This controversial character from the fifth century BCE was a gifted philosopher-poet-physician and something of a charlatan, claiming to revive the dead and control the weather. His powers over nature extended to eliminating the fevers that plagued his region of Sicily. The shady but shrewd Empedocles oversaw the draining of nearby swamps, and along with the bad air went the pesky mosquitoes, which were never associated with the success of the remedy.

Once people figured out that draining marshlands could diminish disease, military strategists latched onto the possibility of reversing public health. Forcing an opponent to occupy noxious habitats could yield victory. Fevers could be deadlier than swords. Not 50 years after Empedocles saved his region from malaria, the Athenian invaders of Sicily were decimated by this disease during their siege of Syracuse.[3] Scholars debate whether the Sicilians maneuvered their defensive line to force the Athenians to establish summer camps amid the marshes or drew their enemy into the wetlands through the ruse of negotiating surrender. In either case, mosquitoes and malaria conspired to break the siege. A century later, the Carthaginian army was wiped out by malaria in the same lethal landscape that had defeated the Athenians, demonstrating why military commanders would be well advised to study history.

The Macedonian strategist Polyaenus provided a detailed account of a particularly villainous, passive-aggressive use of insects.[4] In 306 BCE, a despot named Clearchus seized control of Heraclea, on the Black Sea. Despite being a student of Plato, Clearchus came away with an affinity for tyranny rather than philosophy. He surrounded himself with a circle of brutal henchmen who robbed and raped the locals. When the people appealed to their leader to control his lawless lackeys, Clearchus merely shrugged and offered the ancient version of "boys will be boys." The only recourse that he could offer was for the citizens to build him a walled acropolis to restrain his dishonor guard.

The Heracleans complied, but Clearchus's goons continued their felonious ways. Fearing insurrection among his increasingly fractious subjects, Clearchus arrested Heraclea's democratic council on trumped-up charges. This was a fine first step, but he needed to rid himself of dissidents within the general populace. Using his own men to conduct a political cleansing would surely feed the flames of revolt, so he turned to nature.

Clearchus drafted an army—including his most likely opponents from the citizenry—for a campaign against the city of Astachus. He led the men into western Turkey during the hottest days of summer, but the invasion was a ruse. His purpose was to station his recruits in the marshy lands outside of the city walls. Clearchus and his personal guard established camp on the high ground, leaving the troops to occupy the low-lying areas. From this vantage he had shady trees, fresh breezes—and a fine view down to his condemned army.

Fevers soon swept through the camp, and by the end of the summer his army of potential insurgents was dead or morbidly ill. Clearchus returned to Heraclea, his political victory cleverly masked by a military defeat. His subjects never figured out the depraved scheme of their ruler, and the tyrant never knew that he owed his reign to a legion of disease-carrying insects.

While maneuvering opposing forces—or in the twisted case of Clearchus, one's own troops—into contact with mysteriously but predictably unhealthy landscapes was a viable strategy, this approach to entomological warfare only worked if suitably soggy conditions were available. However, even without biologically hostile environments, insect vectors could be conscripted by savvy commanders without knowing the details of epidemiology. An imprecise concept of contagion was sufficient to deploy entomological weaponry.

The earliest use of germ warfare might have involved history's most infamous pathogen, *Yersinia pestis*, and the world's most legendary booty, the Ark of the Covenant. The historian Adrienne Mayor has suggested that this sacred chest might have been guarded by plague-infected fleas.[5] The story begins with the Philistines' capturing the Ark from the Israelites at the battle of Ebenezer in 1050 BCE. Fearing that Yahweh would be angry with them for trouncing his chosen people, the Philistines hid the booty in their capital, Ashdod. But it's hard to fool omniscient deities: "After they had moved it [the Ark], the Lord's hand was against that city, throwing it into a great panic. He afflicted the people of the city, both young and old, with an outbreak of tumors in the groin" (1 Samuel 5:9). These swellings were probably buboes: bacteria-laden lymph nodes that are a classic symptom of bubonic plague, or the Black Death. Some biblical interpretations refer to these swellings as "hemorrhoids," which is almost certainly a mistranslation and raises the question of whether an outbreak of inflamed rectal tissue would be a suitable divine punishment.

The citizens of Ashdod desperately wanted to distance themselves from the Ark and the associated plague. So they passed the chest on to Gath, where the people also came down with the horrible swellings—and they passed it on to Ekron, where the disease struck again. Seeing the pestilential pattern, the Philistine priests ordered that the Ark be sent back to the Israelites, along with a load of golden statuary to appease Yahweh. With the return of the Ark, the plagues ended and the Philistines learned not to mess with the Israelites and their god. But what had happened in epidemiological terms—was there an entomological booby trap hidden within the Ark?

The Philistines who brought the Ark home from battle might have brought the plague with them, and those who escorted the chest from city to city could have spread the disease. Mayor proposes another explanation. The Philistines knew that peeking inside the Ark was absolutely forbidden, but perhaps the temptation was irresistible. So Mayor wonders, "Does the story of the Ark suggest that the chest might have contained some object, such as cloth, that harbored aerosolized plague germs, or an insect vector that infected the rodents in Philistine territory?" The microbe that causes bubonic plague is transmitted to humans by infected fleas. While there is a pneumonic form of the disease spread by coughing and sneezing, this variant is not associated with the buboes described in the book of Samuel.

The Israelites who set the pathogenic trap might have surmised that fleas from a sick person would ambush Ark peepers, but such an understanding of insect vectors is improbable. More likely, the Ark's protectors armed the chest with a plague victim's clothing, which was thought to be the source of contagion, while actually harboring stowaway fleas. Bacteria cannot survive for long in the environment, but a flea can live for months—the record being 513 days—between blood meals.[6] A few of these insects riding along in the Ark would have made a very hungry and lethal surprise for anyone who tried to access the chest's sacred contents.

An insect-charged booby trap might be viable when protecting a discrete, valuable resource such as a wooden chest, but it is hardly feasible in the course of typical warfare. In normal military conflicts the would-be entomological warrior must have some means of imposing the infected insects on the enemy. But given their lack of scientific knowledge, ancient armies would have had to stumble upon a tactic that fortuitously delivered the insects in the course of pursuing another objective. Such a coincidence of military tactics and entomological luck would seem incredible, but this unlikely convergence triggered an event more than six centuries ago that serves to warn the modern world of the horrific potential of biological warfare.

The conceptual roots of the apocalyptic conflict can be traced to the castle at Thun l'Eveque in Northern France.[7] This was a gorgeous setting in which to conduct a gory experiment in military tactics. An English king, Edward III, laid claim to the French crown, which seems to be a great way of starting a fight—in this case, the Hundred Years War. To follow up his claim, Edward led an expeditionary force to the continent and took Thun l'Eveque castle. He left behind enough soldiers to defend the stronghold, but the troops soon grew restless. For want of entertainment, they took to harassing the residents in the nearby city of Cambrai. The residents were none too happy with the unprovoked raids and sought assistance from the Duke of Normandy. The French duke's forces laid siege to the castle and used "dyverse great engyns" (catapults and launching devices) for heaving rocks and whatnot into the English garrison. In this case, the nature of the "whatnot" is the key to the historical importance of the battle.

Firsthand accounts during the onslaught revealed that, along with boulders, they "cast in deed horses, and beests stynking, wherby they within had great[er] dystresse thane with any other thynge."[8] The motive behind heaving rotting carcasses has been roundly debated. Some scholars suggest that the offal was used to fill in during shortages of geologic ammunition. This explanation seems a bit implausible—were dead horses really more plentiful than rocks? Others propose that the Normans were just disposing of bodies, which also seems peculiar. After all, a dead horse makes a much better meal than it does a projectile—at least before it becomes a "stynking beest." The most compelling theory is that the carcasses were intended to induce illness, and not just the nausea evoked as rotten flesh plopped into the streets. Disease was thought to be promulgated from the stench associated with decomposition (recall that "bad air" was blamed for malaria). In any case, the defenders abandoned the castle after a relief force from England failed to dislodge the duke's troops.

The launching of dead livestock was not a common wartime practice, in substantial part because there were few catapults capable of heaving a thousand pounds of meat, and cutting a decomposing horse into smaller chunks was presumably beyond the call of duty. However, there were some huge trebuchets, the successors of the catapults. These later devices worked by dropping a counterweight—as much as 50 tons—that was attached to the short end of a lever-arm, with the payload placed in a sling at the other end. Such "great engyns" were uncommon, but a more modest trebuchet could launch a sizeable carcass. While decomposing animals were nasty enough, a far more potent package was sometimes available.

The most devastating act of biological warfare in human history relied on sending insect-infested payloads into a besieged city. As with other uses of vectors during this period of history, the assailants were unaware of the instrumental role that the insects played in the success of their military innovation. Nor could the commander who ordered the assault ever have imagined that his entomological weapon of mass destruction would take 200 times more lives than the atomic bomb dropped on Hiroshima six centuries later.

Kaffa was a Genoese seaport, strategically positioned on the Crimean peninsula of the Black Sea.[9] The city was a thriving hub, linking the maritime commerce of the Mediterranean, the overland caravans of the Far East, and the river trade extending up the Don and Volga into Moscow. After the Mongols razed the city in 1308, it was rebuilt and fortified with two concentric walls. Thirty-five years later, the Tartar army again descended on the region and captured Tana, a trading center east of Kaffa. The Venetian merchants fled to the neighboring stronghold with the Mongols hot on their heels. The army was led by Janibeg, the last Khan of the Golden Horde, that portion of the Mongol empire established by Ghengis Khan's grandson, Batu Khan.

The ensuing siege devolved into a brutal stalemate. Janibeg's forces were relentless, but Kaffa's walls proved impregnable—and the Genoese maritime hegemony ensured that the city could be provisioned. After the conflict had been deadlocked for three long years, a conspiracy of fleas, rats, and bacteria tipped the balance in favor of Kaffa. Or so it appeared at first. The Mongol camp was devastated by an outbreak of bubonic plague. The disease most likely came along with flea-infested rats—stowaways in supplies from the Eurasian steppe.

When the khan saw his army melting away in agony, he had no choice but to break the siege. Janibeg sealed his infamous place in history with one final order. Less than a decade after the Normans had catapulted dead horses, the Khan of the Golden Horde upped the ante of biological warfare. An eyewitness, Gabrielem de Mussis, reported the maneuver that changed the course of western history:

> The Tartars, fatigued by such a plague and pestiferous disease, stupefied and amazed, observing themselves dying without hope of health, ordered [human] cadavers placed on their hurling machines and thrown into the city of Caffa, so that by means of these intolerable passengers the defenders died widely. Thus there were projected mountains of dead, nor could the Christians hide or flee, or be freed from such disaster.[10]

The spread of plague from the Mongol camp is unquestioned, but exactly how the disease made its way into the Genoese city is a matter of lively debate.[11] Some scholars contend that flying corpses were not the vectors; instead, they propose that fleas were exchanged among rats in the no-man's-land between the opposing sides. However, fleas do not typically hopscotch between rats. Moreover, the gap between populations would have worked against such mixing. Given 14th-century siegecraft, the distance from the Mongol camp to the city walls was probably at least half a mile, with the front line as close as 200 to 300 yards—within the range of a trebuchet but beyond the range of Kaffa's archers. The typical home range for a rat is only 50 yards, far short of the distance between the human settlements. So exactly what component of the hideous payloads triggered the epidemic?

The infected corpses were a potential source of direct transmission, as the pathogen can be acquired by contact with infected tissue. Bacteria-laden buboes would presumably rupture on impact, providing a rich source of contagion. In the course of clearing these gruesome remains from the streets and rooftops, the residents might have become infected via cuts on their hands. A few people could have acquired the disease in this manner, but epidemiology is a numbers game. Whether the frequency of direct infection would have been sufficient to spark a full-fledged epidemic is doubtful. Plague has a far more efficient means of infection.

The bacteria, *Yersinia pestis*, that cause bubonic plague are transmitted by fleas in an elegantly evolved process.[12] The pathogen is acquired when the insect feeds on an infected mammal. Some 30 species of fleas and more than 200 species of rodents can harbor the bacteria. Rats are most relevant to human outbreaks, with the rat flea, *Xenopsylla cheopis,* and the human flea, *Pulex irritans*, being the most common one-two punch. The bacilli multiply in the stomach of the flea and create the perfect conditions for their own transmission. The microbes form a gelatinous mass in the upper digestive system of the flea. So when the blood-thirsty insect bites a host, rather than being able to suck a liquid meal into its stomach, the insect regurgitates bacteria into the bloodstream of the host. This blockage means that the hungry flea is condemned to repeatedly seek out and bite hosts in futile attempts to feed—while transmitting the pathogen to another victim in the course of each frustrated meal.

Within three or four days of being bitten, a human victim is wracked with fever and excruciating headaches. Then, the lymph glands swell to painful proportions, forming the infamous buboes in the armpits and groin. Conditions deteriorate as the bacteria spread to the blood stream, liver, and spleen. In short order, delirium mercifully gives way to a coma, with seizures and bleeding from various orifices portending the end. In some cases, a secondary infection occurs in the lungs. From here, a pulmonary form of the disease develops, and "pneumonic plague" can then be transmitted by infective droplets. However, we can be quite sure that the bodies being hurled into Kaffa were in no condition for sneezing or coughing.

The only trouble with holding the fleas solely responsible for the epidemic in Kaffa is that these insects don't hang around long after their host's death. However, we can posit that a portion of the catapulted bodies were reasonably fresh, and the insects don't bail out immediately—even a determined flea can't move very quickly within the folds of a person's clothing. Given the efficiency of pathogen transmission, not many fleas would have been needed to spread disease among the city's rats, dogs, cats—and humans. The Mongol khan clearly meant to spread the suffering of his troops into the streets of Kaffa, but what came next was surely unimaginable even to the most vengeful martial mind.

The Genoese evacuated in a desperate attempt to escape the epidemic. Healthy and sick people, along with the usual complement of stowaway rats and their attendant fleas, headed out to sea. Within a year, the ships from Kaffa had sailed across the Black Sea to Constantinople, then along the Mediterranean to the islands of Sicily, Sardinia, and Corsica, and finally to the cities of Genoa, Venice, and Marseilles. At each port, infected people, rats, and fleas disembarked, seeding southern Europe with plague. Disease spread through Italy and France into the heart of the continent, and by 1350 all of Europe was embroiled in the pandemic. The Black Death took two-thirds of the population of Hamburg, half of the population of Florence, and 1.7 million of England's 3.8 million inhabitants. Twenty-five million people, more than a quarter of Europe, became inadvertent victims—collateral casualties, in modern parlance—of Janibeg's parting shot on Kaffa.[13]

We can vilify the Mongol khan, but this is not entirely fair, given that his tactic was repeated in subsequent centuries, albeit without the large-scale consequences. In 1422, Prince Sigismund's forces launched corpses (along with manure and garbage) during the siege of Karlstein, but no epidemic ensued and it is not clear whether the bodies were diseased.[14] The siege of Reval, however, had all but one of the ingredients needed to become a repeat of Kaffa.

At the turn of the 17th century, Karl XII of Sweden was on a kingly tear across northern Europe to acquire as much territory as possible.[15] His acquisitive rampage ground to a halt when the Russians counterattacked and forced a portion of Karl's forces to hole up in the Estonian city of Reval. The Russians were beset by the plague and soon resorted to hurling corpses. As in Kaffa, the tactic worked—plague irrupted within the walls of Reval. However, the dying populace had no way to escape, and the epidemic was contained within the city limits.

By this time, human warfare was changing and so was the role of insects. Although cannons and gunpowder shifted attention to exploding—rather than living (or dead)—payloads, insects flourished with the new military strategies. Vector-borne disease played a decisive role in the course of many human conflicts in the 18th and 19th centuries, although the combatants were largely unaware that six-legged soldiers were often far more deadly than their human opponents. The proboscis proved mightier than the sword—or the cannon. It was as if, having been conscripted and exploited for centuries, the insects declared their independence and fought for themselves. And they didn't lose a battle for two hundred years.

5

THE VICTORIES OF THE VECTORS

Insects have carried disease onto the battlefield for more than two thousand years. For example, during the Peloponnesian War in 429 BCE, flea-infested rats on Greek warships brought plague from Ethiopia and hastened the collapse of the Athenian state.[1] The beneficiaries of this and other such fortuitously timed epidemics were apt to interpret outbreaks as divine interventions. The ancient Hittites and Babylonians were particularly explicit in this regard, paying homage to Irra, the archer-god who could fire arrows of disease into the enemy.

The Greeks had even deeper insights, believing that Apollo decimated enemies by using invisible plague-arrows and sending infestations of rodents.[2] So while the rodent-disease association had been made, fleas remained as invisible arrows. In 396 BCE, the citizens of Pachynus, Sicily, implored Apollo to loose his biological weapons on the approaching Carthaginians. Their prayers were answered when a pestilence—most likely bubonic plague—irrupted on the invaders' ships and the fleet turned back before reaching Sicily.[3]

Such events were relatively isolated, until civilization provided the ecological conditions that painted a target for Apollo's insectan arrows on the backs of soldiers. Several factors conspired to create an unprecedented opportunity for disease vectors. As the human population grew, the scale of armies increased commensurately. In the 18th and 19th centuries, encampments became like small cities with even less sanitation than their permanent counterparts.

With impotent medical interventions and miserably inadequate nutrition, the exhausted troops were sitting ducks for the insect vectors. Surviving on meager rations and drinking polluted water, the soldiers' defenses were pitifully weak against six-legged infiltrators and their microbial comrades. Factor in the lack of bathing and laundering, and the insect parasites were virtually assured of a rout. Moreover, as the devastation of war reached into urban centers overflowing with humans and waste, the beleaguered cities became epidemiological powder kegs.

If all of this were not enough, the modern era also ushered in transportation technologies that allowed armies to cross geographic barriers that had long kept people—along with their insect-borne diseases—isolated from one another. With shipbuilding and maritime navigation racing far ahead of medicine, vectors had a bilateral field day. European invaders of the New World brought with them new organisms—including infective microbes, blood-feeding insects, and adaptable rodents—and were greeted by a fusillade of unfamiliar pests, parasites, and pathogens. Hans Zinsser, author of the 1934 classic in the field of medical entomology, *Rats, Lice and History*, maintained that the famed battles of early modern warfare "are only the terminal operations engaged in by those remnants of the armies which have survived the camp epidemics."[4]

Napoleon Bonaparte has been called the greatest military mind of all time. He was a brilliant tactician, and perhaps no human opponent could consistently get the better of him. But six-legged enemies were another matter altogether. For every one of his soldiers killed on the battlefield, four succumbed to disease—and in most cases the illness was courtesy of a pathogen carried by an insect.

Napoleon's first lesson in the capacity of insects to alter the course of a military campaign came when the Ottoman Empire declared war on France in 1799.[5] Napoleon anticipated that the Turks would attack Egypt, which the French had taken a decade earlier. Figuring that the best defense was a good offense, the French sailed to Syria and unloaded 13,000 troops looking to bring the war to the Turks on Napoleon's terms. As his army easily captured a series of coastal towns, it appeared that the plan to foil the impending Turkish offensive would succeed. However, a stunning reversal of fortune was waiting in Jaffa.

When the French stormed the city, the soldiers became crazed with bloodlust. Over the next three days, Napoleon's men ignored the enemy's attempts to surrender and bayoneted some 2,000 Turks. The slaughter culminated with Napoleon's ordering the execution of 3,000 prisoners. But the fleas of Jaffa would wreak revenge. Within two days of victory, 31 French soldiers were hospitalized with bubonic plague and 14 were dead. The doctors kept their diagnosis secret to avoid creating panic, and the commander visited the sick ward to reassure his troops. But Napoleon could not play nursemaid; he had a war to win (see Figure 5.1).

Figure 5.1. A romanticized recreation of Napoleon visiting his sickened troops in Jaffa, by Antoine-Jean Gros (1799). The French ruler ordered the painting to dispel rumors accusing him of having poisoned his sickest soldiers. Transmitted by fleas, an outbreak of bubonic plague cut short Napoleon's effort to defeat the Turks in Syria. Most of the 2,000 men that died in the campaign were victims of plague, with an incredible 92 percent mortality rate among those showing symptoms. (Art Resources)

The French offensive moved northward, and the plague came along. Having been routed at Haifa, the Turks finally made a stand in Acre. Napoleon's army laid siege to the port city, but their cause was futile. The British navy gleefully supplied the Turkish troops with food and ammunition. For two bloody, feverish months, Napoleon's army relived the suffering of Janibeg's horde. When a last, desperate assault on the stronghold failed, Napoleon asserted that Acre was no longer of strategic value because plague had irrupted within the city. This excuse thinly disguised the real reason for his retreat. The French would soon lose a war of pathogenic attrition unless they could make it to safety.

Napoleon headed toward Egypt with his army continuing to wither. The sickened troops could not make much speed and Napoleon worried that the Turks would catch them and even the score. By the time they made it back to Jaffa, where Napoleon had executed his Turkish captives, the French commander saw no option but to issue a similar order—for his own men.

Napoleon directed that 50 of his sickest men be given an overdose of opium. The doctor protested, but the Turks were within hours of the city and Napoleon could not afford to slow the rest of his troops by burdening them with their dying comrades.

Three months later, Napoleon and the remnants of his army reached Cairo. Two thousand men had died in the expedition with many, perhaps most, succumbing to plague. Some historians maintain that disease was not decisive in the defeat, but it was surely a major factor in the physical and psychological condition of the French throughout the campaign. Napoleon's doctors reported a staggering 92 percent mortality rate among those showing symptoms. The onset of fever was tantamount to facing a firing squad, with the fleas of Jaffa delivering death at point blank range. But this was only an entomological skirmish compared to what the French would face 30 months later and 7,200 miles to the west.

Sans moustiques, les Etats-Unis pourraient être francophones—that is, "without mosquitoes, the United States might be French-speaking." In particular, without a strange convergence of African slavery, Haitian audacity, and French hegemony, American history could have been profoundly different.[6] In 1801, François Dominique Toussaint L'Ouverture—the son of slave parents—declared himself Haiti's governor-general for life. There was only one small problem: France considered itself the owner of this Caribbean island.

L'Ouverture was no fool. After his pivotal role in the slave uprisings of the 1790s, he rapidly consolidated his power base. A savvy politician, L'Ouverture was a fair-weather ally of Spain and France, depending on who had something to offer. The French promoted him to the rank of general, and it was from this military perch that he named himself the island's ruler. Such insolence was too much for Napoleon. He had plans for an empire rooted in French Louisiana and extending far up the Mississippi River. So Napoleon sent his brother-in-law, General Victor-Emmanuel LeClerc, along with 20,000 troops to reclaim Haiti. As in the Syrian campaign, all went well at first.

The French landed in late January of 1802 and seized control of the island after a few short, bloody battles. Had LeClerc been a rapt student of entomological history, he might have predicted that insects would create military havoc in the Caribbean. In the 1650s, the English chose not to invade Cuba when they mistook fireflies along the shoreline for the torches of Spanish troops. And in 1761, termites clandestinely hollowed out the French stockades built to defend the Antilles. When the English gunfire reduced the wooden shells to sawdust, the British easily overran the crumbled bulwarks. In 1802, however, the insects were not so subtle.

Soon after LeClerc's initial victory, many of his soldiers became sick with a high fever. The first wave of 600 debilitated men doubled in a week. And those who had initially fallen ill began to develop horrible symptoms culminating in gruesome deaths. Yellow fever would make the French wish they'd left Haiti to its malcontent governor-general.

Yellow fever is caused by a virus that was imported to the New World from Africa via the slave trade.[7] At first, native American mosquitoes circulated the pathogen among treetop monkeys. But it was not long before the virus was reunited with its African vector, which had a taste for humans. The yellow fever mosquito, *Aedes aegypti*, was a stowaway in water barrels on slave ships, and once it teamed up again with the pathogen, the deadly duo started making history.[8]

From 1693 to 1901, more than 100,000 people died from yellow fever in the United States. Epidemics ravaged the cities of the Gulf Coast and Atlantic seaboard, including New Orleans, Mobile, Charleston, Norfolk, Baltimore, New York, and Boston. Disease-carrying mosquitoes moved inland, triggering an outbreak in Philadelphia that killed 10 percent of the city's population in 1793. A yellow fever epidemic took at least 20,000 lives across more than a hundred towns in the southern United States in 1878.[9] And theirs were not easy deaths.

The symptoms of yellow fever begin within a week of being bitten by an infected mosquito.[10] The individual first experiences a high fever, debilitating headaches, and severe muscle pains. Often, there is a period during which the patient seems to have recovered for two or three days—and then the disease returns with a vengeance. The person's skin becomes jaundiced as the virus attacks the liver. The disease was nicknamed "Yellow Jack" by the British, in reference to the coloration of the skin and the yellow flag flown by ships with victims in quarantine. A bizarre, but not infrequent, symptom involves constant hiccupping so that even when the patient is lucid there is no rest or comfort. Splotches of blue and black appear on the body as vessels rupture. In many cases, the victim vomits what appear to be coffee grounds, but this is actually coagulated blood from internal hemorrhaging, which explains why the Mexicans referred to the disease as "el vómito." As blood continues to seep from damaged vessels and bleeding develops from every orifice, major organs begin to fail. Delirium gives way to coma, with death soon to follow.

Not all patients succumb to the disease. By controlling fever, preventing dehydration, and providing transfusions, today's physicians can save four of five victims, although convalescence may require months or years. On the other hand, without medical care, mortality can be staggering. The British saw a force of 27,000 men reduced to a mere 7,000 by "Black Vomit" on their failed expedition to capture Mexico and Peru in 1741.[11] And in Haiti, LeClerc and his troops were well on their way to proving the old adage: those who fail to learn history are doomed to repeat it.

In April, the rains came to Haiti and the mosquitoes followed. *Aedes aegypti* is what entomologists call a "container breeder"—it lays its eggs in water-filled barrels, troughs, pots, and buckets that humans provide in abundance. There were evidently plenty of breeding sites, because by the end of the month one-third of the expeditionary force was dead. With the French losing 30 to 50 men per day, L'Ouverture saw an opportunity to negotiate the terms of a truce. But he underestimated the skullduggery of LeClerc, who invited the Haitian leader to dinner only to have him arrested and sent to France. Such dishonorable tactics enraged the locals, who resumed hostilities. Through the long, tropical summer, the French continued to suffer horribly. By October LeClerc was down to one-fifth of his initial force, and he finally succumbed to yellow fever himself.

Napoleon sent reinforcements, but for most of these men it would be a one-way trip. Along with fresh troops came a new commander, General Jean-Baptiste Donatien de Vimeur, Comte de Rochambeau. He was brutal, verging on genocidal, in his tactics, which included releasing 700 fighting mastiffs to shred the Haitians. But the mosquitoes of Haiti were far deadlier than the dogs of France. Rochambeau saw another 20,000 soldiers succumb to yellow fever in the year after their arrival. The French had no choice but to capitulate, returning to Europe with a scant 3,000 soldiers and leaving no fewer than 40,000 as mute testimony to the power of insect-borne disease.

Along with Rochambeau's retreat went any hope of France's controlling North America.[12] While the French military was suffering defeat in Haiti, the French government was negotiating the sale of its interest in North America. In 1803, Napoleon sold his country's land in the Mississippi Valley to the fledgling United States for a mere $15 million, and a year later Haiti became the first independent nation in Latin America. If a tiny proto-nation could—with the help of insect vectors—repulse tens of thousands of Napoleon's men, then what could the largest nation on earth accomplish when it unwittingly allied with six-legged mercenaries to engage the Grande Armée of France a few years later?

In the early 19th century, Napoleon looked to the east and saw an opportunity and an obstacle. The opportunity was India, a British colony that Napoleon coveted. But to take this prize, he would need to move his army overland, as Admiral Nelson's fleet would pulverize him on the high seas. Consequently, the campaign would take him through Russia, a tenuous ally who would not tolerate a few hundred thousand French soldiers tromping across the countryside. So Napoleon found an excuse to invade. The Russians had been trading with England, thereby violating Napoleon's decrees prohibiting commerce with France's archenemy.

In late June 1812, the Grande Armée amassed for the invasion.[13] Realizing that 450,000 men would require a tremendous amount of food, Napoleon hoped to stretch his army's supplies by prohibiting the soldiers from touching their rations while crossing Poland. Until they reached Russia, his troops would have to live off the land—and steal from the peasants. The troops met with no resistance from the villagers, who were woefully impoverished and, most important, terribly infested. Upon reaching the Niemen River marking the Polish-Russian border, the marauding soldiers found that they had not only acquired the peasants' stores but their annoying vermin as well. And compared to the devastation soon to be wrought by these six-legged time bombs, the Grande Armée's plundering of Poland would look like a diplomatic mission of mercy.

While spending two weeks in Vilna to rest and recover, Napoleon's men began to develop raging fevers and rashes on their chests and backs. Over the next few days, the debilitating fever persisted and the rash spread to cover a victim's body. After another week of severe headaches and muscle pains, a sick soldier was spared this torment by periods of stupor and delirium. But these respites meant that a patient's heart and brain were swelling with fluids, and death would come soon. If they had but known, perhaps the sick could have gleaned some perverse satisfaction in knowing that typhus was just as lethal to its vector.

We play host to three kinds of lice (order Phthiraptera): pubic or crab lice (*Pthirus pubis*), head lice, and body lice (aka "cooties").[14] The latter two beasts are subspecies of *Pediculus humanus*, differentiated primarily by their preference for living on our scalps or in our clothing. Head lice lay their eggs, or nits, on the hair of their hosts, which give rise to both the next generation of insects and colorful expressions. We have louse eggs to thank when we accuse a person who focuses on minute issues of being a "nit picker" and when we assert that something is examined with a "fine-toothed comb" (a device used to extract the barely discernible nits). Body lice flourish in the warm, moist environment provided by undergarments. The insects crawl from the folds of cloth to grab a blood meal and then retreat to mate and lay eggs, with a female producing as many as 5,000 offspring in just three months. About the size of a grain of rice, these tiny insects don't wander far unless a body becomes inhospitably hot from fever or cold from death—as with a typhus infection.

The pathogen, *Rickettsia prowazekii*, is a rickettsia, which can be thought of as a specialized bacterium that can reproduce only within a host's cells.[15] Oddly enough, although the louse acquires the microbe by feeding on an infected human, the insect does not pass the pathogen via feeding. Rather, the rickettsiae multiply spectacularly in the louse's gut cells until they rupture and release enormous numbers of microbes into the doomed insect's feces. The unwitting human then acquires the fecal-borne pathogen through breaks in the skin caused by vigorous scratching (few conditions itch more than an infestation of lice) or across mucous membranes.

This was not the first time that typhus had altered the course of French military history. In the early 1500s, Charles V rose to power in large part owing to this insect-borne disease wiping out all but 4,000 of a 28,000-man French army. And in 1741, lice proved that they had no nationalistic leanings, annihilating a force of 30,000 Austrians who consequently turned Prague over to the French. But these outbreaks paled in comparison to the toll that louse-borne disease would take on Napoleon's army.

While the ever-retreating Russian army refused to engage Napoleon, the Polish lice were winning a war of attrition. One month into the campaign, the Grande Armée had lost 80,000 soldiers to disease, with dysentery adding intestinal misery to the typhus epidemic. With Moscow still nearly 300 miles to the east, Napoleon pushed onward. By the end of August, he had lost nearly half of his men, and a couple of weeks later his invasion force had deteriorated to 130,000 troops. And still, the Russians refused to fight.

In early September, Tsar Alexander I finally ordered his military to engage the enemy at Borodino, where the French killed 50,000 Russian troops in a one-day battle. Bolstered by this victory, Napoleon made a final thrust into the heart of Russia. When his troops entered Moscow a week later, they discovered that the Muscovites had burned three-fourths of the city and destroyed all of the food stores. The Russians refused to surrender and simply waited until the hungry, louse-ridden invaders had no choice but to retreat.

In mid-October, Napoleon's army was down to 95,000 exhausted men. Reaching Smolensk in early November, they were now brutalized by raging fevers within and bitter cold without. At times, the Grande Armée was losing 250 men per mile on the return to Vilna, where half a million of Napoleon's men had passed through in June. Just 7,000 able-bodied men along with 20,000 stragglers left the city. Typhus raged among the 25,000 who were left behind, and only 3,000 lived to continue their journey home the following year. Napoleon's army made it home in late December, with fewer than one in ten having survived the campaign.

In a matter of half a year, the Grande Armée had lost 400,000 men, with more than half dying from disease (primarily typhus) and many thousands more succumbing to hunger or cold because they were weakened by the louse's lethal microbe. The French would never fully regain their military might. It is ironic, and perhaps fitting, that Napoleon—one of history's shortest generals (standing just five-and-a-half feet)—was finally beaten by one of the animal kingdom's smallest creatures (stretching only a tenth of an inch).

Warfare had become a struggle against human opponents and insect-borne disease. And the Americans would soon affirm Zinsser's interpretation of military history: the glorious battles of the nation's bloodiest war were "the terminal operations engaged in by those remnants of the armies which have survived the camp epidemics."

6

A MOST UNCIVIL WAR

An entomologist started the U.S. Civil War.[1] At 4:30 A.M. on April 12, 1861, at the harbor entrance to Charleston, South Carolina, a cannon shot rang out from Fort Sumter. The man who pulled the lanyard was Edmond Ruffin, the editor of the *Farmers' Register* and a naturalist who had devoted himself to the study and control of grain moths. The bombardment continued for 34 hours, and the next day the United States officially declared war on the Confederacy. It is oddly apropos that an entomologist fired the first shot in a war during which insects would kill far more soldiers than would canons, firearms, and bayonets.

The Civil War showcased forms of entomological warfare that military commanders had been refining for centuries. The bloodiest of American conflicts not only marked a culmination of insect prowess on the battlefield but also provided an ominous glimpse of what was to come in the next century. To understand the place of this conflict in the drama of entomological warfare, we must first appreciate the insect actors and their supporting cast of pathogens.

Of the 488,000 soldiers who perished in the Civil War, two-thirds died of disease—and insect-borne pathogens were among the primary killers.[2] Although lice transmitted typhus, fleas carried plague, and mosquitoes spread yellow fever, these diseases claimed only a couple of thousand victims. Typhus was unknowingly suppressed by delousing, a popular pastime that the soldiers sardonically termed "skirmishing." Like primates grooming one another on the plains of Africa, the men patiently picked the eggs (nits) from each other's hair, a process that they called "(k)nitting work." The vermin were given nicknames alluding to human enemies: "bluebellies" by the South and "graybacks" or "Bragg's bodyguard" by the North. Infested clothes were boiled in salt water or singed over a fire, the sound of extermination being compared to that of popping corn. As for bubonic plague, biologists speculate that the disease never had a good chance to develop in the eastern United States, as the vector favored drier environments. And yellow fever may have been suppressed by a series of fortuitous frosts that limited the life span and range of the mosquito carrier during the war.[3] Rather than these insect-microbial partners that we've already met, the Civil War featured a unique pair of deadly duos: flies carrying enteric pathogens and mosquitoes transmitting malaria.

As for intestinal maladies, the medical records of the Union allow more precise estimates than those of the Confederacy.[4] There were 1,739,135 cases of diarrheal disease among the Federal troops leading to 44,558 deaths, and scholars estimate that the Confederates suffered more than a million cases with at least 30,000 deaths (see Figure 6.1). The various enteric pathogens were commonly transmitted by a retinue of flies that followed the armies and flourished in the detritus of war. Consider that the Army of the Potomac had 56,000 mules and horses, and one begins to get a picture of the tons of feces that littered the camps. Slit-trench latrines were rarely dug, so human excrement added to the mountains of animal dung. Battlefields were often littered with corpses, and even if human remains were buried before becoming flyblown, the shattered bodies of livestock were a maggoty windfall. But in an odd twist of entomological fate, the flies' affinity for decaying tissue sometimes turned these insects into medical saviors.

After major battles, the poorly equipped doctors were overwhelmed by casualties. Often, days would pass as the mangled men waited their turn, and blow fly maggots (family Calliphoridae) would infest the wounds. This would seem like adding horrific insult to injury, but perceptive doctors soon realized that the infested wounds healed faster and led to fewer amputations. Some, such as J. F. Zacharias of the Confederate army, even took the next logical step:

> During my service in the hospital at Danville, Virginia, I first used maggots to remove the decayed tissue in hospital gangrene and with eminent satisfaction. In a single day, they would clean a wound much better than any agents we had at our command [scalpel and nitric acid].[5]

The fly larvae provided two benefits: they consumed the dead and decaying tissue, and they excreted a nitrogenous waste product, allantoin, which accelerated the breakdown of necrotic flesh and promoted growth of new tissue. However, the number of lives saved by the feeding and excrement of maggots was far exceeded by the lives lost to cholera, dysentery, and other such illnesses.

Figure 6.1. If this ward in Washington D.C.'s Carver General Hospital is typical, most of the soldiers are suffering from pathogens transmitted by lice, fleas, flies, or mosquitoes. Of the 488,000 soldiers who perished in the Civil War, two-thirds died of disease—and insect-borne pathogens were the primary killers. Fly-borne intestinal maladies and mosquito-borne malaria accounted for about 5 million cases and more than 150,000 deaths. (Library of Congress)

The various species of flies that feed on carrion, feces, and garbage are called "mechanical vectors." That is, they transmit the pathogens without the microbes reproducing within the insects. However, the offal heaps, dung piles, and cesspools provided plenty of opportunity for bacteria to flourish, so the flies had no difficulty picking up enough microbes to infect the soldiers. These insects operated as a public transportation system for pathogens with regular stops at the fetid latrines and field kitchens. In addition to the enteric diseases carried by flies, another malady has been attributed to these vectors: typhoid. However, Civil War physicians often lumped this disease with the other major, insect-borne disease: malaria.

The Union medical records reported 1,315,955 cases of malaria, with 10,063 deaths.[6] Medical boards thought that malaria and typhoid could transform into one another, so they also reported typho-malarial fever, which accounted for another 57,400 cases and 5,350 deaths. In all, the Confederates probably added another million victims and 10,000 corpses to the tally. Although the physicians believed that typhoid and malaria were interchangeable diseases, the two maladies have little in common other than their symptoms.[7] Victims suffer high fevers and incapacitating weakness, along with diarrhea, stomach pain, and nausea. While typhoid's fever may also cause a rosy rash and malaria's fever is sandwiched between bouts of soaking sweats and bone-rattling chills, these symptoms did not differentiate the maladies in the minds of 19th-century physicians. Both illnesses debilitate a person for weeks, but malaria is far more likely to remain within a victim's body for years.

Typhoid is caused by bacteria that enter via the digestive system and then spread to the bloodstream, bone marrow, and liver. The microbe uses the bile ducts to return to the intestinal tract, where it is passed in feces. The Civil War didn't produce the highest rate of typhoid among 19th-century conflicts. That dubious honor goes to the Spanish-American War in 1898, which saw 369 Americans fall in battle while 1,939 died of typhoid. Medical experts later concluded that the pathogen had been spread by flies that flourished in unsanitary military camps.

Malaria, on the other hand, has a rather more complicated story.[8] The cause of the disease is a protozoan parasite—one of four species of *Plasmodium*—that proliferates in the gut lining of an *Anopheles* mosquito. Once the microbes burst from the insect's cells, the single-celled pathogens migrate to the mosquito's salivary glands. When the vector feeds on a host, the protozoa are injected into the bloodstream. On reaching the liver, the protozoa set up house and reproduce. Their offspring are released into the circulatory system, where they infiltrate red blood cells. Here, the microbe either continues to divide and attack blood cells or it produces gametes—the protozoa's version of eggs and sperm. These male and female gametes are ingested by another mosquito in the course of blood feeding. Once in the insect, the gametes fuse to form a zygote that enters the mosquito's gut lining, and the cycle is complete.

Despite confusion as to what caused malaria, it was the only insect-borne disease for which there was an effective intervention.[9] Quinine could prevent malaria, but prophylaxis often failed due to underdosing; when higher amounts were used for treatment of infected soldiers, quinine provided significant relief. The Union Army consumed more than 19 tons of the drug during the war, and their blockade of the Confederacy drove the price of the quinine sulfate from $5 to $500 per ounce. Smugglers profited handsomely, with one sneaking $10,000 worth of contraband medicine inside a dead mule.

While the shortage of quinine was vexing, the greater problem was the excess of mosquitoes.

The most strategically important aspect of vector-borne disease was the way in which it set the stage for major campaigns. Mosquitoes often played the odd role of peacemakers, leaving opposing forces too sick to fight. The soldiers found that malaria "ate out their vitality, and even those who reported for duty dragged themselves about, the mere shadows of what they had been." Dozens of potential clashes were avoided, delayed, or minimized because of illness, but a couple of cases exemplify the importance of insect-borne disease.

Historians characterize Major General Frederick Steele as "quiet, unimaginative, fairly competent, lacking in drive or initiative and content to comfortably settle down like a police precinct captain"—in other words, a perfect man to stand by as his troops withered.[10] Steele's 15,000-strong Army of Arkansas was supposed to drive the Confederates from the Mississippi River in a campaign that began in August 1863 and quickly bogged down. All the ingredients for a medical disaster were in place: an abundance of disease carriers (many of the Union soldiers had spent the previous summer in the South), a woefully inadequate supply of quinine, and a superabundance of what were, according to a military surgeon, "the largest, hungriest, and boldest" mosquitoes ever seen.[11]

The Army of the Arkansas never had a chance. Medical records showed a malaria rate of 1,287 cases per 1,000 men in the first year of the campaign—an impossibility unless one considers that this is a recurrent disease. That is, the average soldier suffered more than one bout of sweat, fever, and chills in the course of a year. Typically one-half to two-thirds of the men were too sick to answer reveille. And in some units, the losses were even worse; malaria reduced the Sixth Minnesota from 937 men to 79 in a matter of weeks. Although the Union fed 50,000 fresh troops into Steele's army, they could not offset the 178,000 medical casualties in the two years of the impotent campaign. Only once did the major general take the offensive, and the Confederates—who were also wracked by fever, but apparently less so than the Union—repulsed this feeble effort. In the end, Steele lost five times more men to disease, primarily malaria, than to combat. In other campaigns, however, what seemed initially to be a defeat by the insects turned into a victory of sorts.

From April to June 1862, two of the largest armies of the Civil War were poised for a massive conflict at Corinth, Mississippi.[12] The scale of bloodshed promised to exceed that of Shiloh, earlier in April. However, this potentially decisive battle atrophied into a few minor firefights. By the time the Union was ready to attack, 173,315 soldiers were too sick to shoot. Swarms of flies first delivered a cornucopia of intestinal maladies. A colonel from Illinois took time from his bout of dysentery to note in his journal that he'd never seen house flies so thick. Then, by the end of May, the mosquitoes arrived in force and malaria swept through the camps of both sides.

Under the command of General Beauregard, the Confederates retreated down the Mississippi River valley. General Halleck's Union forces gave a half-hearted chase. Although some contend that Halleck feared his human enemy, what he genuinely dreaded was losing his army to disease. In retrospect, Halleck's strategy of not driving deep into the South paid off. By mid-summer the Confederate forces were suffering 179 cases of malaria per thousand men—three times the rate of Union soldiers. It seems that the general had made the critical link between ecology and disease, an association that others would fully grasp and exploit.

The devastating consequences of insect-borne diseases were not lost on the best of the war's commanders. They soon realized that the deadly phantoms could be turned into lethal weapons. This advance in military strategy required a novel version of the old rule of conflict. That is, when it came to entomological warfare, the best defense (against the insects) was a good offense (against the enemy).

No military mind of the 19th century surpassed General Winfield Scott's grasp of the strategic value of vector-borne disease, even without knowing the role that insects played on the battlefield.[13] Scott's knowledge of yellow fever shaped his invasion plan during the Vera Cruz campaign of the Mexican-American War. Although he had no idea that mosquitoes carried "Black Vomit," he recognized that avoiding the hot, wet summer of 1847 was key to minimizing his losses on the march to Mexico City. Unfortunately, the War Department's logistical ineptitude put Scott's amphibious invasion—the first in U.S. history—months behind schedule. As a consequence, his troops were bedeviled by mosquitoes and paid a horrendous price before storming the Halls of Montezuma. In the course of the war, 1,192 American soldiers were killed in action while 11,155 died of disease. Although his efforts at strategic timing had been a bust, Scott knew that the key to beating the enemy was to first avoid losing to disease.

Fifteen years later, when General Scott formulated his "anaconda constrictor plan" for strangling the Confederacy, he again specified that victory was contingent on seasonality. He called for the campaign to take place in November, after "the return of frosts to kill the virus of malignant fevers below Memphis."[14] The general's emphatic argument was not persuasive to the top brass, and the first attempt to take Vicksburg ensued in the summer of 1862. Camped in marshy areas with mosquitoes flourishing, malaria raging, and quinine in short supply, the troops were too sick to mount an attack on the Confederate positions that occupied higher, drier ground. Exemplifying the Union debacle was the Seventh Vermont, which arrived on June 25 with nearly 800 men and by the middle of July had fewer than 100 answer reveille. It didn't help the Connecticut regiment that their assistant surgeon prescribed daily drilling in the broiling sun, confidently maintaining that "if we don't exercise and perspire abundantly we shall get poisoned with malaria and die."[15]

General Scott's plans, had they been implemented, would have protected his troops from insect-borne disease. His insight was soon followed by the next step in the development of entomological warfare: using, rather than avoiding, blood-feeding insects. In this case, a Southern general rediscovered the strategy pioneered by Clearchus.

Beginning in spring 1861, the cry from the Union was "On to Richmond!" To the Northern mind, capturing the capital of the Confederacy would define victory. The first campaign was led by General George B. McClellan, who intended to lead the Army of the Potomac up the Yorktown peninsula and into Richmond.[16] When the Union forces landed below Yorktown in April 1862, McClellan's medical director realized that a large swamp near the camp had the potential of spreading "malarial poison" once the weather turned warm. McClellan sought to avoid the impending epidemic by leading his army up the peninsula, while pushing back the Confederate forces with relative ease. However, as the Union troops closed in on Richmond in the early days of summer, increasingly fierce enemy resistance forced them to encamp near the sluggish Chickahominy River (see Figure 6.2).

Although malaria had been prevalent in this area for a century, the draining of swamps had dramatically reduced its incidence in the years before the Civil War. But war destroys culverts, drainage ditches, and canals while creating trenches, pits, and wheel ruts. As the summer wore on, corpses rotted in shallow, swampy graves and fetid latrines filled with human waste. Meanwhile, the mosquitoes and flies thrived. Malaria, typhoid, and dysentery descended on the Union camps with a vengeance, just as McClellan's nemesis had planned.

General Joseph E. Johnston, the Confederate commander in Richmond, knew exactly what he was doing when he refused to deploy his smaller army

Figure 6.2. The 5th New Hampshire Infantry slogging through the marshes along the Chickahominy River, outside of Richmond, Virginia. With his troops outnumbered, Confederate General Joseph E. Johnston kept the enemy pinned down in the swamps and allowed insect-borne disease, primarily malaria, to win a war of attrition. Thanks to mosquitoes and savvy military tactics, the Union's Peninsular Campaign of 1862 collapsed by late summer. (Library of Congress)

in a direct engagement of the enemy. Through a series of constant withdrawals he'd slowed the Union forces until they were just five miles from the Confederate capital. There, he applied just enough resistance to keep them pinned down along the Chickahominy. By May, Johnston was drawing harsh criticism from his superiors for not throwing his men into a full counterattack. The political leadership saw the Union Army just a few miles from the heart of the Confederacy and thought Johnston's forces were dillydallying with defensive maneuvers. The frustrated general finally retorted, "I *am* fighting, sir, every day! Is it nothing that I compel the enemy to inhabit the swamps, like frogs, and lessen their strength every hour, without firing a shot?"[17]

The shrewd strategist knew that a bedridden enemy soldier was preferable to a corpse. The dead required a burial detail, if time and resources allowed, but the sick and wounded required care—doctors, nurses, beds, medicine, equipment, food, and transportation. Neither the dead nor the ill could fight, but the latter burdened the military machine. For each soldier who fell ill, two more were lost to the associated demands. And McClellan's army was foundering under the burden of 2,000 soldiers who languished in the Yorktown hospital—along with logistical costs of shipping home thousands more to recover or die.

Johnston finally attacked the Union forces on the last day of May in the Battle of Seven Pines. The two-day battle was indecisive in terms of the Peninsular Campaign, but it was a turning point for Johnston and his Confederate Army. The 55-year-old general was badly wounded and had to relinquish his command. He was replaced by a West Point classmate who had also been Winfield Scott's chief aid in the Mexican-American War. General Robert E. Lee proved to be nearly as aggressive as the mosquitoes, and McClellan began to lose his nerve.

The Union troops retreated to Harrison's Landing on the James River at the end of June, where the navy could protect them from Lee's counteroffensive, but nothing could deter the insect onslaught. As summer came, the Union commanders began to fully appreciate the enemy's strategy. General John E. Wool realized that the Confederates were intentionally exploiting the unhealthy environment into which they had first drawn and now driven the Federal troops: "The rebels will do all in their power to keep McClellan where he is with his army, in the hope that death and desertion will so thin his ranks that by fall his army will be reduced by one-half."[18]

When surgeon Jonathan Letterman took over as medical director on July 4, he found the Union forces in a state of near collapse. He reported that "after about 6,000 had been sent away on transports, 12,795 remained," and at least one-fifth of these men were sick.[19] Just two weeks later, another 7,000 soldiers were sent to the rear while replacements put the Union force at nearly 20,000. But these fresh troops were just so many mosquito meals and so much fly fodder. In August, Union leaders evaluated the fast-eroding situation. McClellan was losing a regiment a day between insect-borne disease and combat casualties among his sickly troops. With the epidemic certain to continue into autumn, the Army of the Potomac was ordered to withdraw. The Peninsular Campaign was over—the Confederacy had successfully allied with the insects to crush a larger army.

While large-scale applications of strategies that depended on insectan allies proved effective, only rarely did a commander have the acumen to pull off such a subtle scheme given that science had yet to reveal the intricacies of vector biology. Some of the more localized, tactical uses of insects as weapons were also quite cunning, while others were more like updates of ancient practices—as with the use of angry bees.

The 132nd Pennsylvania Volunteers at the Battle of Antietam demonstrated why commanders dread hysteria among the troops and how insect-induced panic can shape an engagement.[20] Only a month into service, the Pennsylvania regiment was untested in combat until they closed in on the Roulette farm outside of Sharpsburg and the Confederate soldiers provided a baptism of fire. The green soldiers demonstrated that they had more courage than smarts by continuing their advance while seeking cover in the Roulette's bee yard. As they moved stealthily past the rows of hives, a cannon round ripped through the yard. The air was filled with angry bees and hot lead. Some of the men dropped their muskets and dashed into the nearby fields. The slapping, swearing regiment was disintegrating and the Union commanders worried that the panic would spread across the entire front. The Pennsylvania unit was ordered to double-quick march past the Roulette farm, which allowed the troops to escape the bees but left them without cover. The Confederates exploited the opportunity with a devastating volley of musketry. The survivors dropped to their bellies and bravely continued their advance on the enemy, proving that enraged bees have the potential to turn the tide of battle more decisively than lead balls.

Such uses of bees in the Civil War were more a matter of opportunism than planning—with one crafty exception. Faced with overextended supply lines, both Union and Confederate soldiers relied on plundering farms in order to feed themselves—and honey was a golden treasure. A feisty Georgia woman knew well the proclivities of hungry soldiers, and she prepared her entomological defenses accordingly.[21] When the Union soldiers sauntered onto her property, they greedily eyed her beehives. But the men failed to see the cord running from one of the hives, across the yard, and through a hole in the door of her cabin. As they approached their sweet booty, she sprung the trap. Yanking the cord, she toppled the hive, sending the bees into a frenzy. The infantrymen were driven from the yard and some of the cavalry were thrown from their horses as the insects vented their wrath. Once the soldiers had left and the bees had exhausted their fury, she reset her booby trap and went about her business. She reportedly deployed her six-legged bodyguards on several occasions and the soldiers never succeeded in taking her food or supplies.

The tactical uses of bees in the Civil War almost seem quaint within the annals of biological warfare. But there were far more sinister gambits involving

insects—tactics that darkly hinted at what would come as the science of entomology and the practice of warfare forged a diabolical alliance. With a bit of
planning and luck, insects might be used to inflict yet another form of human
suffering: hunger. The protracted course of modern warfare means that supplies are vital to victory. Whether the enemy dies from bullets or starvation
doesn't much matter. During the Civil War, for the first time in history, a
government was accused of having used insects to wage agricultural warfare.
The Confederacy alleged that the Union had intentionally introduced a devastating crop pest from Mexico.

The harlequin bug (*Murgantia histrionica*), a strikingly patterned, orange-
and-black, thumbnail-size insect, has a spectacularly catholic palate and a
penchant for Southern crops. Piercing plants with its elongated mouthparts,
the harlequin bug can destroy fields of asparagus, bean, beet, Brussels sprouts,
cabbage, cauliflower, collards, eggplant, horseradish, kohlrabi, mustard, okra,
potato, radish, and turnip. If none of these is on the local menu, then it will even
attack fruit trees. And the best part—at least from the Union perspective—was
that the harlequin bug rarely ventured north of the 40th parallel.

Although extensive crop damage added to the suffering of the South, there
was never any direct evidence that Northern operatives had seeded enemy
fields with this foreign mercenary.[22] Entomologists now suspect that the insect
probably moved up from Mexico on its own, but the importance of this episode lies not in its ultimate explanation but in what the accusation reveals
about the role of insects as weapons. Whether or not the harlequin bug was
conscripted by the Union, both sides were well aware of the potential for
insects to be used as means of destroying the enemy's agriculture.[23] And awareness of using living organisms to cause suffering behind enemy lines was not
limited to starvation; inducing sickness was also considered.

Although science had not provided the essential knowledge that would
have allowed insects to be weaponized during the Civil War, at least some
military minds were contemplating how mysterious fevers could become part
of a deadly arsenal. A shortage of know-how, rather than an abundance of
morality, prevented insect-borne diseases from being made into weapons. The
best-documented attempt at biological warfare failed, at least in large part,
because the role of insects as vectors was yet to be understood.[24]

In 1863, Dr. Luke Pryor Blackburn attempted to smuggle clothing from
yellow fever victims into the North as a means of spreading the disease. The
Confederate surgeon also sent clothes gathered from yellow fever wards to
President Lincoln in an assassination attempt. Although records indicate that

Dr. Blackburn (also known as "Dr. Black Vomit") was court-martialed for his efforts, his reputation apparently was unsullied, as he was later elected governor of Kentucky. Had the doctor known of the role played by the insect vector, he might well have been able to transport infected mosquitoes from afflicted regions into enemy cities.

It would not be long before scientific knowledge would catch up with military imagination. In the late 1870s, the work of Louis Pasteur and Robert Koch led to the germ theory of disease. In 1889, the little-known Theobald Smith, a medical doctor working for the U.S. Department of Agriculture's Bureau of Animal Industry, was the first scientist to definitively link an arthropod (the tick *Boophilus annulatus*) with the transmission of an infectious disease (Texas cattle fever). With this breakthrough setting the stage, Sir Ronald Ross and his team soon drew the link between malaria and anopheline mosquitoes. In 1900, Walter Reed and his associates discovered that *Aedes aegypti* was the vector of yellow fever. But the entomological breakthrough that would change the course of war in the dawning century involved one of the lowliest creatures, an insect incapable of flight and no larger that a typewritten "l"—as in *louse*.

7

ALL'S LOUSY ON THE EASTERN FRONT

During World War I, the European continent provided history's largest experiment in entomological warfare tactics. For the first time, scientific understanding of insect-borne diseases allowed these agents to be exploited as "passive weapons," demonstrating that the best offense could be a good defense. Rather than forcing the enemy into infested habitats, science provided the means for military leaders to protect their own forces from the ravages of disease-carrying insects that were part and parcel of war. The advantage of metal armor had been known for centuries, but biological armor now transformed the battlefield.

The grand experiment, however unintentional, allowed military historians to compare the course of war when an army was vulnerable versus when it was protected from the ravages of lice. From 1914 through 1918, the Eastern Front was a worst-case scenario for typhus, while the Western Front was relatively vermin free—an utterly unique experience in the annals of entomological warfare.

If an entomologist were to have written a recipe for a typhus epidemic, no finer list of ingredients and instructions for their mixing could be found than those of Eastern Europe:[1]

Begin with a population of weakened human hosts. After Archduke Ferdinand was assassinated in July 1914, Austria declared war on Serbia. The Serbs were an exhausted people, having just finished a war with Turkey—the third major conflict in two years.

Next, take the already vulnerable hosts and pummel them thoroughly. The Austrians bombarded Belgrade and smashed their way through the towns and villages of the north. Civilians abandoned their homes and a wave of refugees poured into the countryside.

To the weakened hosts, add a large dash of new blood, so that the population is brimming with defenseless bodies. The fight was not entirely one-sided. Earlier conflicts had left the Serbs worn down, but the survivors were battle-hardened. They managed to capture some 20,000 Austrian prisoners.

Make sure to thoroughly crush any semblance of hygiene or medical infrastructure. As the offensive continued, hospitals were destroyed and medicines were impossible to find. There had been only about 400 physicians for all of Serbia, and most of these doctors closed their clinics to defend the homeland, leaving the nation essentially without medical care.

Add a heaping amount of lice and microbes, mix thoroughly, and simmer. By late November, typhus began to spread among the refugees. From there, the disease soon infected the weary Serbian army and their POWs. At first there was little alarm, as typhus had been a part of life in Eastern Europe for centuries. But never before had there been such a bountiful mix of ingredients for brewing an epidemic.

Finally, if the stew is not yet boiling feverishly, dump in fatigued bodies until the pot is overflowing. On December 3, the Serbians launched a fierce counterattack. After three days of bloody fighting, the Austrian invasion force was crushed. More than 40,000 prisoners were taken, burdening the Serbs with one POW for every four of their own soldiers. Between the captives and the depleted supplies of food, water, shelter, and medicine, the nation was strained far beyond its capacity.

The epidemic irrupted almost simultaneously from a constellation of filthy camps, ravaged villages, and war-torn cities. Initially, mortality rates were running at 20 percent. But as the scant supply of medicine was depleted, the rate rose to 60 percent. A shortage of grave diggers—along with doctors and nurses—soon added to the grisly conditions. By April 1915, there were 10,000 new cases each day. With one in six people contracting typhus, more than 200,000 Serbs perished, including 70,000 soldiers. Fully half of all of the Austrian POWs died from the epidemic.[2]

The battlefield misery led to a bizarre application of entomology: insects were used to produce self-inflicted wounds and provide a reprieve from the front.[3] Fearing the horrors of war, soldiers on the Eastern Front collected *Paederus* beetles, the insects possessing the potent toxin that so intrigued the Romans more than two thousand years earlier. The weary men pulverized the beetles and applied the powder to minor wounds, mucous membranes, or even their eyes. The severe inflammation that followed was often taken to indicate a raging infection, assuring the victim of a medical ticket to the rear. Insects were also generating unexpected events on a national scale.

Paradoxically, louse-borne typhus protected the nation of Serbia. After repulsing the initial Austrian incursion, the country was absolutely helpless to defend itself against another attack. However, the Central Powers (Germany and Austria-Hungary, along with Bulgaria and the Ottoman Empire) knew better than to invade a land in the midst of a raging epidemic. As such, the Allied Powers did not have to contend with their enemy storming through Serbia and establishing a front with Russia—a strategy that might have substantially altered the course of the war. Russia, however, was not spared from a devastating—albeit nonhuman—invasion.

As the First World War was winding down on the Eastern Front, lice and their microbes were just getting started. The war had been hard on Russia, with famine weakening the nation and refugees spreading lice throughout the countryside. And since the overthrow of the tsar in 1917, essential services had utterly disintegrated. Nearly six times more Russians would die of insect-borne disease in the years after the war than died of battle-related trauma during the war.[4]

In the two decades prior to World War I and the Russian Revolution, the country had suffered about 82,000 cases of typhus each year. This number rose to 100,000 in the early years of the war, climbed to 154,000 in 1916, and thereafter the disease was rampant. Over the next five years, conservative estimates place the number of cases at 20 million, with 3 million deaths. There may have been 30 million infected, with as many as 10 million dead. In 1919, Vladimir Lenin darkly pronounced that "either socialism will defeat the louse, or the louse will defeat socialism." Although the insects lost, the Red Army witnessed the potential of entomological warfare. However, a military must be able to harness a destructive power before it can be exploited. And a critical lesson in this regard came from the trenches of the Western Front.

The British and French knew the phenomenal capacity of typhus to alter the course of war.[5] Just 60 years earlier, during the Crimean War, they had allied with the Turks to fight the Russians. Nearly two-thirds of the 167,755 soldiers who died in the conflict succumbed to disease, and the situation was even more skewed in terms of casualties (i.e., those killed, as well as those injured, wounded, or otherwise incapacitated). Bombs and bullets wounded 197,399 soldiers, while typhus debilitated 767,411. The western European nations were horrified by these losses and set out to understand the cause of the disease.

A French scientist, Charles Nicolle, made the breakthrough in 1909. While serving as the director of the Bacteriological Laboratory at Rouen, he definitively linked lice and typhus—a discovery that earned him the Nobel Prize in Medicine in 1928. Once the vector was known, controlling typhus became a matter of suppressing the insect carrier. The military quickly grasped the importance of hygiene—louse-ridden soldiers were casualties-in-waiting. When World War I broke out, the generals on the Western Front were determined to wage a wholesale assault on insects and were poised to accomplish something unprecedented in European history: cause more deaths by combat than by disease.

Shortly after the opening salvos, the two sides bogged down in the grim conditions of trench warfare. The crowded, filthy conditions were ideal for lice, and infestation rates quickly soared to nearly 90 percent.[6] With the vectors in place, only the microbe was missing. To prevent infected insectan infiltrators from arriving via the Eastern Front, commanders strictly limited the movement of troops from Serbia and neighboring countries. Having cut the disease's supply line, western forces initiated a two-pronged attack on the lice: prevention and intervention (see Figure 7.1).[7]

Special Sanitary Units made sure that the British soldiers were keenly aware of the importance of fighting filth. Indeed, one of the greatest insults to be hurled against a battalion by its replacements came to be "they left a dirty trench." But convincing the troops of the value of hygiene was an uphill

Figure 7.1. Dorsal view of a male body louse; the dark mass inside the abdomen is a previously ingested blood meal. The insect's legs and flattened body are well adapted to avoid being dislodged from their hosts, so during World War I entomologists on the Western Front advised the troops to keep their hair short and faces shaved, change clothing often, keep infested uniforms away from their quarters, and wear silk underclothes—all to deny the insects a reliable foothold. (Photo by James Gathany, courtesy of CDC)

battle. The military aspired to substantially higher standards than much of the civilian population—and incoming recruits were woefully uninformed. Some of London's poor still clung to the old notion that healthy children hosted robust lice infestations.

In a remarkably farsighted move, the British Expeditionary Forces added two entomologists to each of their Sanitary Units. These experts provided simple and effective advice to the soldiers on how to keep their six-legged opponents at bay. Eliminating cover for the enemy was an important tactic, so troops were told to keep their hair short and faces shaved (see Figure 7.2). Because body lice live in clothing and use commando raids to grab a blood-meal, another maneuver was to deny them safe haven. Silk underclothes were recommended, as the fine texture made it difficult for the vermin to gain a reliable foothold. Soldiers were also advised to change clothing as often as possible and to keep infested uniforms away from their quarters. The entomologists had found that the adult lice starved within a week without a blood meal.

Figure 7.2. Typhus shaped the course of the First World War on the Eastern Front, where an epidemic of this insect-borne disease in Serbia served to keep the Central Powers from invading Russia. An understanding that lice were the carriers of this disease prevented major outbreaks on the Western Front. Here, members of the 6th Infantry are seen picking "cooties" out of their clothing near Nantillois, France. Such simple practices prevented the vectors from reaching outbreak levels. (Courtesy of Disabled American Veterans)

The eggs were another matter; brushing and ironing were the two best means of removing and destroying nits. In particular, a hot iron applied to the seams of shirts and pants—the bunkers of body lice—roasted the enemy within their emplacements. But such elaborate assaults were beyond the means of the soldiers in the trenches, so special forces had to be deployed.

Generals had long considered the infantry, cavalry, and artillery as the backbone of a winning army, but the lowly Quartermaster Corps was essential to victory on the Western Front. And among these uncelebrated providers of food, clothing, and supplies, the least assuming units proved to be among the most vital: laundry companies. With clouds of chlorine and mustard gas rolling across Europe, the launderers got into the spirit of chemical warfare. Rather than soap and water, dry-cleaning processes with volatile solvents were found to more effectively wipe out the insectan enemy entrenched in folds and seams. Having beaten the insects on the clothing front, the Allied and Central Powers extended the hygienic battlefield to the human body (see Figure 7.3).

Figure 7.3. A delousing station for soldiers after coming from the lines on the Western Front in World War I. These soldiers are from the 125th Infantry, 32nd Division, near Montfaucon, October 22, 1918. Trenches, dugouts, woods—the entire front was vermin-infested. Although infestation rates among the troops initially reached almost 90 percent, British Sanitary Units—which included entomologists—and the American Quartermaster Corps launched an intensive and effective campaign against the vermin. (Courtesy of Disabled American Veterans)

The account of Private James Brady of the British Army provides a compelling view of the Germans' delousing process:

So far as I recall Nov. 11th 1918 came and went within the dreary confines of Giessen prisoner-of-war camp, without us having the slightest inkling of what was going on in the "free" world outside. . . . Soon after breakfast we were paraded in groups of around fifty men and marched at a hot pace through the camp to the precincts of one of the most comprehensive delousing stations we had ever come across. Fashioned out of some ancient farm-buildings with high-roofed barns on the fringe of the camp, it was manned by a forbidding horde of untidy German soldiery, garbed in long, off-white short-sleeved gowns, each armed with the oddest collection of "toiletry" gadgets—hair-clippers, scissors, razors (safety and otherwise), scrubbers, hand-brushes, loofahs, sponges, rough-haired towels, huge blocks of evil-smelling ersatz soap, and large canisters of equally evil-smelling "disinfectants."

Altogether the joint looked like something designed by a demented Heath Robinson [a British cartoonist with a Rube Goldberg–like sense of humor], peopled by a gang of mentally disturbed sadists intent on inflicting injury to anything in sight. Furthermore, each "torturer" had a horrible grin on his face. We didn't like the look of things one bit. But it turned out to be quite a comedy. Suddenly, a giant of a fearsome-looking Prussian guard-type screamed out one word which we all understood: "STRIP." Then at a signal from the giant, the good-natured torturers descended upon us with something akin to glee—the barbers with their rusty, dull-bladed clippers and shavers first—until, within the swish of a whisker we were reduced to the bald bareness of our birthdays.

The scene was bizarre in the extreme and not lost on those of us with a sense of the humour. But that was only the beginning. A few shouted words of command from the senior NCOs and we were ushered shivering with cold, into the main building and shunted through a badly-lit maze of narrow duck-boarded corridors and cubicles where for a full thirty minutes we were drenched alternately with fountains of hot and cold water assaulting us from every angle, steamed with jets of scalding vapours, scraped, soaked, soaped, submerged in cauldrons of slimy oil, again bombarded with torrents of hot water, battered with rough towels, brushed with canvas sacking, finally propelled head-first into a huge bath of soothing water before being disgorged, pink and panting, into a barn-like room—there to be handed back our very own uniforms, now stiff and hot from dry-heat ovens and stinking of ersatz disinfectant which reminded me of the ablutions at Ripon camp on inspection day.

It may be said that, as we recovered our breath and dressed ourselves in our clean, lice-free uniforms, everybody felt there was a good deal to commend German de-lousing methods. It was the nearest approach to bliss in captivity that we'd ever experienced, and we could but concur when the German orderlies smiled at us and said, "Good, Jah?" We marched back to our billet light of head as well as of foot and empty-bellied, ready to gorge ourselves on our newly-acquired Red Cross parcels.[8]

Had either side failed to hold the lice at bay, the loss of troops to disease would likely have meant precipitous defeat. Although typhus was largely neutralized, another less virulent disease added misery to life in the trenches. Head, body, and pubic lice found a new rickettsial ally.[9] Unknown before World War I, trench fever made its appearance in France and Belgium. Although 800,000 men would contract the disease in the course of the war, there were few fatalities. A victim experiences a sudden fever, loss of energy, dizziness, and headache followed by a rash and severe aching oddly concentrated in the shins, justifying the malady's other name: shin bone fever. The fever, which can reach 105°F, persists for five or six days, then drops for several days only to return in as many as eight cycles.

Trench fever made an encore performance in the Second World War, but it was less prevalent given that soldiers were not massed into filthy ditches for months on end. The disease disappeared for a half-century until irrupting among the homeless population of the United States in what was sardonically called "urban trench fever."[10] Epidemiologists are uncertain of where the pathogen had been hiding until the 1990s. However, given the microbe's apparent capacity to lurk in the environment, and the louse's infamous ability to exploit grubby hosts, we might expect further ambushes by trench fever in the coming years as pockets of poverty expand throughout the world.

Trench fever was a pathogenic seed that germinated amid the privation of battle and continues to sprout in unsanitary conditions. But even more invidious seeds were planted in the First World War in the form of novel ideas rather than new illnesses. Military scientists understood the germ theory of disease, the basics of vector biology, and the rudiments of epidemiology. Having used this knowledge to prevent disease, only a malevolent twist of logic was needed to imagine how an army might conscript insects to induce an epidemic in the enemy.

THREE

BRINGING FEVER AND FAMINE
TO A WORLD AT WAR

QUESTION: Will you describe the methods and the special equipment employed by Detachment 731 for the large-scale breeding of fleas?

ANSWER: The 2nd Division had four special premises for the mass breeding of fleas, in which a fixed temperature of +30°C was maintained. Metal jars, 30 cm high and 50 cm wide, were used for the breeding of fleas. Rice husks were poured into the jars to keep the fleas in. After these preparations, a few fleas were put in each jar, and also a white rat for them to feed on. The rat was fastened in such a way as not to hurt the fleas. A constant temperature of +30°C was maintained in the jars.

QUESTION: What quantity of fleas could be obtained from each cultivator in one production cycle?

ANSWER: I don't remember exactly, but I should think from 10 to 15 grams.

QUESTION: How long did a production cycle last?

ANSWER: Two or three months.

QUESTION: How many cultivators were there in the special section which bred the parasites?

ANSWER: I don't recall the exact figure, but I should say from 4,000 to 4,500.

QUESTION: Consequently, with its available equipment, the detachment could produce 45 kilograms [99 pounds] of fleas in one production cycle?

ANSWER: Yes, that's true.

QUESTION: What was intended to be done with these fleas in the event of bacteriological warfare?

ANSWER: They were to be infected with plague.

QUESTION: And employed as a bacteriological weapon?

ANSWER: Yes, that is so.

—*Testimony of Major General Kawashima Kiyoshi, Chief of the Medical Service of the First Front Headquarters of the Japanese Kwantung Army, to the Khabarovsk War Crimes Tribunal in December 1949*

8

A MONSTROUS METAMORPHOSIS

In 1894, bubonic plague irrupted in Canton, China, and spread to Hong Kong. From this port city, the lethal trio of fleas, rats, and bacteria stowed away on ships heading around the world. As the ensuing pandemic began claiming 12 million lives, two scientists raced to discover the microbe responsible for the Black Death. The winner beat his rival by a matter of days, and the victor's name is now known to every student of microbiology. Alexander Yersin, an eccentric French doctor, shares his name with the plague bacterium, *Yersinia pestis*. And the loser? Kitasato Shibasaburo is no more than a footnote in the chronicles of science. This Japanese microbiologist was, however, a vital link in a chain of events that led to the most diabolical program of entomological warfare ever devised.

Kitasato studied under one of the greatest pathologists in history, Robert Koch. And Kitasato was a fast learner. He worked alongside other Japanese scientists to develop a public health system far ahead of anything in Europe or North America. Then the Japanese converted their science into an unprecedented military breakthrough.

In every conflict up to the Russo-Japanese War, disease had taken a greater toll than bullets and bombs. But at the beginning of the conflict in 1904, the Japanese instituted an elaborate and effective program of hygiene and medical care.[1] When the war ended 18 months later, 1.5 percent of the Japanese troops had been killed in battle, but only 1.2 percent had succumbed to infections. Not only had Japan defeated the microbes and the Russians, but they emerged as an exemplar of wartime compassion.

While the Russians left their wounded behind, the Japanese provided medical care to the enemy, paid POWs a modest salary for their labor, and returned all prisoners—some in better shape than they had been before the war—to their homeland. But within just 20 years, Japan's military would be transformed from a model of morality into a template of depravity. This change would be catalyzed by an individual whose own transmogrification from healer to monster was a microcosm of his country's degeneration.

In June 1892, Ishii Shiro was born into a world of power and privilege (see Figure 8.1).[2] His parents were the aristocracy of the village of Chiyoda-Mura, a couple of hours southeast of Tokyo. Based on centuries-old tradition, the Ishii family received tribute from the peasants, who showed deep respect for their feudal lords. So Ishii Shiro—the youngest of four brothers—grew up being waited on by servants in a stately villa amid verdant bamboo groves and fruit orchards.

Although privilege often spawns indolence, Ishii was an energetic student. Throughout his early education, teachers were amazed at his abilities. By adolescence, Ishii's sense of social entitlement and his formidable intellect had conspired to shape a domineering personality. His size and bearing only enhanced this persona. At 5 feet, 10 inches, Ishii towered over his contemporaries, and his normal speaking voice boomed over the hushed tones of the demure Japanese. He was fanatically loyal to the emperor, having been well served by the stratification of Japanese society. Seeking to satisfy both his brilliant mind and darkening heart, Ishii began to formulate a plan to combine his growing interest in medicine with his access to political power.

In April 1916, Ishii was admitted to the Medical Department of Kyoto Imperial University. Brilliant and arrogant, Ishii breezed through his classes and alienated his classmates. He had no use for them, but the academic patriarchs of this prestigious university could prove most useful, if carefully

Figure 8.1. General Ishii Shiro, the mastermind of Japan's Unit 731. Ishii was responsible for developing biological weapons during World War II, in a program that made extensive use of human experimentation. The breakthrough in terms of operational weapons came when Ishii realized that by using insect vectors, the pathogens would be protected from environmental degradation, provided with the conditions needed to reproduce, and carried directly to the human enemy. (Bulletin of Unit 731, Masao Takezawa)

manipulated. And Ishii's genius extended beyond textbooks; he was a brilliant social climber. Through a paradoxical blend of obsequiousness and brazenness, he became a frequent visitor to the home of the university president. If this affront to cultural norms was not sufficient, Ishii swept away all vestiges of propriety by marrying the president's daughter. He graduated in December 1920, having earned a medical degree and cemented his social standing.

A month later, Ishii began his military training and by summer he was commissioned as a surgeon–first lieutenant. Assigned as a physician to the Imperial Guards Division, he quickly found that medical science was much more to his liking than caring for sick people. As a consummate player in the power game of the Japanese military, Ishii managed to secure a transfer to a research posting at the First Army Hospital in Tokyo in the summer of 1922. There, he acquired a reputation for long nights of debauchery and even longer days of research. The latter caught the eye of his superiors, who assigned Ishii to his alma mater for postgraduate studies.

Ishii arrived at Kyoto Imperial University still seeking the optimal path to glory. Medicine and the military had fueled his journey, but Ishii had not been able to chart a clear course into the future. His clarity of purpose finally came in the form of two events, one experiential and the other intellectual. This pair of epiphanies put Ishii on a one-way road to infamy.

The first of Ishii's signposts appeared when he was sent to the island of Shikoku in 1924.[3] Having devoted himself to studies of pathogenic microbiology and preventive medicine, Ishii was an obvious choice to investigate the outbreak of a mysterious disease. When he arrived, Ishii found the patients gaunt and shaking uncontrollably with chills. Soon, they became unable to move their arms or legs, and the inexorable spread of paralysis culminated in a merciful death. Ishii and his colleagues ascertained that the killer was a previously unknown mosquito-borne virus. The discovery of Japanese B encephalitis was a professional coup for Ishii, but the lasting effect for him was witnessing the disease's sociopolitical repercussions. A sudden, unaccountable malady had killed 3,500 people, swamped the medical infrastructure, evoked terror among victims, induced chaos among authorities—and planted a seed in Ishii's mind as to the capacity of an insect-borne disease. What if such power could be harnessed?

Ishii returned to the university and completed another two years of study and research, earning a doctorate in microbiology. He had begun to establish himself as a preeminent medical scholar, publishing a well-received series of papers in prestigious journals. To stay on top, a scientist must be a voracious reader, and Ishii read with one eye aimed at the cutting edge of his field and the other directed toward military tactics. The latter soon focused on a report of the 1925 Geneva Disarmament Convention.

First Lieutenant Harada, a physician and member of the Japanese War Ministry's Bureau, had attended the Geneva Convention, and his government had dutifully signed the agreement to prohibit biological weapons.[4] Although Harada's report had been largely overlooked, Ishii saw within it the key to his future—and that of the Japanese empire. The prohibition of biological agents in warfare was based on a few straightforward considerations. Poison gases had been brutally effective in the First World War, and military scientists were frantically searching for deadlier agents. These research programs were ineluctably drawn toward biological weapons. And the viability of using disease to wage war had been dramatically enhanced by the development of mass-immunization methods to protect the aggressor from a "boomerang" effect. Ishii reasoned that powerful nations would invest their time only in banning weapons that were likely to be wickedly effective. This logic compelled him to take his case to the highest levels of government.

Using his connections within the Tokyo hierarchy, Ishii finagled his way into the heart of the Japanese War Ministry.[5] There, he made an intelligent and impassioned case for initiating a biological warfare program. But he failed to provoke sufficient paranoia among the Japanese leaders to convince them to support his initiative. They appreciated the hypothetical arguments, but Ishii lacked hard evidence that the rest of the world was preparing such horrific weapons. Whether their hesitation reflected the politics of practicality or the vestiges of honor, they would not launch a biological warfare program without a more compelling case. Ishii accepted the challenge with his typical fervor.

Although he was in the midst of starting a large family, Ishii knew that he had to leave his wife and children if he was to prove that other nations were secretly violating the Geneva Protocol. His future depended on convincing the authorities to fund a biological warfare program and, of course, to put him at the helm. Currying favor from his superiors, Ishii received support for a round-the-world tour of military facilities. Leaving in the spring of 1928, Ishii spent two years visiting more than 20 countries, including the powers of Europe along with Egypt, the Soviet Union, Canada, and the United States.

When Ishii returned to Japan in 1930, he had acquired considerable circumstantial evidence of biological weapons programs in other nations. However, his new information regarding misconduct elsewhere in the world was not nearly as important as the ideological shift that had transpired in his absence.

Ultra-nationalism had infused the political system, and the Japanese High Command had embraced aggressive expansionism. They were primed to believe that the western powers were pursuing nefarious weapons. The passion of the government, the power of the military, and the potency of science made for a supercharged combination. Just four months after returning to his homeland, Ishii was promoted to major and appointed to the Tokyo Army Medical College. In parlaying his ascending status into influence at ever-higher levels of the government, his lobbying began to yield tangible results.

Having caught the attention of Koizumi Chikahiko, Japan's most eminent military scientist and dean of the Army Medical College, Ishii garnered support for establishing a department of immunology—a front for his first forays into biological weapons research. Ishii was put in charge of the Orwellian "Epidemic Prevention Research Laboratory." Although protecting Japanese troops from disease was part of the agenda, the military understood that the laboratory's ultimate goal was to initiate epidemics. The formalization of Ishii's program sent a clear signal to the Japanese government that the military considered biological warfare to be a viable line of research and development.

To be fair, Ishii was not entirely monomaniacal in his pursuit of inflicting disease. While at the college, he made his final, positive contribution to human well-being.[6] Ishii invented a ceramic filter that eliminated the need for boiling water as a means of sterilization, a most difficult proposition in the course of battle. He received generous royalties from the Japanese Army and Navy above the table and lucrative kickbacks from suppliers under the table. The latter practice would continue in various guises throughout Ishii's career and allow him to amass considerable wealth. Most important, his standing in the power structure skyrocketed. With access to the emperor's inner circle, Ishii had the clout to make things happen in a big way.

Having made exciting inroads with animal studies at his Tokyo laboratory, Ishii knew that the next step would require a moral leap for even the most zealous political and military leaders. The preliminary results had to be verified: he needed human guinea pigs. Ishii anticipated that practical concerns would also work against him. The biomedical facility could not ensure containment, so infecting human subjects amid the bustling population of Tokyo would be too dangerous. He needed to conduct such hazardous work abroad.

Only a fool would expect another sovereign nation to willingly put its populace at grave risk to support the Japanese biological warfare effort.[7] Ishii was a visionary but he was no fool. So he'd watched attentively as the Kwantung Army of Imperial Japan provided the perfect location for his dream to materialize—a place where people "could be plucked from the streets like rats" and their government would not so much as murmur a protest.[8]

Manchuria had long been the foster child of Asian powers.[9] Since the 12th century, this northeastern region of China had passed through the control of four dynasties, followed by various warlords in the 19th century. In 1898, the Russians forced the Chinese to grant them a lease covering much of the Kwantung Peninsula on the southern coast of Manchuria. When the Japanese thrashed the Russians in 1905, the victors took custody of the peninsula. Such a bold move might have outraged other world powers, but the United States was busy with its own colonial efforts. In a quid pro quo of epic proportions, Japan recognized the U.S. claim to occupy the Philippines in return for the United States' accepting Japan's suzerainty over Korea, control of the coastal city of Port Arthur, and—most critically—occupation of the Kwantung Peninsula.

The Japanese established the Kwantung Army to cement their foothold in Manchuria. By 1919, the chaotic bureaucracy that masqueraded as a government of the region had been converted into a military organization under the aegis of Japan. The occupiers realized that their resource needs had outgrown their borders, so the peninsula was seen as a stepping stone to the coal, iron, oil, and metals of Manchuria. The United States had been placated, but the Asian powers bristled at Japanese expansionism.

With China trending toward unification under Chiang Kai-shek and Russia flexing its muscles to the north, Japan couldn't simply invade Manchuria without provoking the wrath of these formidable nations. The Japanese needed an excuse. So in September 1931, they blew up a section of their own track on the South Manchurian Railway. Attacking yourself would seem to be an odd ploy, but the Japanese declared the destruction had been the work of Chinese insurgents. Japan had to "defend" its interests, so the army attacked a nearby garrison of sleeping Chinese soldiers—and then kept going. Thousands of troops poured through the door opened by the "Manchurian Incident," and by the end of 1932, the Kwantung Army controlled the entire region.

Manchuria became a puppet state, with every Chinese official having a Japanese "adviser" who pulled the strings. And Ishii had the perfect setting for his program. Here was a land in which the putative government could not object to his work. Moreover, Manchuria was a scientific paradise, with excellent facilities built using natural resource revenues and without the bothersome strictures of Japanese culture. The pièce de résistance was the availability of human guinea pigs, provided by the *kenpeitai*. This military police force used sporadic attacks by the Manchurians' underground resistance as an excuse to arrest virtually anyone.

In 1932, Ishii moved his laboratory and staff to a military hospital in the bustling cosmopolitan city of Harbin.[10] Ishii soon discovered that the hustle and bustle of Harbin made it difficult to keep secret the sinister activities of his unit. Ever the slick operator, he convinced the High Command to approve construction of a new facility outside of the city.

Beiyinhe was an unexceptional village of about 300 homes within a diffuse scattering of settlements called Zhong Ma City by the locals. But there was nothing city-like about the area, which consisted of subsistence farms. Beiyinhe was 60 miles south of Harbin and situated a few hundred yards from the rail line, making it both isolated and accessible. The logistics were ideal for Ishii's unit, and nobody would notice if a few hundred peasants suddenly relocated. Or disappeared.

Late in the summer of 1932, the Japanese army swept into Beiyinhe and torched the entire village, save a large building that was suitable for Ishii's headquarters. Chinese laborers were conscripted to build what became Zhong Ma Prison Camp. They were made to wear blinderlike shields so they could not figure out what they were constructing. Even so, those who worked on the most sensitive area—the inner section of medical laboratories within the prisoners' quarters—were executed once the building was complete to ensure secrecy. In all, the facility included about 100 brick buildings, comprising laboratories, offices, living quarters, dining areas, warehouses, cell blocks—and a crematorium. The camp was surrounded by a 10-foot brick wall topped with high voltage wires. At the entrance, a drawbridge led to twin iron doors.

Ishii's house of horrors could hold 1,000 prisoners, although there were normally about half this many. The average life expectancy of a captive was one month. The inmates were shackled but well fed, not out of any sense of compassion but to ensure useful data from the experiments. Most of the records were destroyed, but documents recovered by the Chinese after the war provide a glimpse into the research at Zhong Ma Prison Camp.

According to one report, Ishii's minions captured 40 mice (perhaps rats) from an area in which plague was endemic near the Manchurian-Soviet border.[11] The scientists collected fleas from the rodents, extracted plague bacteria from the insects, and injected the microbes into three communist guerrillas. The subjects became delirious with fever (the data sheets reveal that one had a fever of 104°F on the 12th day) and were vivisected while unconscious. Ishii converted such initial successes into personal gain, being promoted to lieutenant colonel in the summer of 1935. However, the use of human subjects soon set the stage for disaster.

Less than a year after the prison was up and running, rumors spread among the local villages that inmates were being killed.[12] Eyewitness testimony to the horrors inside Zhong Ma Prison Camp came in the fall of 1936. The Japanese guards had drunk themselves into a stupor in the course of the Mid-Autumn Festival (in an ironic twist, this holiday celebrates the Chinese overthrow of their earlier oppressors, the Mongols). Seizing the opportunity, the prisoners who had enough strength to stagger from their cells made their break. Most were soon recaptured, but a band of partisans found a dozen of the fugitives wandering in the nearby woods and hid them from the Japanese. The escapees' stories of atrocities spread throughout the region. With the real purpose of Zhong Ma Prison Camp no longer secret, the Japanese military was faced with either terminating their biological warfare project or relocating the operation. Ishii might have saved the program on his own, but the Russians gave his superiors a compelling reason not only to continue but also to dramatically expand development of biological warfare.

In the midst of the crisis concerning the prison break, the Japanese military police arrested five Russian spies in the Kwantung region.[13] The infiltrators were not nearly as worrisome as the materials they were carrying: glass bottles and ampulae containing bacteria responsible for dysentery, cholera, and anthrax. The threat of biological sabotage was all that Ishii needed to secure endorsement of his work at the highest levels of the Japanese command. So in 1937 Zhong Ma Prison Camp was obliterated and construction began on a facility that would usher in the darkest days of entomological science.

9

ENTOMOLOGICAL EVIL

Pingfan was an undistinguished cluster of eight or ten villages about 15 miles south of Harbin. That is, until the Japanese army made the residents sell their homes for a pittance, razed the hamlets, and forced 15,000 Chinese laborers to build Ishii's dream.[1]

When completed in 1939, the facility was a bizarre cross between a bio-medical death camp and a resort spa. Within its two square miles, Pingfan comprised more than 150 structures: headquarters building (with a moat), administrative offices, laboratories, barns, greenhouses, a power station, a school, a brothel, recreational facilities (including a swimming pool), housing for 3,000 scientists, dormitories for technicians and soldiers—and a prison for the inmates, along with the requisite crematorium (see Figure 9.1).

Ishii had learned an important lesson from the escape fiasco at Zhang Ma, so Pingfan was surrounded by a 15-foot wall topped with high-voltage lines, barbed wire, and watchtowers. In a stroke of architectural paranoia, most of the new buildings within the facility were kept to one story so that they could not be seen from beyond the walls. The operation was further hidden by being named the "Anti-Epidemic Water Supply and Purification Bureau" but soon came to be known by its infamous moniker: Unit 731. Secrecy even trumped a sacrosanct tradition—while the entrance to Pingfan was devoid of symbols, all other Japanese installations displayed the imperial chrysanthemum on their front gates.

Despite these efforts, nobody could hope to conceal a facility of such magnitude. Needing a cover story, the Japanese told the local people that Pingfan had been converted into a lumber mill. From this absurd effort at deception arose the sickest humor, as the scientists of Unit 731 came to refer to their human subjects as *maruta*, meaning "logs."

Figure 9.1. Most of the Pingfan facility that housed Unit 731 was destroyed by the Japanese in an effort to cover up evidence of their biological and entomological warfare program and human experimentation. The ruins of the power plant remain as mute testimony to the enormous scale of research and development efforts—and the suffering and depravity that took place. (Photo by M. Ziegler)

While waiting for his demonic Shangri-la to be completed, Ishii played politics in Harbin, enjoying a life of luxury with his wife and seven children. His schmoozing paid off in the form of a phenomenal annual budget of 10 million yen (an office clerk, working 90 hours a week in Tokyo, earned five yen a month). Unit 731's budget would eventually rival that of the Manhattan Project in the United States.

Ishii and company occupied Pingfan in the fall of 1938. Staffing such an enormous facility was one of his most important administrative duties. There had been enough military doctors to support the relatively modest work at Zhang Ma, but Pingfan was an enormous facility. Ishii routinely spent three months a year in Japan, enticing top scientists with promises of unfettered research in unparalleled facilities and valued service to the nation. The military spawned Ishii's fiendish program, but it could never have flourished without the complicity of the medical and scientific communities.

The scope of Unit 731's work included toxins, plant pathogens, vaccines, and a gruesome range of projects for which human testing was deemed necessary, including studies of frostbite, high-altitude decompression, and poisonous gases. But the bread and butter of Ishii's research program was human disease: finding a microbe and a means of delivery that would constitute a lethal weapon system. Unit 731 initially adopted a shotgun approach. The list of pathogens known to have been tested reads like a who's who of human disease: from anthrax, brucellosis, and cholera to typhoid, venereal diseases, and whooping cough. Not satisfied with their own inventory, the Japanese tried to secure pathogens from other sources, including the Americans.

When Naito Ryoichi arrived in New York on February 29, 1939, nobody was expecting him.[2] However, he seemed to be a respectable scientist on a credible mission. An assistant professor at the Army Medical College in Tokyo, Naito presented a letter of introduction from the Japanese Embassy in Washington. The document explained that Naito was a medical researcher seeking samples of yellow fever virus for vaccine development. He was directed to the office of Dr. William A. Sawyer, director of the virus laboratories at the Rockefeller Institute of Medical Research.

Sawyer was immediately suspicious. Yellow fever had little potential to afflict Japan. Moreover, to prevent the spread of the disease, both the League of Nations and the Congress of Tropical Medicine had explicitly prohibited the importation of the virus into Asian countries for any reason. A Japanese scientist should have known about these restrictions, but Naito feigned naivete. Sawyer gave his regrets, sent Naito on his way, and wrongly assumed that the matter was closed.

Three days later, one of Sawyer's technicians was stopped on the street by "a man with a foreign accent." The man, almost surely Naito, offered the technician $1,000 for a sample of the Asibi strain of yellow fever—an extremely virulent form of the virus. The man explained that a professional rivalry kept Sawyer from providing the needed sample. When this appeal failed, the bribe was tripled. Becoming alarmed, the technician fled from the increasingly desperate man and reported the incident to the Rockefeller Institute. The information was passed to the State Department, where the strange event was dutifully filed. Although the incident caught the eye of the Army Surgeon General's office, the U.S. military had largely pooh-poohed biological warfare. Fortunately for the Japanese, the Americans were nearly as naive as Naito pretended to be.

Although unable to secure the most virulent strains of some diseases, the screening process at Pingfan soon narrowed the list of pathogens.[3] Based on operational considerations and test results, the scientists in Unit 731 focused their studies on two agents deemed to have the greatest potential for weaponization: cholera and plague. Early tests at Pingfan concentrated on various means of spreading bacteria via sprays and bombs. These direct approaches had the advantage of being simple, but the disadvantage of being ineffective. The animal and human subjects did not become infected at nearly the desired rate. However, wartime scientists rarely have the time to perfect their creations before they are deployed by the military.

In the summer of 1939, the Japanese engaged in their first major border clash with the Soviets. Skirmishes near the village of Nomonhan had rapidly escalated, and the Kwantung Army was getting thrashed. The precarious position of the Japanese was deteriorating and they needed a tactical advantage—some secret weapon that the better-equipped Russians lacked. So they called in Unit 731.[4] Ishii saw the Halha River, which roughly divided the armies, as the key to his plan. He dispatched two teams of commandos, who paddled rubber rafts to the Soviet side of the river and poured six gallons of a salmonella and typhoid concentrate (not as promising as cholera, but worth a shot) into the water. His unit also provided the Kwantung Army with 2,000 warheads filled with plague bacteria for shelling the Russian troops. Unit 731 had fired its first biological shots of the war. But there is a big difference between shooting and hitting something.

Undeterred by the epidemiological evidence, Ishii presented the campaign as an unqualified triumph. As for the river assault, Ishii pointed to the Soviet losses to the waterborne diseases dysentery and cholera, while carefully avoiding the troublesome details: neither of the relevant pathogens was poured into the river, Japanese troops suffered similar losses to these diseases, and the only sure victims of the attack were the 40 Japanese who accidentally contracted typhoid fever in the course of handling the jugs of microbes. And the bacterial bombs? Ishii reported that plague had taken a toll on the Soviets at Nomonhan. Of course, there was no sense in his pointing out that the Japanese forces had been similarly sickened, which would suggest that either Unit 731's microbes had drifted back into the Japanese lines (not likely) or the biological attack had not inflicted disease on the enemy (very likely).

Ishii and his scientific staff fully understood that simply dumping pathogens into a moving river and exploding shells laden with bacteria were unlikely to trigger disease outbreaks. So why risk operational failure? In 1939, Unit 731 was on an upward trajectory but clever research would not impress the Japanese hierarchy. To build his empire, Ishii needed tangible results—such as a cleverly scripted battlefield drama. Nomonhan was the perfect stage, and Ishii knew that his audience was both clueless and desperate—clueless as to how to assess the performance of the microbial actors and desperate for any reason to cheer. When Ishii lit up the "Applause" sign for biological warfare, the Japanese leadership gave him a standing ovation. In 1940, Emperor Hirohito decreed a substantial increase in funding for Unit 731. But Ishii knew that his production had been a flop and directed his scientific staff to ensure that the next performance would be a legitimate blockbuster.

The researchers understood that biological warfare had run into two fundamental problems. First, human pathogens are well suited to living in host tissues but poorly adapted to living outside. Heat, cold, desiccation, and ultraviolet radiation quickly destroy the microbes. And second, bacteria cannot move in the environment. Following release from a spray nozzle or bomb casing, only an infinitesimal minority of microbes happen to drift passively into human contact. Even these fortunate few must be inhaled or ingested—and then avoid the host's immunological defenses. There had to be a more reliable delivery system.

The answer finally came to Ishii and his staff: they'd been too clever by half.[5] Rather than forcing human ingenuity at every step of disease transmission, the key was to exploit what millions of years of evolution had painstakingly developed: vectors. Insects solved the problems that had confronted Unit 731 scientists. Fleas, flies, and their ilk protected fragile microbes from the harsh environment while carrying the bacteria directly to the target. And, as an added bonus, some vectors support microbial reproduction in their tissues, effectively amplifying the pathogenic payload.

Delivering infected insects to a military target, however, required genuine innovation. The initial approach was to fully mimic nature, releasing flea-infested rats behind enemy lines. The researchers packaged the animals in parachute-delivered paper containers.[6] As an added twist, the containers self-ignited after releasing their contents, thereby destroying evidence of a biological attack. But the rodent payload required complicated handling and logistics that precluded large-scale attacks.

Eliminating the rats, the Japanese began to adapt an existing delivery system.[7] The Uji bomb was originally conceived to carry a slurry of bacteria, and its ten-quart compartment could hold a lot of fleas (see Figure 9.2). But the high explosive used to rupture the steel casing and aerosolize the enclosed pathogens killed most of the insects. Glass casings were tried, but their fragility made loading a risky business. Then Ishii hit upon the perfect material.

Exploiting Japan's ceramic heritage, he commissioned village artisans to fashion bombshells. Never suspecting the payload of their ceramic containers, the craftsmen soon perfected the casings. At this point, Ishii standardized production and moved fabrication within the walls of Pingfan.

Early trials involved loading the ceramic bomb with 3,000 to 6,000 fleas, with a few rats aboard to provide an in-flight meal for the insects. The rats did

Figure 9.2. An Uji bomb designed by Japan's Unit 731 for carrying ten quarts of bacterial slurry. This basic design was modified to disperse thousands of plague-infected fleas over a target. The disease vectors were packed in small porcelain bulbs set within a ceramic bomb casing. A modest, timed charge exploded the ten-quart payload and thereby released fleas from an altitude of about 500 feet. These devices were used against Chinese targets, but the Japanese soon turned to direct spraying of fleas from airplanes. (Photo by M. Ziegler)

not survive the impact, which was fine, given that the fleas were supposed to be seeking human hosts. In any case, the rats were soon abandoned as they added extra weight, and the fleas did fine without a snack on the way to the target.

In the "new and improved" Uji bomb, the plague-infected fleas were packed into small porcelain bulbs set inside the larger bomb casing. When a modest charge exploded the casing, its fragments shattered the thin bulbs. With an effective delivery system in place, the next challenge was to produce pathogens and vectors in massive quantities.

Pharmaceutical companies and breweries served as the models for industrial-scale production of microbes. Despite some novel challenges (nobody worries if a few yeast cells escape from a keg of beer, but a vat of deadly bacteria is another matter), it was not long before Pingfan was culturing more than 1,500 pounds of microbes every month. Although microbial production methods were well known, nobody had mass-reared fleas.

Unit 731 developed increasingly effective methods for breeding enormous quantities of fleas.[8] At first, the insects were simply produced using human hosts. Ishii's scientists housed a group of ten prisoners in an isolated shed. The men were dressed in heavily padded clothes and seeded with fleas. The human incubators were expected to meet their daily production quota by harvesting a hundred fleas. While 1,000 fleas a day might have been sufficient for research needs, this level of production could not meet the demands of an operational weapon system. So the Japanese devised a process that yielded a phenomenal stockpile of infected vectors.

To start the production cycle, feral rats were caught and chloroformed, and boys were employed to pick fleas from the rodents. The insects were placed into test tubes that were then upended over the shaved bellies of anesthetized, plague-infected rats. Once the fleas had fed and acquired bacteria, the insects were transferred to incubators stocked with uninfected rodents and fleas. This method allowed the continuous production of infected vectors. So great was the demand for rodents to fuel the furious rate of insect production that a four-story granary (an exemption to the one-story constraint for secrecy) was built to feed and house the colony.

As each generation of fleas matured, semi-nude workers harvested and packaged the infected insects. By dressing in loin cloths, the men ensured that a flea landing on bare skin could be detected and brushed off before it had a chance to bite. Major General Kawashima Kiyoshi, a physician and chief of the Medical Service, described the scale of production:

In the detachment's 2nd Division there were specially-equipped premises capable of housing approximately 4,500 incubators. Three or four white mice were put through each incubator in the course of a month; these mice were held in the incubator by means of a special attachment device. There was a nutritive medium and several kinds of fleas in the incubator [both rat fleas and human fleas were produced]. The incubation period lasted three to four months, in the course of which each incubator yielded about ten grams of fleas. Thus, in three to four months the detachment bred about 45 kilograms [99 pounds] of fleas suitable for infection with plague.[9]

Given Kawashima's testimony and the fact that a kilogram of fleas consists of about 3 million individuals, at peak production the Japanese could produce more than *half a billion* plague-infected fleas per year. Such a biological capacity surely delighted Ishii, but he also needed to develop a psychological capacity among his scientists for inflicting suffering.

If Unit 731 was to refine entomological warfare, the researchers had to be desensitized to human suffering.[10] After thousands of repetitions, the killing of animals becomes mundane. From here, one need only begin to speak, and then think, of humans as laboratory animals to callous the soul. In this way, the experiments at Pingfan eroded the vestiges of moral constraints among the scientists.

Within the walls of Pingfan, no structures had a more sinister purpose than the Ro and Ha buildings—the prisons that housed the human subjects.[11] Each building was 120 feet long and 65 feet wide, with Ro dedicated to males and Ha containing both sexes, along with children and infants. The structures could house 400 prisoners, but they typically operated at half capacity. To obtain reliable experimental results, the prisoners were kept in decent conditions: the buildings had central heating and cooling, the cells had flush toilets, and the inmates were provided with nutritious food.

Most of the *maruta* were acquired from Harbin. Han Chinese were most numerous, but experimental subjects included Mongolians, Koreans, White Russians, and Jews. From a processing center in the city, they were transported to Pingfan in freight cars or trucks, deposited at the facility's administrative building, and given numbers. The Japanese assigned code numbers up to 1,500 and then began over again. This allowed sufficient differentiation for purposes of scientific record keeping but confused any attempt to determine the fate of an individual or reconstruct the extent of experimentation. Despite the coding system, it is evident that from 1940 to 1945, the researchers at Pingfan used at least 600 human subjects per year, and some estimates put the number closer to 2,000. What is certain is that nobody who left the administrative center via an underground tunnel to either the Ro or Ha building lived to tell the tale (see Figure 9.3).

Figure 9.3. The former administrative building at Pingfan now houses a museum with dioramas, artifacts, and other displays of the work done by Unit 731. Although thousands of people died within the walls of the facility, the Japanese used an ambiguous numbering system to make it extremely difficult to trace individual prisoners. Only 277 names are known today, and these are engraved on this memorial. (Photo by M. Ziegler)

Human subjects were the key to the rapid development of biological weapons at Pingfan. Various modes of infection were tested—injection, inhalation, contact wounds, contaminated shrapnel—but nothing was more promising than plague-infected fleas. A postwar report based on interviews with Unit 731 scientists detailed the findings of particular experiments:

> The fleas were mixed with sand before being filled into the bomb. About 80 percent of fleas survived the explosion which was carried out in a 10-meter square chamber. . . . Eight of the 10 subjects received flea bites and became infected and 6 of the 8 died.[12]

The most heinous aspect of this research came from the medical scientists' compulsion to precisely monitor the course of infection, an infatuation that led to the practice of vivisecting human subjects. The bizarre rationale was offered by one of the medical technicians—along with a graphic account of the process:

> The results of the effects of infection cannot be obtained accurately once a person dies because putrefactive bacteria set in. Putrefactive bacteria are stronger than plague germs. So, for obtaining accurate results, it is important whether the subject is alive or not. . . . As soon as the symptoms were observed, the prisoner was taken from his cell and into the dissection room. He was stripped and placed on the table, screaming, trying to fight back. He was strapped down, still screaming frightfully. One of the doctors stuffed a towel in his mouth, then with one quick slice of the scalpel he was opened up.[13]

The scientists considered the Chinese to be an inferior race but suitable as laboratory animals. Indeed, the racism of the Japanese was global in scope, stretching to include Americans and Europeans. In previous times, military honor would have stood between POW camps and Unit 731, but Ishii convinced his underlings to extend their studies of human subjects to "white rats."

The POW camp at Mukden was 300 miles southwest of Pingfan.[14] With the war in the Pacific intensifying, the Japanese fully expected to use biological weapons against Caucasian troops, and the scientists thought it important to determine if soldiers of European descent would respond differently to the pathogens. Although the nature and extent of experiments remain matters of debate, the best estimates place the number of POWs used as guinea pigs between 200 and 1,500. Americans seem to have been the principal subjects of experimentation, although British, Australian, New Zealand, and Dutch prisoners were also used. In a particularly gruesome move, a Japanese scientist announced his plan to use a portion of liver excised from an American POW to develop a poisonous bait for bed bugs (family Cimicidae). We may never be certain how many POWs were converted into laboratory animals, but we can be sure that knowledge of human experimentation extended far beyond the walls of Pingfan (see Figure 9.4).

Ishii was eager to reveal his work to his superiors.[15] Films of human experiments were screened by high-ranking officers, but Ishii needed support beyond the military. So he proudly showed his documentary to the prime minister, Prince Takeda (Hirohito's cousin), and Prince Mikasa (the emperor's youngest brother). There can be little doubt that Hirohito himself knew that human experimentation was ongoing in Manchuria. And Unit 731's "open secret" extended beyond the halls of the military and government—Japan's scientific community was fully aware of the research being conducted at Pingfan.

Ishii gave regular presentations at army medical colleges, civilian universities, and scientific conferences. For those who missed his lectures, the nature

Figure 9.4. A rare photo documenting the work of Unit 731. Although no higher-quality version of the image exists, the photo is sufficient to show a doctor standing in front of a pile of bodies of Chinese prisoners who had been used as human guinea pigs. Corpses were disposed of using crematoria in the Japanese biological warfare laboratories. As a result of experiments, field tests, and attacks with biological weapons during World War II, the Japanese killed a total of 580,000 Chinese—slightly more than three-fourths by entomological weapons.

of Unit 731's work was so thinly veiled in scientific journals as to constitute a veritable confession. Consider Ishii's publication on epidemic hemorrhagic fever in the *Japan Journal of Pathology*.[16] The methods section of the paper described how infected ticks (suborder Ixodida) were macerated in a saline solution, the mixture was injected into monkeys, and the consequent symptoms were monitored. In violation of scientific standards, Ishii did not name the species of "monkey" used in this research, which should have been a tip-off that something was amiss. Furthermore, the body temperatures of the experimental animals reached 104.4°F, and even the sickest monkey never attains such a high fever. Only one primate could register such a temperature—a very ill human. In all, Unit 731 scientists published or presented more than 100 papers, eventually becoming so bold and crass as to refer to the experimental subjects as "Manchurian monkeys" (there are, in fact, no such creatures).

Ishii's demented playground soon grew too large for the walls of his fortress. Although the hub of his empire remained in Pingfan, Unit 731's spokes stretched across Asia.[17] The biological warfare network grew to include at least 10,000—by some accounts more than 20,000—physicians, nurses, microbiologists, entomologists, plant pathologists, veterinarians, and other scientists. Some two dozen satellite facilities—some disguised as Red Cross units—formed the support structure for Unit 731.

Lying 70 miles north of Pingfan, just two hours by train, Anda Station served as Unit 731's proving ground. A typical experiment conducted at this remote airstrip was described by Kurushima Yuji, a medical orderly. A square-mile grid of 1,000 boxes lined with sticky paper was laid out, after which an airplane dropped a bomb, which exploded about 300 feet above the grid and rained uninfected fleas over the site. The range and pattern of falling insects provided data needed to set the operational parameters of Uji bombs. However, to fully endorse the device for military use required testing with infected fleas and human targets. According to General Kawashima:

> In the summer of 1941, experiments were performed at Anda Station on the use of the Ishii porcelain bomb charged with plague fleas. . . . The persons used for these experiments, fifteen in number, were brought from the detachment's inner prison to the experimental ground and tied to stakes which had been driven into the ground for the purpose. . . . Flags and smoke signals were used to guide the planes and enable them to find the proving ground easily. A special plane took off from Pingfan Station, and

when it was over the site it dropped about two dozen bombs, which burst at about 100 or 200 meters [500 feet] from the ground, releasing the plague fleas with which they were charged. The plague fleas were dispersed all over the territory. A long interval was allowed to pass after the bombs had been dropped in order that the fleas might spread and infect the experimentees. These people were then disinfected and taken back by plane to the inner prison at Pingfan Station, where observation was established over them to ascertain whether they had been infected with plague.[18]

They had. Some died in just two days; others lingered for ten days or more.

In the early 1940s, a combination of drought and wartime destruction led to a famine in China and spawned Detachment 100.[19] Fascinated by the scale of death—hunger had killed nearly 3 million Chinese—Ishii advocated exploiting this vulnerability through biological warfare. To develop methods for the destruction of crops and livestock, the Japanese chose a site near Changchun, about 150 miles south of Pingfan.

The historical record is frustratingly silent as to whether insects were developed for crop destruction, although it seems likely that they were considered, given their importance as pests of rice and other Asian crops. For example, the brown planthopper (*Nilaparvata lugens*) causes "hopper burn" (the rice plants become brown and crisp owing to the insects' feeding) and the green leafhopper (*Nephotettix virescens*) is a vector of tungro, a devastating viral disease that can wipe out thousands of acres of paddies in a single season. As for vector-borne animal diseases, tick-borne piroplasmosis, a debilitating ailment of horses caused by a single-celled parasite, was investigated by Detachment 100.

Another major biological warfare installation rubbed salt in China's bloody wound. Nanking—a thriving city, resting amid lush forests and stunning mountains in the heart of the country—had been the target of Japan's most brutal conquest. In retribution for having resisted the Imperial Army and served as Chang Kai-shek's capital, Nanking was beaten into submission. During a horrific eight-week period beginning in December 1937, the Japanese turned the city into a scene of depravity: 20,000 women were raped and 200,000 men were slaughtered. A year later, Unit Ei 1644 opened for business.[20]

Ishii appointed his childhood friend Matsuda Tomosada to head Ei 1644. A staff of 1,500 needed a large facility, and—adding further insult to injury—the unit commandeered a hospital in the heart of the city. In short order, the building was surrounded by the requisite ten-foot brick wall topped with barbed and electric wire and patrolled by guard dogs. To a visitor, the first floor of the converted hospital would have appeared to be a conventional medical laboratory. The second floor, however, would have looked like some sort of diabolic zoo, with cages of mice, rats, and ground squirrels; containers of fleas and lice; and flasks of cholera, typhus, and plague. The third floor was a chamber of horrors. The recollections of a soldier assigned to Ei 1644 provide a terrifying tour:

> One had to pass through the main offices in order to get to the third floor, where the cages were. . . . Inside the door, the room was about ten by fifteen meters [30 by 50 feet] with cages all in a row. Most of the *maruta* [up to 100, but usually 20 to 30] in the cages were just laying down. In the same room were oil cans with mice that had been injected with plague germs, and with fleas feeding on the mice.[21]

The primary purpose of the Nanking facility was to support the work of Unit 731, with 100 rearing chambers being dedicated to the production of fleas. However Ei 1644 also had a unique program in which lice were mass produced and infected with typhus. There are no records indicating if, when, or where these weaponized creatures were released, so we might presume that they were not terribly effective against the enemy.

The other auxiliary units in support of the Japanese biological warfare effort sprung up in response to opportunity and need.[22] For example, Unit 673 in Songo was largely dedicated to the study of the epidemic hemorrhagic fever named after this Manchurian city. Songo (the disease) is caused by a virus that is transmitted to humans by ticks that have fed on infected rodents.

With a network of thousands of researchers and technicians stretched across eastern Asia, the Japanese Army understandably demanded a tangible return on its investment. Jars of body parts, cleverly designed bombs, reams of data, plumes of greasy smoke from incinerators, and small-scale tests were all very fine. But if Japan was to win the war, the real killing would need to begin. Ishii and company were delighted to comply.

10

JAPAN'S FLEAS AND FLIES

The scale of 20th-century conflicts led to civilians becoming strategic targets. The morale of a populace, the industrial output of a city, and the agricultural production of a farming district were all vital to protracted, large-scale warfare. The horrific toll on noncombatants from the German Blitzkrieg, Allied bombing, and V-1 rocket attacks in the European theater was not lost on the Japanese, who needed no excuse to attack the Chinese populace, but welcomed the implicit acceptance of such tactics by the international community and the opportunity to further test the moral waters.

The Japanese began using poison gas against the Chinese in 1937 as a military tactic that also served as a probe of political sensitivities.[1] When compelling evidence of chemical warfare was brought to the League of Nations in 1939, nothing was done. Japan had already resigned from the League to protest the body's condemnation of the Manchurian occupation, and the international community's attention was focused on German aggression in Europe. With the rest of the world turning a blind eye, Japan turned to entomological weapons.

In the summer of 1940, plague broke out in the city of Xinjing following what may have been the first attack using flea-charged Uji bombs.[2] However, there is only fragmentary information on the complicity of the Japanese or the scale of suffering. The role of Unit 731 in subsequent disease outbreaks became unambiguous, as Ishii documented that his six-legged soldiers could reliably deliver death with an even more efficient delivery system (see Figure 10.1).

By modifying aerial spraying equipment, the scientists found that aircraft could directly release clouds of fleas over enemy targets. This method was used at Chuhsien, where plague irrupted a month after the attack.[3] The outbreak developed more slowly than hoped, but 21 people eventually died

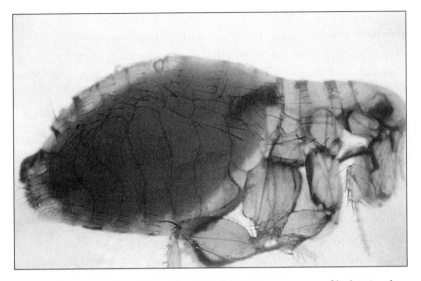

Figure 10.1. The oriental rat flea is one of the insect vectors of bubonic plague, and it was the key to Unit 731's entomological weapons program. This photograph shows a mass of bacteria in the insect's digestive system (the dark mass just in front of its blood-filled gut). This blockage prevents the hungry flea from ingesting a meal and forces the insect to regurgitate bacteria into the host, thereby spreading the pathogen in its futile efforts to feed. (Photo courtesy of CDC)

and many times that number were afflicted. Such losses were sufficient for sabotage operations, but the High Command would want more bodies in a full-fledged attack. And in short order Unit 731 delivered.

The raid on Quzhou demonstrated the potential of infected fleas to inflict serious and sustained damage.[4] The attack in the fall of 1940 triggered an outbreak that continued for the next six years—the city was still losing people after Japan had lost the war. Perhaps some of the death toll was a consequence of the populace being completely unfamiliar with plague. Their first experience with the disease, courtesy of Unit 731, cost 50,000 lives. Entomological warfare was proving to be wickedly effective, but Japan's finest scientists thought they could do better than relying on bloodthirsty parasites.

Believing that technological cleverness could trump evolution, Ishii's staff devised a method for protecting and distributing pathogens without the assistance of vectors.[5] They loaded up three planes with granules containing plague bacteria. On contact with water, the pearly-white grains were designed to swell, rupture, and release their lethal payload. Millions of granules rained down over Kinghwa in November 1940—and then the researchers waited and waited . . . and waited. There were no reports of disease in the city. The microbiologists at Unit 731 grudgingly concluded: no fleas, no plague.

The attacks escalated toward wholesale entomological warfare. For example, the stockpile at Hangzhou for an impending raid included 11 pounds of cholera bacilli, 150 pounds of "typhus" (whether this represented pure microbes or infected lice is not clear, but the latter seems likely given the weight of the payload), and 15 million plague-infected fleas.[6] As Ishii's confidence grew, he began to allow his minions a greater role.

Ishii entrusted Colonel Ota Kiyoshi with the raid on Changteh.[7] To ensure that his master would not be disappointed, the colonel used a bit of overkill. Ota supervised more than a hundred men in producing, loading, and releasing 100 million infected fleas. Within days, an 11-year old girl had died of plague and an outbreak was underway. The disease swept through the city and into the surrounding villages. That first fever-wracked season left 500 dead, and another 7,000 would succumb before the epidemic subsided.

Over the next two years, Unit 731 would attack more than a dozen villages, towns, and cities, causing more than 100,000 casualties. Although the exact number of targets and victims of infected fleas will never be known, some firsthand accounts of the suffering and panic inflicted by entomological warfare have survived.

Archie Crouch, an American missionary, was stationed in Ningbo, a bustling port city with all the requisites of a good target: sultry weather, a thriving colony of wharf rats, and a dense population of humans. Crouch was in the unusual and unfortunate position of being perhaps the only western eyewitness of an entomological raid by the Japanese.[8] His diary provides a haunting view of what transpired that fateful day:

> October 27, 1940. . . . It was unusual to have an air-raid alarm that late in the day. . . . I heard nothing until the plane was over the city. It was flying very low, and that, too, was unusual, since the bombers usually came in groups of three, six, or nine. As this lone plane circled over the heart of the city a plume of what appeared to be dense smoke billowed out behind the fuselage. I thought it must be on fire, but then the cloud dispersed downward quickly, like rain from a thunder head on a summer day, and the plane flew away.[9]

Seeking an explanation of what the plane had dropped, Crouch learned that "the gossip around the city that morning was that the plane had dropped a lot of wheat, so much in some streets that the people were sweeping it up for chicken feed." In fact, the payload had included millions of fleas along with a generous portion of sorghum, wheat, and rice. The purpose of the grain was to attract rodents to the drop site, so that the fleas would have plenty of opportunity to find a blood meal.

The full meaning of the strange aerial deposit would become evident to the people of Ningbo. A week after the raid, Crouch wrote:

> We would soon learn that fleas carrying bubonic plague can cause more civilian and economic destruction than squadrons of planes carrying bombs . . . when the first bubonic plague symptoms appeared among people who lived in the center of the city.[10]

The first wave of sickness swept up 20 people, a modest start but sufficient to catalyze an epidemic. Along with the spread of disease came another outbreak, every bit as important to the Japanese goal of economic and industrial disruption—terror. Unless the authorities took decisive action, panicked people would flee the infected zones and spread plague throughout the city and into the countryside. Understanding the gravity of the situation, the Chinese organized a Herculean quarantine program. According to the journal of the duly impressed American missionary:

> Armies of brick masons were organized to build a fourteen-foot-high wall around the six square blocks in the center of the city where plague was concentrated. The plan was to burn that section of the city as soon as the wall was completed and the people evacuated. . . . No one who lived in the area enclosed within the wall was allowed to leave except through the decontamination sheds.[11]

As soon as the evacuation was complete, the Chinese laid trails of sulfur throughout the walled-off area, ignited the powder at strategic points, and watched as "fires from the burning sulfur raced through the maze like sparkling snakes." The people of Ningbo knew that having lost 97 of their neighbors to plague was a virtual victory relative to the scale of suffering that would have taken place without drastic intervention. What they did not know was that some infected rats and fleas survived the conflagration and that plague would return to the city in 1941, 1946, and 1947, with lesser irruptions until 1959. They also could not know why the death rate was so high: the Japanese had used a particularly virulent strain cultured from human subjects. And finally, given the importance of the Confucian concept of reciprocity to the Chinese people, perhaps they suspected—but could not have known—that insects have the capacity for poetic justice.

Handling millions of infected fleas was almost sure to produce accidents, as described by Ishibashi Naokata, a civilian employee of Unit 731:

> Once, during a transfer, the fleas got loose and got all over the airport. There was a scare that everyone working in there would become infected, and a lot of commotion followed. We sprayed large quantities of insecticides over the airfield, and because of it extensive areas of grass died and turned a bright red.[12]

A few losses to "friendly fire" did nothing to alter the path of entomological warfare, which was on a trajectory toward ever more deadly operations (see Figure 10.2).

With the Japanese having lost any qualms about waging war with insect-borne disease, the scale of biological attacks was limited only by military logistics. The attack on the Zhejian region combined conventional troops with microbial and entomological weapons in a massive, coordinated campaign.[13] Unit 731 assigned more than 300 men to support the 14,000 Japanese infantry in

Figure 10.2. Wang Binhong was a victim of Unit 731's attack with plague-infected fleas at Congshan, where a third of the 1,200 residents succumbed to the disease. He was sent away from the village to avoid the outbreak, but fell ill shortly after he returned. His father had heard of a man who survived the disease by consuming alcohol, so the 15-year-old boy was fed nothing but alcohol. Wang Binhong survived both the treatment and the disease. (Photo by M. Ziegler)

a retributive and strategic offensive. Colonel Doolittle's audacious bombing raid of Tokyo, Yokohama, and other cities had enraged the Japanese, and they knew that the American pilots had received sanctuary among the villages of Zhejian. The Japanese were seeking both to punish the Chinese and to ensure that the region would not become a host for airfields in support of U.S. bombing runs.

The Japanese troops were ferocious in meting out punishment to the settlements where American airmen had received aid, and Unit 731 was able to drive home the lesson with their special brand of suffering. Infected fleas rained down from the planes, and pathogenic bacteria flowed from barrels and nozzles. Despite these tactics, the Japanese biological campaign was largely ineffectual. While the military leaders were in no mood for philosophizing, the scientists understood that sometimes we learn more from our failures than from our successes. Although outbreaks of plague had laced the region, the use of cholera had been a disaster. Pouring and spraying the bacteria put the attackers at greater risk than the defenders—the Japanese suffered 10,000 biological casualties and 1,700 fatalities, while few Chinese contracted the disease.

Ishii's staff realized that unless a better means of dispersing the bacteria could be found, cholera was a loose cannon. But its potential to devastate an enemy was simply too great to abandon weaponization. Until now, cholera had been viewed as a water-borne disease, so dumping bacteria into wells or spraying microbes over ponds and streams seemed to be sensible tactics. The problems were that handling the concentrate often infected your own troops, that even large volumes of culture were quickly diluted in most water sources, and that the bacteria required some rather particular conditions to survive, let alone flourish, in aquatic systems. There had to be a better way of sparking an epidemic.

The scientists of Unit 731 knew that, once under way, cholera spreads like wildfire. The disease produces a flood of watery diarrhea that is swimming with bacteria. Death from dehydration occurs when a person loses about 15 percent of available body water. This lethal volume works out to be about ten quarts—and with a quart of liquid feces per hour, death from shock, kidney failure, and circulatory collapse may occur in less than half a day. Once the flow of contaminated excrement gains momentum within a human population, the disease proliferates rapidly as victims contaminate the water supply. The problem for the Japanese military was how to get the cycle started.

Merely dumping bacteria into the enemy's water hadn't infected enough people to trigger an outbreak. A new approach was needed—or the adaptation of a tried-and-true tactic. If insects were effective in delivering plague, perhaps these carriers could be conscripted for cholera. From this small leap of ento-mological logic came the greatest military success in the modern annals of biological warfare.[14]

Japanese epidemiologists realized that the key to triggering the initial wave of infection was to put high numbers of bacteria in intimate contact with even a relatively small fraction of the target population. What Unit 731 sought was a cholera carrier with a strong affinity for humans.

House flies (*Musca domestica*) get their name for a very good reason—they flourish among human habitations.[15] To be more precise, these insects are what entomologists call "filth flies," for it is not our houses but our garbage and sewage that the adults and maggots find so tasty. With respect to enteric disease transmission, flies acquire pathogenic bacteria while feeding on excre-ment from infected individuals. Studies have found 4 million bacteria per fly in insects collected from slums, and almost 2 million in flies from apparently clean neighborhoods.[16] The hairs on the fly's body function as microbe mag-nets, while the sponging mouthparts of these insects are peppered with tiny pores and fine hairs ideally suited for picking up pathogens. To make mat-ters even worse, because house flies can consume only fluids, they regurgitate prior to feeding on solid food to initiate the breakdown of their meal. The insect vomits up a soup of enzymes, including a portion of its last snack, onto the prospective meal—a bit of food on a person's plate, cup, hands, or lips. And given the ease with which house flies could be produced en masse, the Japanese had three of the essential ingredients to trigger a cholera epidemic: bacteria, vectors, and people. There was just one problem.

Unfortunately for the Japanese, the Chinese were fastidious, not particu-larly prone to producing breeding grounds for flies around homes, schools, and factories. Introducing bacteria-laden flies would create an outbreak only if public hygiene could be undermined. Fortunately for the Japanese, war is messy—and with a bit of timing, the consequent filth can be a powerful pre-lude to entomological warfare. With this final piece in place, the Japanese were ready to launch an attack the likes of which the world had never seen.

Yunnan Province had become a thorn in the side of the Japanese. This region hosted an Allied supply line into China, providing Chiang Kai-shek's Nationalist forces with the supplies and arms they needed to resist the Japanese. The route leading from Burma through the city of Baoshan and into southern China made this tropical region one of the most strategically vital areas in the war. On May 4, 1942, a wave of 54 Japanese bombers descended on Baoshan, dropping tons of explosive and incendiary bombs. The city was decimated:

10,000 people died in the raid and more than three-fourths of the buildings were destroyed. Mixed in with the conventional ordnance were a number of ceramic-shelled bombs.

At first, these bombs appeared to be duds—the casings had burst open without exploding. But the nature of these special devices was soon evident. Lin Yoyue, a retired elementary school teacher, described the bizarre contents as being a "yellow waxy substance [with] many live flies struggling to fly away." He had discovered Unit 731's brainchild, officially called the Yagi bomb, but known among its developers as the "maggot bomb." It was divided into a section packed with a gelatinous slurry of bacteria and a compartment loaded with flies—not larvae, as the device's vile nickname would suggest, but winged adults. On impact, the casing burst and the insects were splattered with a slimy coating of cholera bacteria. Released from their confinement, the flies dispersed into the decimated city. The populace had no chance to ponder the unusual invaders, as the Japanese were not done with their dastardly plan.

The planes returned for three more bombing runs on May 5, 6, and 8. Rather than simply moving the rubble around, these attacks had a purpose unique in the annals of aerial bombardment. The goal was to move the people. Sickening Baoshan was a fine start, but the supply route of the Allies could just be moved to bypass the diseased city. The Japanese sought a regional epidemic, and the series of bombings was intended to drive the infected people into the countryside—carrying cholera along with them. The refugees unwittingly complied, taking along their churning intestines.

By June, cholera had spread into more than half of the counties in Yunnan Province. Villages were ravaged as far as 125 miles from Baoshan, with 25 to 50 percent mortality being typical. Some 60,000 of the city refugees died of the disease, with more than twice this number succumbing throughout the region. The final tally reached 200,000 victims across an area equal to that of Pennsylvania—a scale of death that provided the Japanese with an unqualified victory. The Allies' supply line was utterly contaminated. Moreover, with this epidemic raging, the Chinese Nationalist Army could not base troops in the region. By creating this diseased no-man's-land, the Japanese were free to divert thousands of soldiers to other fronts. Any strategy that worked this well was worth repeating.

In August 1943, the Japanese duplicated their "decimate-and-contaminate" ploy in the northern province of Shandong. Because Japanese troops were in the area, a new twist was added—their soldiers were vaccinated against cholera to prevent losses to "friendly fire." Once again, an epidemic swept through the region afflicting towns and villages and spreading into parts of adjacent provinces.

Cholera's final score from the maggot-bomb campaigns: China 410,000, Japan 0.[17] Yunnan and Shandong became the Hiroshima and Nagasaki of China, with flies and microbes taking as many lives as atomic bombs took in Japan.

By now, the frequency and characteristics of disease outbreaks—along with eyewitness accounts of falling fleas and fleeing flies—left no doubt as to Japan's tactics. General Chiang Kai-shek communicated the nature of these attacks to Winston Churchill and Franklin Roosevelt.[18] In 1943, Roosevelt threatened to retaliate against the Japanese if entomological raids persisted. The Japanese were undaunted and, more important, they were becoming desperate. Threats of future retribution coming from across the ocean hold little sway when a nation sees an enemy massing in the present, across its border.

11

JAPAN'S PLEAS AND LIES

The imminent conquest that Japan faced in 1943 had been unimaginable just a few years earlier. Indeed, 1941 had been a heady time for the Axis powers. In June, Nazi Germany invaded the Soviet Union, and Japan was anxious to prepare for its part in the offensive. From the highest levels of the Japanese military came the order for Unit 731 to accelerate its work on plague. The plan was to initiate massive epidemics within the Soviet Union, softening the enemy in preparation for a conventional offensive from the east, as the Third Reich invaded from the west.[1]

At 4:45 A.M. on June 22, 1941, 4 million German, Italian, Romanian, and other Axis troops poured over the Soviet border. The offensive became a mad dash into the heart of Russia, with the Soviet troops in utter disarray. So paltry was the resistance that Panzer divisions outraced the infantry. By late fall, the Germans had surrounded Moscow and laid siege to Leningrad—and winter came.

Unable, unwilling, or unprepared, Japan failed to launch an attack from the east, allowing the Soviets to focus their might on a single front. They counterattacked that winter, and the following summer saw a series of bloody battles with neither side gaining much ground. By fall, it was clear that the Axis offensive had lost all momentum. The tide was turning.

The Soviet counteroffensive in the winter of 1943 unleashed a juggernaut that buckled the German lines and opened an unstoppable surge to the west the following summer. With the Red Army reaching the Baltic States and punching into Poland and Hungary in 1944, Japan's biological warfare plan shifted from optimistic offense to desperate defense. As the war on their western front became a rout, the Soviets turned their attention to the Manchurian border.

By this time, Ishii had been reassigned to Nanking as the chief of the First Army Medical Department.[2] But the Kwantung Army needed him to oversee the enormous increase in weapons production that was viewed as the best hope for repulsing the impending Russian invasion. Ishii was summarily transferred back to Unit 731 and given the responsibility of preparing a biological arsenal sufficient to stop the Soviets.

Ishii's plan called for using 300,000 rats as the basis for producing several hundred million infected fleas. A Japanese officer recalled that in the summer of 1945, "the squad [of soldiers dressed in civilian clothes] was brought up to 30 men and the mass trapping of rats intensified."[3] When the military could not spare any more soldiers to hunt rodents, the Japanese began paying a bounty to children for every live rat they could bring to gates of Pingfan. Having soon depopulated the countryside of rodents, the locals were encouraged to start their own colonies. The farmers quickly realized that rearing rats generated significant income, and the peasants began to provide Pingfan with a steady stream of live rodents, never knowing—or perhaps wondering about —the fate of their squeaking livestock.

With an initial capability to generate 130 pounds of fleas in each three-month production cycle, Ishii exhorted his staff to labor unceasingly toward a goal of 200 pounds. Once the frantic scientists and technicians approached this output, Ishii pressed even harder. Referring to the high appraisal and enormous faith that the Imperial Army expressed for Unit 731, he called on his staff to redouble their efforts. Ishii drew up plans to produce annually more than 5 billion infected fleas—more insects than there were people on the planet.

Such a mind-boggling quantity of fleas demanded an enormous amount of blood. To feed and infect so many insects would have required 50,000 incubators—far more than Unit 731 could possibly maintain. Circumstantial evidence indicates that the Japanese were pursuing a more productive system, perhaps one in which huge numbers of fleas could be fed on blood through an artificial membrane. Whether or not this technology was fully developed, the limiting step in production became acquiring sufficient volumes of infected blood to feed the fleas. The Japanese turned to one of the largest pools of blood at their disposal, their human livestock—the *maruta*.

In Nanking, Ei 1644 began keeping Chinese prisoners in kennels—rows of 40 or 50 animal cages about four feet on a side.[4] These *maruta* did not need to be decently fed or housed, for their bodies were nothing more than living incubators. They were inoculated with bacteria and the disease was allowed to progress. Once a victim was deemed ready for harvesting, a technician dragged the delirious prisoner from his cage and administered chloroform or restraints as needed. Then, the *maruta* was taken to surgery where his femoral artery was severed.

Although this massive buildup of biological arms was motivated by the expectation of a Soviet invasion of Manchuria, U.S. forces were also moving toward Japan. If entomological warfare could repel the Red Army, might these weapons also halt the Americans' methodical advance across the blood-soaked islands of the Pacific? The United States had ignored and discounted the danger of Japanese biological weapons for most of the war. Racist perceptions assured the Allies that their uncivilized, heathen enemy would not be capable of unleashing a weapon that required scientific sophistication and technological prowess. On several occasions, only the whims of war kept the Americans from learning a brutally hard lesson as to the hazards of bigotry and arrogance.

The first near-hit came in March 1942, when the Japanese planned to mobilize their biological warfare units against American and Philippine troops defending the island of Bataan.[5] Pending the outcome of a conventional battle, the plan called for releasing 200 pounds of plague-carrying fleas—about 150 million insects—in each of ten separate attacks. However, when the Japanese forces summarily defeated the enemy in early April, the entomological assault became unnecessary.

In the summer of 1944, the United States attacked Saipan, one of the Mariana Islands of immense strategic value in the war.[6] Japan knew that if the airbase fell into the hands of the Americans, it would become a staging ground for the bombing of their homeland. After 24 days of horrific fighting, the Japanese defense finally collapsed on July 9. Ishii proposed that a commando team could sneak onto the island and introduce plague-infected fleas to sicken American forces and prevent them from mounting air raids. The Japanese leaders approved the plan and dispatched a squad of 20 men under the command of two army medical officers. The U.S. troops might well have paid the painful price of their leaders' underestimation of the enemy, except for a lucky torpedo. The ship carrying the insectan warriors was sunk by an American submarine. Having again dodged the biological bullet, the Americans remained blissfully unaware of the impending entomological assault on Okinawa.

For 82 days beginning in April 1945, Allied and Japanese forces battled for control of Okinawa. To lose this colonial possession was to allow the Americans a foothold in Japan itself. Once again, Ishii plotted to repel the invaders with plague.[7] But as with Bataan and Saipan, the entomological weapons apparently never made it to the target. In the case of Okinawa, however, it is not clear what prevented their use. Perhaps the Japanese drew the line at spreading disease on their own land, or maybe they saw that the war was almost certainly lost and did not want to provoke the wrath of the Americans. At least this latter consideration played a role in what would have been the most memorable entomological attack of the war—a raid on the American mainland.

In 1944, U.S. interrogations of POWs left no doubt that the Japanese were capable of waging biological warfare (although the prisoners claimed that Pingfan was intended only to counter the Soviet potential). And by the end of the year, Americans came to realize that the enemy also had a delivery system capable of reaching the United States. That November, Japan launched more than 9,000 hydrogen balloons armed with incendiary payloads. The release was carefully timed and positioned so that the jet stream would carry the nasty surprise packages to North America (see Figure 11.1).[8]

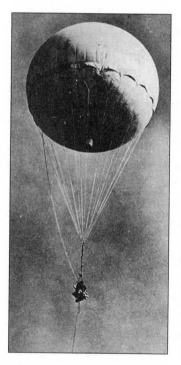

Figure 11.1. A "Type A Fugo," or balloon bomb, used by the Japanese in World War II. The hydrogen-filled balloon was about 100 feet in diameter and could lift a 400-pound payload up to 32,000 feet for its journey on the jet stream across the Pacific Ocean to the west coast of North America. At least 50 Fugos were seen over the state of Washington on a single day. Although the bombs were equipped with incendiary charges, the Allies were concerned that payloads could include biological weapons.

About 300 of the balloons made it to the shores of the United States and Canada. There was some concern that such attacks could serve to bolster Japanese morale, but there was little worry that the firebombs could cause much damage to the wet forests of the Pacific Northwest. American officials became far more alarmed with the possibility that the next wave of balloons could carry pathogens—and even their vectors.[9] However, the Americans hadn't a clue that the Japanese planned to launch their biological attack from beneath the seas rather than from the air.

In the waning months of the war, Ishii managed to convince his superiors that desperate times called for desperate measures. He worked with other military commanders to develop a plan for attacking the U.S. mainland in what came to be called Operation Cherry Blossoms at Night—a lyrical name for a wicked plan.[10] The operation called for one of Japan's unique plane-carrying long-range submarines to sneak into the waters off the coast of California. At nearly 400 feet in length and with crews of 144, these were the largest submarines to ply the depths until the nuclear ballistic missile-toting vessels of the 1960s. Once near shore, the Japanese sub would surface, the fold-up planes (there were three on board) would be assembled for flight within minutes, the plague-infected fleas would be loaded aboard, and the kamikaze pilots would release their pathogenic cargo over San Diego. This attack was to be supplemented with a boatload of commandos launched toward the harbor to spread plague and cholera along the waterfront. The plans were finalized on March 26, 1945, and the components were readied for deployment. At the last moment, the operation was scrapped by the chief of general staff. General Umezu Yoshijiro found his moral courage by tapping into a sense of cultural shame. He realized that such an attack on the United States could trigger retaliation in kind, which might spiral into "an endless battle of humanity against bacteria [for which] Japan will earn the derision of the world."

Despite Umezu's sense of honor, Japan remained committed to waging entomological warfare against the Soviets. And the race was on to prepare the weapons before the Red Army descended. While the stockpile of infected insects was building in 1945, the Japanese command hurried to assemble the necessary field battalions for using the biological weapons. Still clinging to a pathetic vestige of secrecy, these were called "Water Purification Units." The top brass devoted military manpower commensurate with the enormity of the production effort. There were 13 units of up to 500 men and more than 40 units each with about 250 personnel.[11] Despite the frenetic effort to prepare for entomological warfare, only so many *maruta* can be exsanguinated, so many rats caught, and so many fleas produced. In the end, the biological calendar of the Japanese turned slower than the military clock of the Soviets and the atomic clock of the Americas.

On August 6, 1945, the first atomic bomb fell on the Japanese city of Hiroshima. Three days later, the second bomb fell on Nagasaki—and the Red Army poured across the border of Manchuria. The Japanese were utterly overwhelmed. Ishii was stunned by his country's surrender on August 14, but he quickly emerged from his stupor and commenced the most important project of his career: erasing Pingfan and hiding its secrets.

As the Russian troops rolled toward Pingfan, Ishii issued his last commands. First, he ordered the slaughter of the laborers and prisoners. Then, he had critical files shipped to his most devoted subordinates in Japan, where his minions would obediently receive the documents and bury the treasure in the garden of his Tokyo home. Next, Ishii gathered his staff. He needed to squelch them as effectively as he'd silenced the Chinese laborers, but mass murder wasn't an option. Rather, the carrot-and-stick approach seemed most effective—starting with the stick. Akamo Masako, a nurse in Unit 731, recalled:

> It grew dark. Ishii came over to us carrying a big candle and said, "I am sending you all back home. When you get there, if any one of you gives away the secret of Unit 731, I personally will find you, even if I have to part the roots of the grasses to do it." He had a fearful, diabolical look on his face. My legs were shaking. And not just at me—at everyone. "Even if I have to part the grasses . . ."[12]

And then, the carrot. Ishii had amassed a fortune through embezzlement, kickbacks, legitimate royalties, and other means. Moreover, he had been so open with the work of Unit 731 that virtually all of his superiors were implicated in the war crimes. So he was well positioned to secure substantial investments from high-ranking officials to buy the silence of his subordinates with payments of up to 2 million yen.

Ishii's final order was among his most malevolent. The infected rats were released into the surrounding countryside. As Ishii boarded a train that would take him to a waiting plane, the rodents reveled in their newfound freedom. By the time Unit 731's commander arrived in Japan, a wave of plague arrived in China. The ensuing epidemic spread into 22 counties and killed more than

20,000 people.[13] One might imagine that adding a few thousand more victims might have tipped the scale of justice toward the condemnation and prosecution of Ishii and his inner circle. But the other pan on the scale was rapidly offsetting the value of human life: the value of science to the American military. For three long years after the war, Ishii and his lieutenants engaged the American military and their scientists in a convoluted series of investigations, feints, threats, pleadings, and offers that culminated in a Faustian deal that isn't found in the history texts.

After interrogating Ishii and his lieutenants for countless hours, the American investigators arrived at two conclusions.[14] First, the Allied Supreme Command became convinced that the Japanese scientists could be of significant value to the U.S. military's biological warfare efforts. And second, the International Prosecution Section—the organization conducting war crimes trials—determined that the Japanese had conducted human experimentation and disseminated plague-infected fleas on the Chinese people. So it was that General MacArthur's desire for military intelligence came to loggerheads with the chief prosecutor's concern for justice (see Figure 11.2).

The rapidly developing Cold War added fuel to the militarist-moralist fire. The battle against the fascists was over, but the United States saw a potentially greater enemy emerging from the rubble of war—the communists. A coordinating committee of government and military officials cast the matter in starkly pragmatic terms:

> Since it is believed that the USSR possesses only a small portion of this technical information, and since any "war crimes" trial would completely reveal such data to all nations, it is felt that such publicity must be avoided in interests of defense and security of the U.S.[15]

While politicians and generals debated the pros and cons of the information-for-immunity deal through 1947, Ishii preserved his value to the Americans. He began unearthing his buried treasure, providing his captors with 8,000 histology slides, reams of experimental protocols, three massive autopsy reports, hundreds of drawings, and several large cases of documents. This windfall fed the Americans' lust for knowledge and power. The report by the American researchers to their military superiors made clear that ethical constraints were not a major concern to the scientific community:

Figure 11.2. The International Military Tribunal in Tokyo, where Japanese general Tojo Hideki (fifth from the left on the second row) is accused of war crimes. Tojo, prime minister from 1941 to 1944, was hanged in 1948. General Ishii Shiro, the commander of Japan's biological and entomological warfare program, avoided prosecution through a deal with the Americans. Ishii exchanged clemency for the secrets of Unit 731, which turned out to be of little value to U.S. scientists. (Photo by AFP)

> Such information could not be obtained in our own laboratories because of scruples attached to human experimentation. These data were secured with a total outlay of ¥250,000 to date, a mere pittance by comparison with the actual cost of the studies.[16]

In 1948, circumstances finally called the Americans' hand. The Tokyo War Crimes trials were coming to an end and the window of opportunity for presenting new evidence was closing. Attempting to sustain a vestige of legal integrity, the U.S. prosecutors claimed that their case against Ishii and his crew was ultimately too weak to pursue. Capitulating to political pressure, they opted for de facto immunity via nonprosecution.

Having lost this early round of the Cold War to the American biological warfare community, the Soviets came out swinging in the next round.[17] On Christmas Eve of 1949, the Russians formally charged their Japanese captives—including 12 officers and enlisted men from Unit 731 and its sister facility in Nanking—with war crimes. News of the trial alarmed the U.S. State Department, where officials worried that the Soviets would use their proceedings to publicly reveal the immunity-for-intelligence deal that had been struck by the Americans and thereby score a major international propaganda victory. Although the testimony presented in the six-day trial suggested that Ishii had been the kingpin of the operation and that Allied POWs, including U.S. soldiers, had been subjected to experiments, the Japanese-American quid pro quo remained a dirty little secret.

The findings of the Khabarovsk trial—named for the grungy industrial city where the proceedings were held—filled reams of court transcripts, detailing the production system for insect vectors, specifying the various incidents in which insects were used, and providing explicit detail on the rationale for the development of entomological weapons by the Japanese:

> From the materials of the Court investigation it is evident that detachments 731 and 1644 utilized fleas on the basis of the "theory" of the ideologist of bacteriological warfare, Ishii Shiro, from which it followed that in this case the fleas served to protect the plague germs from the action of external environmental factors. According to this "theory," the germs being within the organisms of the fleas, thereby acquired, as it were, a living, protective integument. Attacking human beings when in search of food, the plague-infected fleas bit them and thereby infected them. Thus, the fleas were intended for the purpose of preserving the germs, of carrying them, and of directly infecting human beings.[18]

In the knockdown, drag-out fight between communism and capitalism, the Soviets racked up some major points with their high-profile trial in 1949 and righteous indignation over the nonprosecution of Ishii. Then, when nobody was looking, they resorted to the tried-and-true tactic of hypocrisy.[19] Although the Soviets sentenced the accused to labor camps for periods of 2 to 25 years, when world attention turned to other matters, all of the convicts were repatriated to Japan in 1956. Such leniency was part of a Soviet-Japanese deal in which the prisoners exchanged information beyond that of the trial testimony for clemency in an arrangement highly reminiscent of the United States' bargain with its captives.

If the Soviets' experience was anything like the Americans', then the superpower ended up with the short end of the stick.[20] William Patrick, who worked at Fort Detrick for three decades and served as chief of the Product Development Division, had the opportunity to review the contents of seven steamer trunks sent from Japan. He discovered that the Japanese secrets fell far short of constituting valuable science. For example, to test the infectivity of anthrax spores, the Japanese scientists placed a few milligrams of a wet paste into the ear of a subject, but the dose was not quantified so one couldn't repeat the experiment (not to mention the bizarre mode of application—few people would end up with anthrax paste in their ears during a biological warfare attack).[21] In the end, much of the data was purely anecdotal and many of the experiments were fatally flawed in their designs. But by the time the U.S. scientists figured out the deplorable state of the "science" for which they'd traded their moral integrity, the Japanese were busy trying to live happily ever after.

Ishii enjoyed his retirement, receiving a generous military pension.[22] He continued to consult with the Americans and may have traveled secretly to the United States to provide lectures to biological-warfare scientists. Many of his staff also did very well for themselves, rising to prominent positions in academics, government, and business.[23] One became the chief of the Entomology Section of the Health and Welfare Ministry's Preventive Health Research Laboratories, another emerged as surgeon general of Japan's armed forces, yet another rose to the presidency of Japan's Medical Association, and every direc-

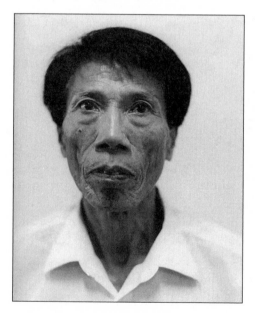

Figure 11.3. Zhong Shu is a Chinese plaintiff in the lawsuit against Japan for damages caused by entomological weapons in World War II. His grandmother died from bubonic plague, following an attack by Unit 731 with infected fleas. Zhong Shu seeks neither an apology nor money from the legal action, and he bears no hatred for the Japanese. Rather, "It is a question of honor. I was the oldest grandson, and my grandmother loved me a lot." (Photo by M. Ziegler)

tor (with one exception) of the Japanese National Institute of Health between 1947 and 1983 had served in a biological warfare unit, half of them having conducted human experiments. It is no wonder that the Japanese continue to deny the atrocities of Unit 731—the cultural embarrassment would be devastating.

Amid rumors that former members of his unit had evened a longstanding score, in 1959 Ishii died—most probably from throat cancer rather than assassination.[24] His story spanned three centuries. He was born at the end of the 19th century, his depraved program of science stained the 20th century, and his final tally was determined in the 21st century. In 2002, an international symposium of historians convened in Changde, China, to arrive at what has become the accepted figure for the death toll from Japanese biological warfare.[25] As a result of attacks, field tests, and experiments, a total of 580,000 Chinese were killed—slightly more than three-fourths by entomological weapons (see Figure 11.3).

The ghosts of Japan's Unit 731 would haunt the Cold War for decades to come. But the Japanese were not the only Axis power to alter the nature of warfare in the 20th century. While Japan mastered the use of insects to transmit disease, Germany devoted its formidable scientific expertise to waging war on crops.

12

BEETLE BOMBS

If Adolf Hitler had not forbidden offensive research on biological warfare, the Germans might have surpassed the Japanese in entomological weaponry. Scholars speculate that Hitler's aversion to unconventional arms may have stemmed from having been gassed in the First World War. Others note that among Hitler's eccentricities was a phobia of bacteria, so producing pathogens by the ton might have been too much for him to contemplate.[1] The Führer, however, was not entirely in control of his military.

Although Hitler's prohibition impeded progress, the Nazis were able to surreptitiously pursue offensive research under the auspices of developing defensive capabilities. Germany's research and development program was centered at the SS Military Medical Academy at Posen, under the supervision of Professor Kurt Blome—Germany's version of Ishii Shiro, but without the megalomania. As part of Blome's network, an Institute of Entomology was established within the Waffen-SS by order of Heinrich Himmler.

The entomological research initially focused on vector-borne diseases. Medical scientists tried to weaponize typhus-infected lice, using prisoners at Natzweiler, Dachau, and Buchenwald as experimental animals. Another program investigated methods for triggering plague outbreaks in enemy ports. And intrigued by the outbreak of malaria in war-torn Greece, Blome discussed whether it might be possible "to spread malaria artificially by means of mosquitoes."[2]

Nobody can be surprised that lice, fleas, and mosquitoes attracted the attention of Nazi war planners; these disease vectors had caused millions of deaths in previous wars. But few could have guessed that the linchpin of the German biological warfare effort would be a lethargic—albeit hungry and fecund—thumbnail-size black-and-yellow striped herbivorous insect, the

Colorado potato beetle (see Figure 12.1). This dumpy insect's military career was launched by virtue of its being in the right place at the right time.

The beetle's early years among people were rather unexceptional.[3] *Leptinotarsa decemlineata* (the species' name being a reference to the ten lines or stripes on its body) was identified in the early 19th century, living quietly along the Rocky Mountains where it fed on prickly nightshade—a poisonous, spine-covered plant. But the nightshade family also includes species of agronomic value, including the insect's namesake (the leaves of which are toxic). So when potatoes were brought to the western frontier, the beetles flourished. The insects also found related plants quite delectable, including egg plant, peppers, tomato, and tobacco.

While pioneers followed wagon trails westward, the beetles followed a food trail eastward. In 1869 the insects reached the lush farmland of Ohio, and by 1874 they made it to the East Coast. Seeing the devastation wrought by the pest, the Europeans banned the importation of American potatoes. The wholesale destruction of potato fields struck a raw nerve on the other side

Figure 12.1. A Colorado potato beetle was the mainstay of the German entomological warfare program in World War II, although the French and Americans also pursued weaponization of this crop-feeding pest. The Nazis estimated that 20 to 40 million beetles would be needed for a major attack on English potato fields. Whether such a coleopteran assault transpired is a matter of debate, but millions of beetles had been stockpiled by the summer of 1944. (Photo by Scott Bauer, USDA/ARS)

of the Atlantic. Not 25 years earlier, the Great Hunger had spread across the continent. Beginning in Ireland, where the suffering was most severe, a fungal disease turned potatoes into black rotting lumps. Between 1846 and 1850, a million people died from hunger and disease owing to the blight, and another two million became refugees. People do not soon forget such misery.

Europe's potato crops stayed beetle-free until World War I, when the influx of troops and supplies from the United States made the quarantine impossible to sustain. When the beetle established a beachhead near Bordeaux, France, the echo of the Irish potato famine became the drumbeat of entomological warfare.

In 1938, an eminent British scientist, John Burdon Sanderson Haldane, published a paper entitled "Science and the Future of Warfare," which launched the military career of the Colorado potato beetle. This influential scientist's entomological predictions were much more accurate than his other conjectures. Haldane dismissed the possibility of developing more powerful explosives, pooh-poohed the notion of finding deadlier chemicals, and soundly renounced the idea of producing aerosols of pathogenic microbes. But he viewed insects as having the potential to play a significant, if not decisive, role in coming conflicts:

> On the other hand, it is reasonably probable that some biological methods will be used. It would be very surprising, for example, if insect pests, such as the potato beetle, were not introduced into this country by hostile aeroplanes in the course of a future war. The potato beetle would not cause a famine, but it would cause a certain amount of trouble and keep a certain number of people busy who could be used for other purposes. . . . The Germans may drop potato beetles on us or the French may drop them on the Germans.[4]

Naming the combatants seemed a particularly bold move. But Haldane surely knew that the increasing beetle populations in western Europe would evoke agricultural anguish—and military temptation. Whether he was a brilliant prognosticator or simply an informed scientist connecting the dots, Haldane's forecast turned out to be uncannily close to the mark. France became the European center of entomological warfare in the late 1930s.[5]

The use of insects as weapons was broached by the French in May 1939 during a meeting of the Veterinary Surgeon's Commission for Disease Prevention in Modern Warfare. Despite the organization's apparent focus on animal health, the insects of interest were plant pests. Lieutenant Colonel Guillot proposed developing a system for dropping beetles from the air onto enemy potato fields. In light of the importance of this crop to the Germans and the demonstrable damage caused by natural infestations of the beetle, the commission approved the plan. By that September, France's program to weaponize the Colorado potato beetle was under way.

Under the leadership of Professor Paul Vayssière, the entomological warfare program began to develop methods for producing and dispersing the beetles. Details are sketchy, but at least some progress was made, as field tests were conducted in Cazaux to assess release methods.[6] How close the French came to having a viable weapon system is not clear, as the Nazi invasion cut short the program in 1940.

Despite their best efforts to destroy evidence, the French left behind enough clues for the Nazis to infer that a biological warfare program had been under way. Professor Heinrich Kliewe of Giessen University's Diagnostic Laboratory for Infectious Diseases was a member of the inspection team, and he knew that the captured equipment and confiscated records were cause for alarm. Kliewe figured that if the French were developing biological weapons, then surely the British and Americans also had active programs.[7] Based on evidence from France and the Fatherland's vulnerabilities, the Germans were somewhat concerned about anthrax, plague, and rinderpest (a disease of cattle). But their deepest worries pertained to foot-and-mouth disease—and potato beetles.

The Germans had been preparing for the Colorado potato beetle well before the discovery that their enemy was weaponizing the insect. In the 1930s, Germany took a two-pronged approach to defending the nation's food against this pest. First, they aggressively pursued the development of new insecticides.[8] This research led to the discovery of organophosphorous compounds, versions of which included chemicals that were quite good at killing insects—and absolutely astounding at killing humans. German chemists had stumbled onto the nerve gases—poisons based on the same general structure as two of today's commonly used home-and-garden insecticides, malathion and diazinon. With relatively minor tweaking of the basic chemistry, German scientists produced the terrible trio of tabun, sarin, and soman—a tiny droplet of which caused horrific convulsions and death within minutes. Although the Nazis had a monopoly on the nerve gases, these chemicals were not used in warfare, largely because erroneous intelligence led the German High Command to believe that the Allies had similar weapons.[9]

In the second prong of the defensive effort, the Germans convened an international committee to recommend countermeasures against the insect menace. By the time the French scheme was discovered, Germany's Potato Beetle Defense Service had 632 personnel engaged in monitoring more than 2 million acres of potato fields for natural infestations. Kliewe used this organization to investigate suspicious incursions of Colorado potato beetles. Major outbreaks in Bavaria and Thuringia were blamed on enemy raids, but the evidence was highly circumstantial. Soon, however, the Germans had the "smoking gun" that they needed to shift their entomological warfare program from defensive measures to offensive tactics.

On April 30, 1942, Kliewe's office received a disturbing report from England.[10] According a reliable German spy, an American B-24 Liberator had delivered an unspecified number of Texas ticks (probably *Amblyomma spp.*) and 15,000 Colorado potato beetles to the British. Shocked by this information, the German Army High Command demanded an immediate risk assessment. Kliewe drafted a report in which he dismissed the ticks as posing a threat to Germany. He was much less certain about the potato beetles. A few thousand of these insects could not be used for mass release, but he warned that their use "by agents and saboteurs has still to be taken into account."[11] The military leaders concluded that Germany could no longer rely solely on repulsing an attack. While continuing to develop defenses against insect invasions, the Nazis decided that they needed the capacity to retaliate in kind, and they put Kliewe in charge of the effort.

The Germans faced two immediate obstacles in creating an entomological weapon. Hitler had banned offensive biological warfare, but Kliewe neatly sidestepped this prohibition by maintaining that the Germans could not hope to adequately defend themselves unless they had knowledge of the sorts of weapons that the Allies might be devising. So the offensive program was cast as a simulated venture into what the enemy might be plotting. With the political problem solved, Kliewe could turn to the practical problem; he needed an infrastructure for research and development.

Kliewe transformed the Potato Beetle Defense Service into the Potato Beetle Research Institute, a subtle change in name that camouflaged a major redirection of the program.[12] Under his leadership, the laboratories were converted to facilities for weaponizing, rather than controlling, insect pests. To wage entomological warfare, the scientists had to address three critical questions: which insect would make the best weapon; how could it be mass produced; and what methods would allow its dissemination into enemy lands?

As for choosing an agent, the Colorado potato beetle's record of damage gave the species a leg up on its rivals. But alternatives were proposed and preliminary experiments were conducted to explore untapped military potential. Between 1941 and 1944, at least 15 species of aphids (family Aphidae), beetles (order Coleoptera), bugs (order Heteroptera), flies, and moths (order Lepidoptera) were evaluated for their capacity to decimate asparagus, corn, pasture grasses, pine trees, potatoes, rapeseed, turnips, and wheat.[13] Several plant diseases and weeds were also considered. However, little progress was made with any of these agents, so the Colorado potato beetle became the focus of research. The next challenge was large-scale production.

At the inaugural meeting of the Blitzableiter (Lightning Rod) Committee in March 1943, the scientists and strategists discussed how many Colorado potato beetles would be needed to inflict serious damage on the enemy's agriculture. One of the key scientists, Dr. Bayer, had done his homework and delivered the discouraging news: "England probably has about 400,000 hectares [1 million acres] of potato fields, for whose destruction about 20–40 million beetles would be necessary. At present the production of such great quantities is impossible."[14]

But, being a good scientist, Bayer converted adversity into opportunity. He claimed that the problem of production could be solved with a bit of funding and research—with his laboratory taking the lead, of course. He was clearly not a man to make vacuous assertions, for a month later Bayer informed the committee that the mass breeding of Colorado potato beetles had begun. How Bayer accomplished this remains a mystery. The first entirely successful artificial diet for this insect was not reported in the scientific literature until 2001,[15] so we can surmise that the Germans used fields and greenhouses to provide a year-round supply of potato leaves.

Whatever system Bayer and his associates invented, it was tremendously productive, for he expected that enough beetles to mount an attack would be on hand by the summer of 1944. Bayer calculated that if the insects were released over the potato fields of eastern England, "[the] destruction of the fields will reduce food calories by 6%."[16] While falling short of famine, this deprivation would add substantially to the suffering of an already beleaguered nation. With the insects in the production pipeline, the third question became a pressing matter: how to disperse tens of millions of beetles?

The Germans believed that a low-tech means of dissemination would be effective if the enemy were unprepared to quickly find and suppress incipient infestations. And this was precisely the state of affairs in war-torn Britain.

So the Blitzableiter Committee approved field trials to evaluate direct aerial releases.

The best documented release experiments took place near Speyer, in October 1943.[17] In the first trial, 40,000 potato beetles were marked and dropped from a plane. Fewer than 100 beetles were recovered. A subsequent aerial release of 14,000 beetles yielded just 57 recoveries. These results meant that either the Germans were bad at recapturing beetles or the beetles were good at dispersing. The latter interpretation was favored. However, these sorts of field experiments were diminishing the stockpile of insects for the raid on England, so the Germans switched to using inanimate models in experiments involving the airdrop of as many as 100,000 wooden surrogates.

Interestingly, the Germans never seemed to recognize the most obvious advantage of dropping fake beetles. Releasing tens of thousands of live pests over Germany to test an entomological weapon system had the same downside as aiming a gun at one's own head and pulling the trigger to see if it is loaded. The beetles falling on the German countryside in the name of military science had no allegiance and were more than happy to bite the hand—or the fields—of those who bred them. When a major outbreak developed in southern Germany, Hermann Göring accused the Allies of waging an entomological attack, notwithstanding Kliewe's research program having released thousands of beetles in the afflicted region the year before.[18]

While wooden beetles were falling on German farms, live beetles were reproducing like mad in Nazi laboratories. By June 1944, the German High Command was informed that "the use [of the Colorado potato beetle] is possible at any time."[19] Despite all the preparations, there is only limited evidence that the beetles were ultimately released. German records provide no clear documentation, although there was considerable incentive to destroy files pertaining to biological warfare in the closing days of the war. All we have is a single accusation from a British naturalist.

In an *International Herald Tribune* article, Richard Ford recounted potato beetle attacks.[20] He claimed that cardboard box–bombs armed with 50 to 100 insects were dropped over English fields. The earliest incident took place in 1943 on the Isle of Wight, where teams of evacuee children—pledged to secrecy—were dispatched to the infested sites and put to work rounding up the beetles and dropping them into vats of boiling water. Ford reported that another attack took place in Sussex, although "how many of these Colorado beetle bombs were dropped in England, I do not know."[21] The veracity of the retired British Museum official's testimony is diminished by the absence of physical evidence (none of the boxes were preserved), the incongruous timing with the German program (beetles were apparently not ready for full deployment until 1944, although small-scale trials on enemy lands may have been conducted before then), and the lack of corroboration by other beetle-bomb eyewitnesses.

At least the story of beetle invasions was more plausible than the bizarre rumor of entomological warfare that swept Britain in 1940. After the evacuation of Allied troops from Dunkirk came fantastical stories that the Germans had created an omnivorous strain of grasshoppers (family Acrididae) that would be used to starve England into surrender.[22] Perhaps their own military history gave the British a reason for believing this wildly imaginative scuttlebutt—after all, they had used a naval blockade in the hopes of starving the Germans in World War I. Whatever its origins, British concern for entomological warfare became an ironically self-fulfilling prophesy.

On December 6, 1941, Prime Minister Winston Churchill received a "Most Secret" memo from a member of his War Cabinet. Lord Maurice Hankey, head of the British biological warfare program, sounded an ominous warning:

> I would not trust the Germans, if driven to desperation, not to resort to such methods [as biological warfare]. It is worthy of mention that a few specimens of the Colorado beetle, which preys on the potato, were found in some half a dozen districts in the region between Weymouth and Swansea a few months ago: although these are not important potato districts and no containers of other suspicious objects were discovered, there were abnormal features in at least one instance suggesting that the occurrence was not due to natural causes.[23]

The beetle outbreaks were almost certainly not the dastardly work of the enemy, given that the infestations arose more than a year before the Germans began mass-producing Colorado potato beetles. But Hankey's concerns were genuine, as his memo also requested permission to organize defensive measures against entomological attacks. Churchill authorized such a program on January 2, 1942. And as part of developing this effort, the British had Colorado potato beetles flown from the United States—the shipment that the German spies discovered and interpreted as evidence of offensive preparations.[24] So it was that the Allies inadvertently catalyzed the Nazi's program of entomological warfare.

While the British and Germans were unwittingly goading one another into escalating efforts to prepare for a deluge of Colorado potato beetles, the Americans were apparently plugging away with their own experiments.[25] Although there is some evidence of a beetle-breeding project, nothing suggests that these insects were deployed against the Nazis. The targeting of postwar Germany is another matter, at least according to some sources.

The East Germans, along with the Czechs and Poles, periodically alleged that the Americans were using potato beetles to starve communist citizens and induce "economic collapse."[26] The most explicit accusation was made in 1950, when the East German State Secretary of the Ministry of Agriculture and Forestry formally charged the United States with scattering beetles over potato crops. The Czech Minister of Agriculture joined the fray, claiming that "Western imperialists this year again are spreading the Colorado beetle in our fields, this time as far East as Slovakia."[27] Eastern European media published photographs of "potato bug containers" allegedly attached to parachutes and balloons, and Polish and East German children were regularly dispatched to the Baltic coast in search of beetles. The East Germans took to calling the insect *Amikäfer*—a clever combination of the German words for "American" (*Amerikanischer*) and "beetle" (*Käfer*). The government also produced a series of beautifully rendered posters showing potato beetles with the black and yellow stripes replaced with red and white ones (along with white stars on a blue thorax) falling from airplanes and marching across a map of Germany (see Figures 12.2 and 12.3).

Cold War paranoia combined with World War II history during the 1969 Geneva disarmament conference, at which a British representative expressed his nation's fears of entomological warfare: "[It] is possible to envisage the use in war of biological agents which are not microbes; hookworm, for instance, or the worm causing bilharzia, or even crop-destroying insects such as locusts or Colorado beetles."[28]

The United Nations also waded, albeit awkwardly, into the issue of Colorado potato beetles being used as weapons. A 1969 UN report on chemical and biological weapons initially asserted that this insect would not be a practical weapon.[29] But in a consummate example of diplomatic double-speak, the report then described how weaponized beetles might be used with impressive success:

To use it for this purpose, the beetle would have to be produced in large numbers and introduced, presumably clandestinely, into potato-growing regions at the correct time during maturation of the crop. In the course of spread the beetle first lives in small foci, which grow and increase until it

Figures 12.2 and 12.3. Propaganda posters warning that the Americans were using Colorado potato beetles were distributed in the early days of the Cold War. Although once widely distributed, these are now rare historical artifacts and high-quality images are difficult to acquire. As seen in the left-hand poster, the East Germans took to calling the Colorado potato beetle "Amikäfer"—a blend of the words for American (*Amerikanischer*) and beetle (*Käfer*). The flag-bearing version of the pest is a clever adaptation of the spotted (if not starred) thorax and the striped hind wings (see Figure 12.1). The poster on the right admonishes people to battle the insectan incursion in the name of peace. European concerns about the possibility of using these pests as entomological weapons were expressed at the 1969 Geneva disarmament conference, and as recently as 1999 Russian military leaders have implied that clandestine releases by the Americans were still taking place. (Both courtesy of the Kloss Museum)

becomes established over large territories. The beetle is capable of astonishing propagation: the progeny of a single beetle may amount to about 8,000 million in one and a half years. Since beetles prefer to feed and lay their eggs in plants suffering from some viral disease, they and their larvae may help transmit the virus thereby increasing the damage they cause. The economic damage caused by the beetle varies with the season and the country affected, but it can destroy up to 80 per cent of the crop. Protection is difficult because it has not been possible to breed resistant potato species and the only means available at present is chemical protection. Were the beetle ever to be used successfully for offensive purposes, it could clearly help bring about long-term damage because of the difficulty of control.[30]

Nor does it appear that the final chapter of the potato beetle's military escapades has been written. Lieutenant-Colonel Valentin Yevstigneyev, Russia's deputy chief of Radiological, Chemical, and Biological Defense Forces, made a most remarkable insinuation during a 1999 interview. When asked about the logistics of dispersing biological agents, he noted that although some insects cannot be effectively scattered from the air, "we do have suspicions about the mass emergence of Colorado [potato] beetles in Russia."[31] But then, given the Soviet interest in entomological warfare that began in World War II, perhaps a guilty conscience was at the core of his suspicions.

13

WAKING THE SLUMBERING GIANTS

With 3 million lice-ridden Russian corpses and another 27 million people afflicted by typhus after the First World War, the Soviets could not have missed the potential of entomological warfare. In 1928, the Revolutionary Military Council initiated the weaponization of typhus. A top-secret institute was founded in the town of Suzdal under the control of OGPU, the forerunner of the KGB.[1] In light of the risks associated with studying human diseases, the facility was transferred to a more isolated site in 1936.

In a move reminiscent of the Japanese occupation of Pingfan, the residents of Vozrozhdeniya Island in the Aral Sea were given six hours to evacuate their homes, and the Soviet biological warfare program set up shop. Typhus might have catalyzed the Soviet program, but interest in other pathogens was formalized in a project code named Golden Triangle, an allusion to its three elements: plague, cholera, and anthrax.

While diseases with the potential to be carried by insects were central to the Soviets' efforts in unconventional warfare, the earliest success came as a serendipitous chemical discovery rooted in entomology. In a tale eerily similar to that of nerve gas in Germany, Soviet scientists searching for better insecticides happened upon a poison that was tough on cockroaches and hell on humans.[2] Phosgene oxime was appropriately called "nettle gas" in light of the first symptom of exposure—a stinging, searing sensation on the skin. If inhaled, the blistering agent lethally burned the victim's lungs. Chemical munitions may have been the first entomologically based weapons out of the blocks, but live insects were not far behind.

In the early years of the Second World War, both chemical and biological agents were favored weapons of partisans. Supplied by their Soviet allies, Polish guerrillas killed hundreds of Germans using poisons, pathogens, and insect infiltrators. In 1943, Russian saboteurs planted typhus-infected lice among German troops occupying the Karachevo region.[3] The results were impressive: 2,808 soldiers were debilitated by disease. But as clever and effective as these covert operations were, the Soviets had much bigger plans for entomological weapons.

History provides a fuzzy picture of most Soviet research on biological warfare. There are whispers of human experimentation at various prisons and detention camps. The strongest evidence was provided by a deserter who recounted a near-disaster in Mongolia.[4] Near the city of Ulan Bator, the Soviets set up a primitive laboratory where they performed experiments on political prisoners and Japanese POWs. To evaluate the military potential of insect-borne diseases, the Russians chained their captives in tents that held cages of plague-infected rats infested with fleas. In the summer of 1941, one of the flea-bitten prisoners escaped and triggered an outbreak that spread to nearby villages. Alarmed by the possibility of an epidemic, the commander ordered air strikes on the afflicted settlements. Once the villages were leveled, the Soviets burned the corpses of 3,000 to 5,000 Mongols, both to ensure that the outbreak was terminated and to destroy evidence of the biological blunder.

Various sources reveal that Soviet research pursued other arthropod-borne diseases, but historians lack tangible proof that entomological weapons were used. There is, however, rather compelling circumstantial evidence of human complicity in insect-vectored diseases amid the bloodiest battle of World War II.

Outside of Stalingrad in the summer of 1942, the German army fell prey to a debilitating malady.[5] The symptoms began with an ulcerating sore, with the nearby glands soon becoming swollen. As the sickness progressed, the soldiers were wracked with fever and chills, pounding headaches, and malaise. Although not many died—probably less than one in ten—the illness brought the campaign to a halt. The disease then spilled over into the surrounding countryside, with more than 100,000 people eventually falling ill. The culprit was tularemia.

There had never been such a major outbreak of this disease in the region, although one of the most common ways to contract tularemia is from the bite of an infected deer fly, horse fly, or tick. The Soviets had been investigating tularemia in their biological warfare program, and at least one modern expert contends that, "This unprecedented spike in tularemia casualties was not likely to have occurred naturally."[6] Not only did the Soviets disavow any knowledge as to the origin of the epidemic, but they also accused the Germans of taking a parting entomological shot at Stalingrad.

The origin of the typhus outbreak that developed as the German army retreated cannot be ascertained with certainty, but human villainy may have been instrumental. The outbreak started within the German forces. The Soviets—claiming to have no idea how typhus had irrupted among the enemy—were positive that the Germans deliberately spread the disease among Russian civilians and troops. Just as flea-ridden bodies had been catapulted over the walls of Kaffa, the Soviet Extraordinary Commission claimed:

> The fascists imposed distinctive, epidemiological diversities [diversions?] aimed at injuring our troops: they threw across the front line the lice-ridden victims of spotted fever [typhus] and prior to their back off dissolved the camps of war prisoners and civilian populations infected by spotted fever.[7]

The Soviet accusations of the Germans during the battle of Stalingrad rang as hollow as their condemnation of the Japanese during the Khabarovsk trial. But the hypocrisy of condemning your enemy's use of insects as weapons while operating your own entomological warfare program was standard fare for the victors. Some critics contend that no country was more disingenuous in its righteous indignation than the United States. By not being a party to the Geneva Protocol, the Americans were able to pursue biological—and entomological—warfare without formal censure.

When confronted with a horrible scenario, a common psychological response is to deny the severity—or even the existence—of the situation. Throughout the 1920s and '30s, denial was the United States' position on biological warfare. The U.S. Chemical Warfare Service, in which any development of biological weapons would have occurred, steadfastly argued that living organisms were not a viable means of waging war. In a 1926 reply to a League of Nations initiative on the subject, the Chemical Warfare Service maintained that "the only method presenting a certain danger would be that of dropping from aeroplanes, glass globes filled with germs."[8]

This American position was further entrenched by an influential analysis conducted by the chief of the Medical Section for the Chemical Warfare Service in 1932. Major Leon A. Fox was considered to be a gifted physician and a brilliant scientist, so his scathing critique became the U.S. Army's official position:

> Bacterial warfare is one of the recent scare-heads that we are being served by the pseudo-scientists who contribute to the flaming pages of the Sunday annexes syndicated over the nation's press. . . . Certainly at the present time practically insurmountable difficulties prevent the use of biologic agents as effective weapons.[9]

In Fox's estimation, microbes could not survive prolonged exposure in the environment, and they certainly could not withstand being blasted from bomb casings. And if the pathogens were infective, the resulting disease would likely boomerang and afflict one's own troops (unless they had been immunized, in which case the enemy would also have secured this protection). This scientific authority was no military strategist, having failed to consider that civilian populations were important wartime targets. However, the Fox Doctrine became the de facto position of the U.S. military. When Ishii Shiro read Fox's article in the journal *Military Surgeon*, he found the analysis deeply flawed.[10] What good fortune to discover such naivete and vulnerability in one's adversary.

Later in the 1930s, a thin crack emerged in the American policy of denial. Intelligence briefs indicated that Japan and Germany had begun to prepare for biological warfare. Using these hints as a reason to revisit the Fox Doctrine, Lieutenant Colonel James S. Simmons of the U.S. Army Surgeon General's office argued that the country was vulnerable to attack.[11] He posited several scenarios, including the enemy's releasing swarms of yellow fever–infected mosquitoes along American shores to weaken the U.S. war effort. Simmons's 1937 report was received with only passing interest, but it put him in a position from which he could subsequently argue for a substantive change in policy.

Two years after Simmons issued his report, the U.S. State Department was notified that a Japanese Army physician (Ishii's minion, Naito Ryoichi) had tried to obtain the yellow fever virus from the Rockefeller Institute for Medical Research. The strange incident yielded no official response, and it might have passed entirely without notice had Simmons not been primed to react. Linking this event to other intelligence, he wrote the secretary of war, Henry L. Stimson, and strongly suggested that the War Department initiate a research program to develop both a defensive response to biological attack and the capability to respond in kind. The momentum was beginning to shift.

In the early fall of 1939, the Chemical Warfare Service released an important but uninspiringly named document to select personnel. "Technical Study No. 10" marked a major turn in military thinking.[12] The report identified nine diseases to which the United States was vulnerable. Six of the illnesses were of particular concern because they were carried by insects and did not require "existing skin lesions nor a co-agent [other than a vector] in order to enter the human body." The report warned "that attack by airplane dissemination of infected insects and other bacteriological materials, is a possibility not to be ignored."

By the early 1940s the U.S. military had replaced the Fox Doctrine with the belief that, ready or not, biological warfare was coming—and America was not ready. Secretary Stimson asked the National Academy of Sciences to appoint a civilian panel to assess the status and future of biological weapons. The WBC (either an acronym for War Bureau Consultants or the result of an effort to fool spies by transposing the abbreviation for the Biological Warfare Committee) was diddling around with a search of the scientific literature when the U.S. Army received a report of the Japanese plague attack on Changteh—the release of 100 million infected fleas had triggered an outbreak that initially killed 500 people and eventually led to 7,000 deaths.[13] Just days later, the WBC and the rest of the nation learned that the Japanese had bombed Pearl Harbor. What had begun as a scholarly study became a deadly serious venture.

Ten weeks after the United States entered the war, the WBC issued its first report. The tome included a staggeringly thorough analysis of biological warfare, covering a range of entomological tactics. So taken was the committee with the potential of insect vectors that they recommended "studies be made to determine whether mosquitoes can be infected with several diseases simultaneously with a view to using these insects as an offensive weapon."[14] In its second report, the WBC pulled no punches, concluding that "the best defense for the United States is to be fully prepared to start a wholesale offensive whenever it becomes necessary to retaliate."[15]

Secretary Stimson was convinced that the United States was facing a ruthless enemy, who had a head start and would use biological weapons if faced with defeat or strategic opportunity. In April 1942, he advised the president to adopt the WBC report. Roosevelt accepted his secretary's recommendation, and Stimson appointed George W. Merck, the president of the giant chemical manufacturer Merck & Co., Inc., as director of a civilian War Research Service.[16] The WRS would replace the WBC and oversee biological warfare preparations.

Placing a poorly funded, nonmilitary committee in charge of research virtually guaranteed that substantive work on biological warfare would go nowhere. Four months after its creation, the WRS figured out that nothing much could be learned without an actual weapons development program. When they pled for assistance, the National Academy of Sciences appointed a cleverly code-named group—the ABC Committee—whose mission seemed the same as its predecessor and whose abbreviation remains a mystery.[17] Although in no position to harm anyone with biological weapons, America was winning the war of acronyms.

Finally, late in 1942, President Roosevelt authorized an expenditure of $250,000 from his Special Emergency Fund to jump-start substantial research leading to production of biological warfare agents.[18] Roosevelt's decision allowed the U.S. Army's Chemical Warfare Service to begin a series of studies, but he remained largely clueless about the nature of what he had funded. A year later, the president was perplexed upon finding a Department of Agriculture request for $405,000 that the Bureau of the Budget could not explain. The War Department had instructed their collaborating agency to keep the project secret, and the frustrated commander in chief asked his special assistant, "Why is it so confidential to destroy insect pests?"[19] Two days later his underling found that the scientists were preparing to defend the nation from a biological attack, but it might have been more revealing if he'd asked either the British or Canadians for a briefing on entomological warfare.

Biological warfare had been on the British agenda since the early 1930s when, in the great round-robin game of suspicion, they received intelligence reports that the Germans were pursuing microbial and entomological weapons. England's initial response was to begin stockpiling insecticides. While Colorado potato beetles were seen as the most likely nemesis at home, the British also had concerns for their colonies. Haldane issued an even more dire prediction:

> It has been suggested by many students of tropical diseases that if mosquitos carrying yellow fever once got a foothold in India, there would be an epedemic [sic] of yellow fever overrunning the whole country. . . . From what we have seen of the behaviour of certain Powers in recent wars, I do not think it would be beyond them to introduce an epidemic of that kind if they wanted to paralyse the Indian Army.[20]

Haldane's speculations led the British to stockpile yellow fever vaccine to protect their far-flung troops. Meanwhile, the Germans launched their own campaign of accusations. In the weeks leading up to the outbreak of World War II, Joseph Goebbels, the Nazi minister of propaganda, claimed that the British were attempting to introduce yellow fever into India by collecting infected mosquitoes from West Africa and releasing them from aircraft over cities.[21] Why the British would have attacked their own colony wasn't entirely clear, but the Germans were evidently worried about the Allies' capacity to wage entomological warfare.

The British press countered with reports that the Germans had released innocuous bacteria in the London Underground and Paris Metro as "practice attacks" to determine the susceptibility of these cities.[22] These incidents were almost surely a hoax, but England was not satisfied with simply waiting for the enemy to attack.

In 1940, the British government built a secret biological warfare laboratory at Porton Down.[23] The facility grew rapidly into a scientific behemoth that featured the largest brick building in the nation. Porton Down concentrated on bacteriological weapons, although some work was conducted on using house flies to vector *Salmonella*—a nasty intestinal pathogen that people who have suffered food poisoning can respect, if not appreciate. As the British government's program was cranking up, the Canadians were getting started in a rather different manner.

Sir Frederick Banting, who shared the 1923 Nobel Prize in Medicine for his discovery of insulin, believed that the coming world conflict would be "a war of scientist against scientist"—politicians, at least in Canada, were proving useless.[24] Banting had been stymied in getting his government to grasp the potential of biological weapons. There was some concern that Vancouver, with its questionably effective sanitation and unquestionably large rat population, was vulnerable to plague. But Banting could not convince the Canadian leadership that a war could be fought with living organisms. Undeterred, he struck out on his own. With connections to industry magnates such as the president of the Canadian Pacific Railway, the chairman of T. Eaton Co. (a major department store chain), and the head of the Seagrams liquor empire, Banting managed to raise more than a million dollars from private sources to underwrite his biological warfare research program.

The Canadian scientists met with their U.S. counterparts late in 1941. The Americans hadn't made much progress on biological weapons, so they were eager to learn about the work of their northern neighbors. The two groups struck a deal—the Canadians shared the results of their research on converting yellow fever mosquitoes into biological weapons, and the Americans opened their classified files on botulism, malaria, plague, psittacosis, tularemia, typhus, and yellow fever. The happy quid pro quo might have blossomed had the British not interrupted with incredible news.

Porton Down had been working on various anti-crop and anti-livestock weapons with an eye to destroying their enemy's food supply.[25] In the summer of 1942, British scientists had successfully tested an anthrax bomb against livestock on Gruinard Island in the Scottish highlands. This was only the tip of the iceberg—anthrax, particularly when inhaled, is lethal to humans. The British had hit upon an ideal biological warfare agent, but there was a snag. England lacked the industrial capacity to produce the number of bombs that would be required to turn this pathogen into a full-blown weapon system. The Americans, on the contrary, could manufacture enormous quantities of war materials.

A plan to produce 500,000 anthrax-filled bombs a month put the U.S. military in the big leagues of biological warfare. The Americans might have been the last of the major belligerents to begin substantive work in this field, but the program would grow to employ thousands of civilian and military workers. With $60 million in support, biological warfare would come to rival the Manhattan Project for talent, if not economic resources.[26]

Roosevelt's initial funding had started the ball rolling at Edgewood Arsenal, but the British push for large-scale production left no doubt that a larger facility was needed. The U.S. operation put down roots in an obscure tract of military land in rural Maryland. Detrick Field—which was soon renamed Camp (and later Fort) Detrick—would be the cornerstone of biological warfare.

The military anticipated a cost of $1.2 million dollars to build Camp Detrick into a premier research and development center.[27] They were off by a factor of ten. The facility boasted 245 structures, including laboratories and production plants, housing for 5,000 workers, hospital, fire house, chapel, theater, recreation halls, and—of course—incinerators. Although no humans were killed in experiments, the scientists racked up an incredible tally of animals: 598,702 mice, 32,339 guinea pigs, 16,652 rats, 5,222 rabbits, 4,578 hamsters, 225 frogs, 166 monkeys, 48 canaries, 34 dogs, 30 sheep, 25 ferrets, 11 cats, 5 pigs, and 2 roosters. The production of anthrax was the catalyst for this remarkable venture, but the scientists explored a panoply of biological agents. And among its other units, Camp Detrick housed a flourishing Entomological Warfare Department.

Although some experts saw airborne dissemination as the ultimate means for waging biological warfare, others argued for the use of vectors.[28] In that great American way, the entomologists and microbiologists competed to make their systems operational. Techniques for mass-producing fleas, lice, and mosquitoes were developed, while methods to produce clouds of infective microbes were studied in a special bomb chamber. And both groups of scientists soon received a boost in support, thanks to the enemy.

When transoceanic balloons floated over the west coast of the United States, American scientists managed to squeeze more political value out of the aerial attacks than did their Japanese counterparts. Claiming that the raid "scared them stiff," U.S. researchers warned political and military leaders that Japanese B encephalitis—the disease that had catalyzed Ishii's passion for biological warfare—could be the next payload. According to Colonel Murray Sanders, a leading military microbiologist at Camp Detrick,

> Mosquitos were the best vectors—and we had plenty of those in the States—and our population had no defenses against B encephalitis. We had no experience of the disease in this country. . . . And four out of five people who contracted it would have died, in my view.[29]

Sanders's estimate of mortality was rather exaggerated, but not out of line with other efforts to incite the government. Although claiming that a quart of the virus, formulated as a freeze-dried powder, "would have put America at Ishii's mercy," the scientists knew full well that there was no chance that the pathogen in this form could have been picked up by mosquitoes. But they also understood how little their superiors knew about biology—and the researchers used this knowledge gap to enhance their own prestige and funding. With an infusion of support, a most impressive monument to American industry and ingenuity grew, with poetic appropriateness, in the nation's heartland.

In the Second World War, the town of Vigo, Indiana, stepped up to the plate and hosted one of the least desirable wartime enterprises: an ordnance plant. Perhaps the logic was that had there been an accident, the population center of Terre Haute County would be six miles from the explosion. The good folks of Vigo probably thought that life couldn't get much more dangerous than having an ordnance factory in their backyard. They were wrong. The U.S. military converted the plant into the largest bacterial production facility in the world.[30] A series of failed safety tests (harmless bacteria used to simulate anthrax kept escaping) prevented the Vigo plant from coming on line. But this glitch in the production side of biological warfare didn't keep researchers from moving ahead—and learning their own lessons about safety.

The American scientists were coming to the same conclusions as the Japanese. While containment of pathogens and vectors in the laboratory required stringent but familiar protocols, field testing demanded geographic isolation to minimize the risk of accidental infections. What the Americans needed was a remote location—like an island.

About seven miles from the coast of Mississippi lies Horn Island, a nondescript spit of land about a mile wide and 10 miles long. In the spring of 1943, this homely hunk of sand was transformed into a biological proving ground, with two little problems.[31] The planners had failed to consider that the island was situated in prime fishing grounds and that the winds blew toward the mainland for more than two-thirds of the year.

The biological warfare workers scrambled to find a more remote test site, settling on the Granite Peak Installation, a 250-square-mile tract of wasteland on Utah's Dugway Proving Grounds. Plans were drawn up for a massive complex in the desert, but the facility would not be ready for nearly two years. So the scientists had to squeeze their work into the hours when the boats and breezes were absent in the Gulf of Mexico. With the hazards of releasing clouds of pathogens upwind from Biloxi, the microbiologists were constantly frustrated. Not so the entomologists.

The military could acquire valuable information on the operational potential of vectors without having to release infected insects and endanger unsuspecting fishermen and urbanites. Although Camp Detrick's upper echelon was partial to airborne dissemination of pathogens, the Canadians' progress with rearing and disseminating insect vectors could not be dismissed.[32] Entomologists from the two countries collaborated on a series of field experiments ranging from the banal to the bizarre.

Horn Island saw releases of house flies and salt-marsh mosquitoes (presumably *Aedes solicitans*), and then things got a bit wacky. The Americans were intrigued by the Canadians' work on mass producing fruit flies.[33] Although these insects are not natural carriers of any disease, the potential to rear them in enormous quantities and their affinity for overripe fruit opened the possibility that they could be used to contaminate an enemy's food supply. Tests with these insects were followed by experimental releases of screwworm flies (*Cochliomyia hominivorax*) and "wool maggots" (an assortment of bottle and blow flies in the family Calliphoridae). These were also not normally disease vectors, but at least these flies could inflict injuries to livestock—a more valuable wartime asset than rotting bananas.

These trials had only "limited success," which dampened the entomological enthusiasm. Critics complained that development of biological weapons was being slowed by testing "unconventional modes of dissemination such as the use of insects."[34] However, one element of the American war effort reveled in wildly imaginative methods of killing, and this operation saw flies as the ideal conscripts in defeating the Nazis.

Before the outbreak of World War II, American intelligence operations were a fragmented collection of programs conducted by the State Department and the various branches of the armed services. With the totalitarian regimes rising to power, Roosevelt needed a coherent intelligence service and someone to arbitrate the squabbles among the turf-conscious agencies. Some things never change. In 1941, Roosevelt appointed one of his classmates from Columbia University, William J. Donovan, as Coordinator of Information—a post functionally indistinguishable from President George W. Bush's Director of National Intelligence.

General William "Wild Bill" Donovan earned his nickname either for the gutsy plays that he made as the star quarterback for his college football team or for the fanatical training that he gave his men before pushing them to their limits in combat. Donovan was one of the few men to rise from the enlisted ranks to become a general, and upon retirement he applied his experience to climbing the ladder of politics. He was appointed U.S. District Attorney in New York but lost his run for governor, being a Republican candidate at the wrong time and place. Undeterred, Donovan launched his meteoric ascent into national prominence by being named Roosevelt's Coordinator of Information. Just five months later, the Japanese bombed Pearl Harbor and the president needed intelligence like never before. So Donovan was in the ideal position to head the new Office of Strategic Services (OSS), which would later evolve into the Central Intelligence Agency.

Donovan hired a Boston chemist and business executive, Stanley P. Lovell, to head up the Research and Development Branch of the OSS. Lovell played the real-life role of Q in the James Bond movies, working with his staff to invent spy gadgets ranging from tiny cameras to explosives disguised as dinner. Nothing was too absurd—not even, as we shall see—six-legged commandos. Lovell was at the center of a plot to undermine the Führer's charisma by injecting his vegetables with estrogen to cause his mustache to fall out and his voice to become soprano. His branch developed a capsule of mustard gas that was to be dropped into a flower arrangement during a high-level meeting of the enemy, as part of a plan to blind the German High Command, upon which the Pope would announce that God had punished the Nazi leaders, thereby causing the Roman Catholic soldiers among the Axis forces to lay down their arms.[35] Really. Then came the mission that called for a weapon so bizarre as to make these earlier boondoggles seem downright reasonable.

In February 1942, General Rommel's Afrika Korps pummeled U.S. forces in North Africa, and the Americans became worried that their defeat would encourage fascist Spain to join the Axis alliance. Moreover, the Germans were amassing troops in Morocco, in preparation for cutting off the railroad from Casablanca to Algiers—the sole supply line for Allied forces. A covert operation was needed to debilitate the German troops, break the momentum of the Axis, and save the Allied lifeline. But this required something far beyond feminizing the Führer or blinding German generals—this called for incapacitating thousands of soldiers without being detected. This called for flies.

The plan was to weaken the enemy forces by using flies to spread a witch's brew of pathogens.[36] Given the agency's inability to rear an army of flies, Lovell decided to conscript the local vectors. He didn't know it, but North American house flies couldn't have held a candle to their desert kin. Modern biological warfare researchers have considered using a Middle Eastern strain of these insects to disseminate anthrax, in light of the insects' merciless pursuit of moisture from the eyes, noses, and mouths of humans. But today's tactician would face the same logistical problem as confronted the OSS: How do you contaminate a few million flies with a pathogen?

Lovell was a chemist, but he'd been out of the laboratory often enough to know that flies love dung. And with a bit of research, he discovered a key demographic fact: There were more goats than people in Morocco—and goats are prolific producers of poop. Lovell now had the secret formula: microbes + feces + flies = sick Germans. Now all he needed was a few tons of goat droppings as a carrier for laboratory-cultured pathogens.

The OSS collaborated closely with the Canadian entomological warfare experts to launch one of the more preposterous innovations in the history of clandestine weaponry: synthetic goat dung. Of course flies are no fools; they won't be taken in by any old brown lump. So the OSS team added a chemical attractant. The nature of this lure is not clear, but a bit of sleuthing provides some clues.

Allied scientists might have crafted a chemical dinner bell by collecting and concentrating the stinky chemicals that we associate with human feces (indole and the appropriately named skatole). While these extracts would have worked, the more likely attractant was a blend of organic acids, some of which had been known for 150 years. Two of the smelliest of these are caproic and caprylic acids, which, by no coincidence, derive their names from *caprinus*, meaning "goat." Etymologically as well as entomologically astute, Lovell named the operation Project Capricious. So with a scent to entice the flies, Lovell's team then coated the rubbery pellets in bacteria to complete the lures.

All the Americans had to do was drop loads of pathogenic pseudo-poop over towns and villages where the Germans were garrisoned, and millions of local flies would be drawn to the bait, pick up a dose of microbes, and then dutifully deliver the bacteria to the enemy. Lovell worried about keeping the operation clandestine.[37] The Moroccans had to be persuaded that finding goat droppings on their roofs the morning after Allied aircraft flew over was a sheer coincidence. Presumably a good disinformation campaign can dispel almost any suspicion, or, as Lovell intimated, if the plan succeeded there would be very few people in any condition to raise annoying questions about fecal pellets on rooftops.

Lovell's need for secrecy pertained not only to the enemy but also to his own agency. Operation Capricious was not revealed to Donovan, who, being a "soldier's soldier," would likely have nixed the use of biological weapons as being dishonorable. In the end, however, Lovell didn't have to worry about getting caught by either friends or foes, as the secret weapon was never deployed. Just as the OSS was gearing up to launch the sneak attack, the German troops were withdrawn from Spanish Morocco.[38] They might well have preferred to take their chances with pathogen-laden flies, given that Hitler was sending them to the bloody siege of Stalingrad.

Such opportune turns of the war, rather than moral principle, kept the Americans and British from using biological and chemical weapons. Churchill was prepared to defend his nation with poison gas had the Germans landed on the British shores. And Truman's willingness to use atomic bombs against Japan, even when the United States was headed to near certain victory (albeit at a horrific cost in terms of American lives), leaves little doubt that he would have authorized biological weapons had the situation been sufficiently dire. Although the Americans were never desperate enough to use insects against their enemies, the U.S. military battled against insects in a series of encounters that changed the course of the war.

Military history shows that the best offense is possible only when hostile insects have been neutralized with a good defense. This is why the world's largest employer of medical entomologists for the past half century has been the U.S. Department of Defense.[39] In World War II, the Americans won pivotal entomological battles in both the European and Pacific theaters.

In the summer of 1943, Allied forces fought a bloody, month-long battle to take Sicily, a strategic stepping stone to Italy. By September, the Italians had surrendered, but the occupying German forces were preparing to make the Allies pay a dreadful price for every acre of Italian real estate. When the

Americans landed in the port city of Naples, they figured that the campaign was winnable—if the fight was between them and the Germans. However, the enemy had fortuitously secured the assistance of an unwitting ally. The U.S. troops would never defeat the Nazis unless the American medical units could first conquer the lice.

When the Americans arrived, the Italian people were in desperate straits.[40] Many had taken shelter from air raids in filthy, crowded caves. Within the towns and cities, squalid conditions prevailed: sanitation systems had collapsed, medical care had disappeared, and food was scarce. It was a louse's dream come true. With the number of typhus cases growing at an alarming rate, an epidemic seemed inevitable—and the Americans would be walking right into the medical minefield. Ravaged by disease, the U.S. troops wouldn't have a chance in the bitter fight against the waiting Germans. But the lessons of history had not been lost on the Allies.

In anticipation of the role that insect-borne disease could play in the war, the U.S. Typhus Commission had been formed in December 1942. Although an early typhus vaccine was available and used throughout Europe, the medical corps doubted its efficacy. The military understood that the key to winning the battle against typhus was not medicine but insecticide. The villain of the 1960s was the hero of the 1940s—DDT would save untold numbers of civilians and soldiers in southern Italy and spell the difference between victory and defeat in the coming military campaign.

The commission assembled a delousing program of unparalleled proportions, with various units swinging into full operation within two months of the Americans landing in Italy. Some units were charged with finding targets, such as crowded shelters and diseased neighborhoods. Other squads pursued "tactical delousing"—dusting people, bedding, and clothing with insecticide through a program involving 40 stations, staffed by 439 men. Altogether, they had the capacity to handle 100,000 people per day. The process was quick, easy, and effective, as described by one of the officers:

> [The procedure] consisted essentially of forcefully blowing powder, by hand dusters or power dusters between the layers of clothing worn by the individual and between the innermost layer of clothing and the skin of the body. This was accomplished by a uniform technique, inserting the nozzle of the duster up the sleeve, down the neck (both front and back), around the waistline and into the crotch area of clothing. Hair and any cap or hat were dusted thoroughly. An infested person properly dusted is no longer a menace to others and will remain so for a period of at least two weeks, at the

end of which he should be redusted. Approximately 1 to 1 ½ oz of powder per person is sufficient to insure [*sic*] the thorough dusting of all clothing worn.[41]

In just a few months, the U.S. army had dusted 3,265,786 people with 127 tons of DDT powder (see Figure 13.1). The U.S. Typhus Commission had saved the lives of thousands of Italian citizens and—more to the point in military terms—had provided the Allied commanders with a healthy army ready to take on the Nazis. But just as American troops finally broke through into northern Italy, a mysterious malady began to sap their strength.

Q fever was barely known to science when the United States was drawn into the Second World War, so it is not surprising that the American medics struggled to diagnose the illness that wracked the troops.[42] The men were suffering a range of nonspecific symptoms: severe headaches, skyrocketing fevers, bone-rattling chills, and pneumonia. Along some parts of the front,

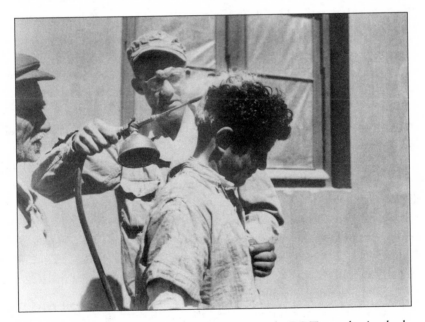

Figure 13.1. A soldier uses a power duster to apply DDT powder in the battle against louse-borne typhus, most likely in Italy during 1943. The Typhus Commission dusted more than 3 million people, saving thousands of civilian lives and ensuring that a healthy Allied force could continue the push into northern Italy. Thanks to DDT spraying in the Pacific theater, the malaria rate among Allied forces dropped by 98 percent from 1943 to 1945. (Courtesy of the Centers for Disease Control and Prevention)

one-third of the units were put out of commission. The rickettsiae that cause the disease are commonly transmitted to agricultural workers from infected farm animals via milk and airborne particles, but ticks are also vectors. We can surmise that eight-legged arthropods probably carried the disease among the troops, given that the soldiers had little exposure to livestock. With supportive therapies, sheer grit, and dint of will, the American offensive pushed on as the soldiers battled microbial and human enemies. While ticks made life miserable in Europe, their relatives were inflicting thousands of casualties among Allied troops in the Pacific.

Much to the relief of the 41 soldiers and sailors who landed on a spit of land off New Guinea in 1944, South Bat Island was uninhabited. But although there were no Japanese lying in wait, the island's nonhuman inhabitants were most unfriendly.[43] An alliance of mites (*Trombicula*) and microbes nearly repulsed the Allied invasion. Within days of their landing, 26 of the men fell ill with severe headaches, fevers, chills, and swollen lymph nodes. The medics became desperately worried as some of the victims broke out in a rash that developed into crusty black scabs. The island's chiggers[44] carried the rickettsia responsible for an exotic disease: tsutsugamushi fever (also called scrub typhus). This tale would be repeated on Guadalcanal and many other islands. One of the worst outbreaks struck a few weeks after troops hit the beaches of New Guinea; the 6th Infantry Division suffered 931 cases and 34 men died. While mites took a toll, another vector was far worse, putting five times more U.S. troops out of commission than did wounds incurred from battle. Chiggers were diabolical, but mosquitoes were hell.

In May 1943, General Douglas MacArthur declared, "This will be a long war, if for every division I have fighting the enemy, I must count on a second division in the hospital with malaria and a third division convalescing from this debilitating disease."[45] Malaria killed only about a thousand American troops, but in the cold calculus of battle, sometimes a general prefers a dead soldier to a sick one. The former requires a burial while the latter consumes medical staff, bed space, rations, and fuel. And 2 million cases of malaria cost the U.S. military dearly in terms of manpower and supplies.

MacArthur was frustrated by his officers' failure to enforce "malaria discipline"—basic procedures to minimize the likelihood of contracting the disease. But then it is hard to blame a soldier for thinking that the armed humans on the other side of a marsh were a far more imminent risk than the clouds of mosquitoes (*Anopheles*) hovering over the swampy ground. As a high-ranking officer on Guadalcanal succinctly noted, "We are out here to fight

Japs and to hell with mosquitoes."[46] But unless someone fought the insects, the officers would not have had enough soldiers to fight the Japanese.

By 1944, the U.S. military had established a formidable medical infrastructure in the Pacific theater, including dozens of malaria-survey units and more than a hundred vector-control programs.[47] Between specialized training and the advent of DDT, the anti-mosquito operations were phenomenally successful. In 1943, the malaria rate among Allied forces in the Pacific was a crippling 208 cases per 1,000 troops, but by 1945 the rate had plummeted to just 5 cases per 1,000 troops.

Through the combined efforts of entomologists and engineers, the U.S. military invented the standard weapon against the insects: a five-pound canister that used a propellant to force an oil-based insecticide through a fine nozzle. This device was adapted by the pest-control industry and became the basis for household aerosol cans that filled store shelves for decades. Of course, the military also waged war on the mosquitoes with more impressive delivery systems, including B-25s and C-47s equipped with 625-gallon tanks loaded with insecticide. Indeed, the success of these systems for dispersing DDT—a notoriously stable chemical—convinced the U.S. military to abandon the long-held belief that chemical weapons had to be volatile. While insecticides were spawning new thinking in chemical warfare, mosquito-borne diseases were causing the military to rethink its position on biological warfare.

Early in the war, yellow fever was viewed with increasing concern. Not only had a Japanese agent tried to acquire a particularly virulent strain from the Rockefeller Institute, but the medical community had sounded the alarm concerning the vulnerability of Hawaii to biological attack with, among other agents, infected yellow fever mosquitoes. The War Department ordered the military to guard food and water, undertake rat and mosquito control programs, and—most fatefully—vaccinate the troops. The military had long held that inoculation could neutralize some of the most dangerous biological weapons. A yellow fever vaccination program was implemented throughout the Pacific theater, but a large portion of the vaccine was contaminated with hepatitis B, and some 330,000 U.S. servicemen were exposed to this virus.[48] The "easy" solution to a biological warfare threat was not without its own risks. But syringes were not the most worrisome vectors.

Mosquitoes transmitted a menagerie of diseases in addition to malaria and yellow fever. On Samoa, mosquitoes carried filariasis, a horrific disease caused by a tiny worm that damages tissues in the lymphatic system and causes enormous amounts of fluid to accumulate. Without treatment, the legs and groin of the victim can swell to grotesque proportions (the testicles can expand to 6 inches in diameter and a half-gallon of lymph may accumulate in the scrotum) and the skin becomes thick and cracked, yielding the aptly named condition of elephantiasis. Mosquitoes spread dengue fever in Saipan, where nearly one-third of the troops contracted the illness in a three-month period in 1944.[49] The symptoms are not as dramatic as those of filariasis, but the writhing elicited by the extreme muscle and joint pains have earned the disease its common name of "breakbone fever." And, finally, Okinawan mosquitoes greeted U.S. troops with Japanese B encephalitis.

Between the Americans' Faustian deal with Ishii and the U.S. military's experience with insect-borne diseases in the European and Pacific theaters, the stage was set for the next act in the diabolic drama of entomological warfare. As the curtain descended on World War II, the Soviets pressed for the death penalty for Nuremberg defendant Hans Fritzsche, on the grounds that he had instigated the Nazis' biological warfare program. The British and Americans insisted upon his acquittal. Hypocrisy has its limits, and the western allies knew full well that they had invested vastly more time and resources into their programs than had the Germans. Moreover, the western nations predicted that biological weapons would surpass nuclear arms as a means of waging war under many scenarios in the near future, a future in which the new enemy was communism. But few strategists could have guessed that this prophesy would be so spectacularly fulfilled within five years. For it was the dropping of insects, not atomic bombs, over Korea that sparked an international firestorm in the early days of the Cold War.

FOUR

COLD-BLOODED FIGHTERS
OF THE COLD WAR

According to the Air Observer Corps, two American planes invaded the Liaoyang area [Korea] at 6 P.M. of that day and again at 6:30 P.M. on March 27 [1952]. At the time when Lu Li-tsun heard the noise of the airplanes, Jen Wan-ku, a militiaman of Pei-chia-ch'ang Village was on his way to the 4th group of inhabitants on patrol duty. He saw about 160 meters [175 yards] away on the southeast a red object of the size of a thermos bottle dropping from the air above the houses of Chang Chia-feng, Wang-Ch'ang and Huang Yü-ch'eng. The object exploded when it was about 3–4 meters [11½ feet] above the roof of the houses producing a feeble noise and an offensive smell . . . Wang Hua-ming, a member of Wang Wen-ch'ang's family saw, through the window, the red object falling in front of their gate when he was sitting on his kang (brick bed). He rushed out of his room but the red object had already disappeared. He went back to his room again and lighted a lamp and saw numerous insects on the outer surface of the window pane . . . Up to March 28th, these insects were found in 36 villages and towns . . . The area in which these insects were found covered 30 kilometers [22 miles] from east to west and 20 kilometers [12 miles] from north to south.

—Report of the International Scientific Commission for the Investigation of the Facts Concerning Bacterial Warfare in Korea and China

14

KOREA'S HAILSTORMS OF HEXAPODS

The case presented by the North Koreans and Chinese in 1952 provides either irrefutable evidence that the United States engaged in the most comprehensive and systematic program of entomological warfare in modern times or compelling evidence that the communists had the most coordinated and insidious program of propaganda in memory. Or something intriguingly in-between. All that we know with absolute certainty is that the Korean War produced the most sensational accusations of the use of insects as weapons of war in the last half of the 20th century.

The nations of the world were bloodied and exhausted at the end of the Second World War, and neither the capitalists nor the communists were particularly anxious to start another shooting war. Although tensions ran high, an uneasy stalemate developed as the spoils of the beaten Axis powers were divided among the victors. In Germany, this meant splitting a nation into east and west portions. In Korea, the 38th Parallel separated the north and the south.

The United Nations appointed a commission to oversee the affairs of Korea, a political tinderbox.[1] The UN commission was dominated by the United States and therefore lacked credibility in the eyes of the communists. American influence in the region was augmented by an increasingly cozy relationship with Japan, which became the west's "unsinkable aircraft carrier."

Meanwhile, the Soviet Union was under Stalin's iron-fisted authority and Mao Zedong was leading the Communist Party to control of China. The nationalist government of Chiang Kai-shek was exiled to Taiwan, leaving the United States with a single beachhead on the Asian continent—Korea. Back home, Joseph McCarthy's communist witch hunts were unfolding, adding fuel to a political fire that had all the signs of becoming a military conflagration. And when the flashpoint came, the U.S. military had to be ready—with all the weaponry that might be needed to fight an ugly war against a godless enemy.

The possibility that the United States waged entomological warfare might have been summarily dismissed by military historians and political analysts if the Americans had not been so obviously pursuing the weaponization of insects. And whether or not a rain of insects fell from American planes, there is little doubt that western nations were laying the groundwork for entomological warfare in the years leading up to the Korean War.

The 406 Medical General Laboratory of the U.S. Army's Far East Medical Section became the linchpin for research and development of insect vectors.[2] Spawned in a warehouse near the Agsugi Air Base in Yokohama in 1946, Unit 406 was originally tasked with providing health services to U.S. soldiers and fostering public health among civilians. But this seemingly benign, even laudable, purpose underwent "mission creep." The Dr. Jekyll to Mr. Hyde transmogrification might have been predictable, given that the unit was the brainchild of Brigadier General James S. Simmons—the man who promoted biological warfare within the Army Medical Corps during World War II. But Simmons was by now too high ranking to actually head the venture.

Lieutenant Colonel W. D. Tigertt was an ideal first commander of Unit 406. He'd conducted extensive research on insect vectors of Japanese B encephalitis, the disease that planted an evil seed in the fertile mind of the young Ishii Shiro in 1924. Under Tigertt's direction, the laboratory grew into a full-scale R&D program with 309 personnel distributed among the departments of bacteriology, entomology, epidemiology, and virology.

The Americans sought expertise from all quarters, including associations with former Unit 731 personnel. Although the scope of their involvement is not fully known, Japanese scientists assisted the U.S. military during an outbreak of an arthropod-borne disease in Korea. Treating uniforms with insect repellent quashed the disease, but beneficiaries were far from thankful. Rather, they expressed their suspicions of Japanese-American collaboration by insinuating that the disease was new to their country.

When Colonel Richard P. Mason took over the helm of Unit 406 in 1951, entomological warfare became a major focus. To provide essential support for the scientific staff, enlisted personnel were given a course in medical entomology beyond that of most major universities. Research initially concentrated on mosquito-borne diseases, but the military scientists eagerly expanded their work to include ticks, mites, lice, fleas, and flies, with particular attention to the breeding and biting behaviors of black flies and midges found in Japan and Korea. The entomology department explained that these insects were of particular interest "because of the many reports of their biting humans and their great potentiality as vectors of disease."[3] Soon, the research agenda outgrew the original facility, and Unit 406 set up a branch laboratory in Kyoto. With the entomological laboratories running at full throttle, the Americans had little choice but to employ Japanese nationals. This logistical necessity became a grievous political mistake.

Among these workers were a number of communist infiltrators, who were delighted to expose the nefarious work of the American military. In 1952, the Japan Peace Council, a leftist organization, distributed a pamphlet entitled "American Bacteriological Base Is Located in the Center of Tokyo." The awkwardly worded leaflet claimed that

> closely wrapped packages of insects laden with germs of infectious diseases such as plague, cholera, scarlet fever, dysentery and meningitis, are being regularly transported there [Unit 406] along with instructions for experiment from Deterric [*sic*] Research Center of the United States. Then Detachment [Unit] 406 immediately begins work of mass cultivation.[4]

The extent to which Unit 406 moved from defensive to offensive research is not clear, but this venture was only the beginning of the American entomological warfare efforts in the years leading up to the Korean War. Tigertt had led the Army Medical Unit at Camp Detrick, and he linked the flourishing biological warfare program in the United States with the nascent entomological warfare initiative in Japan. Under Mason, the flow of knowledge began to run in the other direction, with Unit 406 apparently catalyzing a keen interest in weaponizing insects back home. And the big guns of biological warfare on the home front were not to be outdone.

Theodor Rosebury, the director of the Research and Development Department at Camp Detrick in the late 1940s, published a lengthy report on various delivery systems for pathogens. He rated insect vectors as highly effective carriers and noted that "technical developments discussed as possibilities in this paper have already become realities"[5]—perhaps an allusion to the early work of Unit 406. But the scientists at Camp Detrick did not depend solely on the research from this upstart, outpost laboratory. Preparing for entomological warfare with the communists would require putting the nation's best minds and finest facilities into the effort.

A clear, overall picture of the program is difficult to reconstruct, but glimpses into the workings of Camp Detrick reveal a pattern of research in which entomology was being taken very seriously.[6] Such evidence includes a

"Research and Development Project Listing" of the Chemical Corps dated October 30, 1951, which revealed that $160,000 had been devoted to studies of "arthropod dissemination," and Project No. 411-02-041, which allocated $380,000 to Johns Hopkins University for investigating mosquito vectors of encephalomyelitis viruses. Even Projects that appeared to be defensive in nature turned out, in at least some cases, to have a darker side. For example, by its title, Project No. 465-20-001, "Mechanism of Entry and Action of Insecticidal Compounds and Insect Repellents," would seem to have been concerned with pest management. However, the research synopsis revealed another agenda:

> Information on the mechanism of action of insecticides is applicable directly to problems involved in both the offensive application of and protection against insect dissemination of biological agents. Under project 465-20-001, insect strains resistant to insecticides are being developed. These represent a potentially more effective vehicle for the offensive use in BW of insect borne pathogens.[7]

Nor was the United States the only western nation interested in conscripting insects to defeat the communist menace; the Canadians were every bit as aggressive in their pursuit of the perfect entomological weapon. Dr. G. B. Reed was in charge of the Defense Research Laboratory at Queen's University in Ontario, and by the late 1940s he had devised a remarkably simple means of using insects as weapons.[8] Earlier work in collaboration with U.S. scientists had provided methods for mass-producing vectors, but Reed understood that rearing enough insects to wage a war would be an enormous challenge. Although the Canadians had devised a 500-pound bomb capable of delivering 200,000 infected flies to a target, nobody had figured out how to stockpile millions of live insects.

Reed's breakthrough originated in his laboratory's "media unit," a group of researchers dedicated to producing insect foods that could be laced with various pathogens to infect a range of vectors, including house flies, fruit flies, various biting flies, mosquitoes, chiggers, ticks, and fleas. The most efficient substrate for producing infected flies was found to be canned salmon, which served as both a nutrient-rich environment for the pathogens and a savory diet for the insects. The Canadians took the first step toward Reed's ultimate innovation in arming experimental bombs with live flies and contaminated salmon. When the device ruptured, this dual payload allowed the containerized flies, along with any of their wild brethren who might be buzzing about, to avail themselves of the smelly bait.

Reed realized, as had Ishii, that scientists can be too clever in trying to engineer what nature already provides. The Canadian scientist found that most habitats supported plenty of naturally occurring flies, so the laboratory-reared insects could be eliminated, along with the problems of production and stockpiling. Soon he was developing payloads of house-fly baits laced with pathogens and enhanced with chemical attractants—perhaps synthetic goat dung had not been so absurd after all!

With entomological weapons rapidly becoming viable, western governments had to develop policies and guidelines for the conditions under which such unconventional warfare would be waged. If the Cold War became hot, the difference between having a weapon and using a weapon would be vitally important to the United States and the rest of the world.

At the start of 1950, the United States' position on biological warfare was deeply conflicted. If the Berlin Blockade had led to all-out war a year earlier, the Americans intended to use unconventional weapons. And the U.S. Joint Chiefs of Staff still maintained biological warfare in their emergency war plans. Perhaps limiting the military's option was the National Security Council Directive that stated that "chemical, biological and radiological weapons will not be used by the United States except in retaliation."[9] Yet others maintained that biological warfare would adhere to the established practice concerning the use of nuclear arms: presidential discretion. Whether the U.S. policy was one of militarily constrained retaliation or presidentially ordered first use, the Americans were certainly getting ready.

On the last day of June 1950, the report of the Stevenson Committee was delivered to the power brokers in Washington, D.C. Earl P. Stevenson, the well-connected president of a major engineering firm, had chaired a group of scientists, industrialists, and bureaucrats who had been commissioned to assess American preparedness for biological warfare.[10] The committee's recommendations were unambiguous, if not entirely unbiased. (Stevenson's firm had received lucrative military contracts for developing bacteriological delivery systems.) The experts excoriated the government for having allowed such an important element of the military arsenal to dwindle to near impotency after the Second World War.

In response to the report, the U.S. Department of Defense increased funding for biological weapons development from $5.3 million to $345 million over the next three years.[11] The phoenix of biological warfare began its rise from the ashes on June 30, 1950—and the timing could not have been more suspicious for those who would accuse the United States of using entomological weapons. Just five days earlier, the Korean peninsula had become the stage on which communist and western nations would play out their ideological conflicts in terms of blood.

The Korean War began on June 25, 1950, when 38,000 North Koreans, with the support of 50 Soviet tanks, crossed over the 38th Parallel and advanced on Seoul, the capital of South Korea.[12] The next day, the UN Security Council branded North Korea as the aggressor and called on member nations to unite in repulsing the invaders. President Harry S. Truman, already stinging from the communist successes in Eastern Europe, vowed that he would not lose Korea and ordered American forces to join with the South Koreans under the UN banner.

For months the war seesawed, until the two sides settled into a brutal stalemate. The Americans achieved air superiority while the North Koreans constructed impregnable fortifications. As the deadlock wore on, the western forces initiated "Operation Strangle" to sever the communists' supply lines and break the enemy's will to fight. The war was becoming very ugly.

By the end of 1950, the U.S. Joint Chiefs of Staff expressed their support of the president's stand to consider nuclear warfare, if this tactic was deemed necessary to avoid defeat in Korea. With the atomic bomb on the table, the door was presumably open for other unconventional arms, such as biological weapons. That fall, Camp Detrick had deemed five living agents feasible for military use, including three pathogens deliverable by insect vectors. But it is abundantly clear that the Americans were not the only ones thinking about biological warfare.

With the spring thaw of 1951, both the North and South Koreans found themselves battling microbes as well as each other. Smallpox and typhus irrupted throughout the region, and the Chinese media alluded to the possible role of the Americans in these outbreaks. Newspaper and radio reports reminded people of the biological warfare program that Ishii had masterminded and of the dastardly association between the Japanese war criminals and the American military. In March, Peking radio charged UN forces with manufacturing biological weapons, thereby either expressing a genuine concern for their soldiers and allies or building the foundation for a propaganda campaign—or both.[13]

The North Koreans also began testing the war-crime waters. On May 8, the minister of foreign affairs, Pak Hen Yen, sent an official cable to the president of the UN Security Council alleging that U.S. forces operating in concert with the United Nations had spread smallpox virus in and around Pyongyang.[14] The UN commander adamantly denied the charge, and the Security Council accepted that naturally occurring diseases worsen during the course of a war without there being anything evil afoot. With the international community dismissing the charges, the U.S. military took their game to the next level.

In September, a top secret memo sent by the U.S. Joint Chiefs of Staff hinted that the frustration of being stymied by the North Koreans was eroding the military's traditional reticence to use unconventional weapons:

> National security demands that the United States acquire a strong offensive BW [biological warfare] capability without delay. A sound military program requires the development of all effective means of waging war without regard for precedent as to their use. . . . The adoption of a positive military policy to the effect that the United States will be prepared to employ BW whenever it is militarily advantageous would serve to stimulate Service interest in the BW field and accelerate its development.[15]

American strategists figured that disease could create panic among the Chinese and North Korean populace supporting the front-line troops. Moreover, insect vectors had the ability to find their way into tunnels, caves, and fortifications that had proved resistant to conventional bombing. The position of the Joint Chiefs swayed the U.S. Secretary of Defense, Robert Lovett, who ordered the military to devote the resources necessary to ensure readiness for waging biological warfare at the "earliest practicable time."[16]

Who could order the use of these weapons and under what conditions remained equivocal. This ambiguity may have been designed to allow American leaders to plausibly deny responsibility for biological warfare. And if the pathogenic cat got out of the diplomatic bag, then the U.S. government could always fall back on not having ratified the 1925 Geneva Protocol. Although the rationale behind the United States' nonpolicy on biological warfare is murky, the American government would need diplomatic bulwarks. For the upcoming political battle over entomological warfare proved to be unprecedented in its ferocity and tenacity.

On February 22, 1952, the curtain rose on what was to become an international tragicomedy of epic proportions. Bak Hun Yung, North Korea's foreign minister, issued an official statement to the UN Secretariat alleging that the United States had engaged in entomological warfare. The North Koreans charged that

> the American imperialist invaders, since January 23 this year [1952], have been systematically scattering large quantities of bacteria-carrying insects by aircraft in order to disseminate infectious diseases over our front line positions and rear. Bacteriological tests show that these insects scattered by the aggressors on the positions of our troops and in our rear are infected with plague, cholera and the germs of other infectious diseases. This is irrefutable proof that the enemy is employing bacteria on a large scale and in a well-planned manner to slaughter the men of the [Korean] People's Army, the Chinese People's Volunteers, and peaceful Korean civilians.[17]

Two days later, the foreign minister of the People's Republic of China not only lent his country's support to the charges but also expanded the accusations. Zhou Enlai claimed that during a one-week period, the United States had sent 448 aircraft on at least 68 occasions into northeast China to airdrop contaminated insects.

The Chinese rapidly assembled the People's Commission for Investigating Germ Warfare Crimes of the American Imperialists to report on the extent and nature of the raids.[18] If the Chinese and North Koreans had simply conducted their investigation and taken no further action, the charges could have been interpreted as mere propaganda. But whether it was part of an enormous charade or whether the communists were sincere in their fears, they not only "talked the talk" of having been attacked, they "walked the walk."

The Chinese and North Koreans initiated a massive defensive response to the reported biological attacks. A telegram from the Central Epidemic Prevention Committee in Beijing in March 1952 stated, "the enemy has furiously employed continuous bacterial warfare in Korea and in our Northeast and Qingdao areas, dropping flies, mosquitoes, spiders, ants, bed bugs, fleas . . . thirty-odd species of bacteria-carrying insects . . . in a wide area."[19] The Central Military directed the Chinese army in Korea to undertake a sweeping epidemic-prevention campaign to protect troops and civilians.[20] Within days, medics administered the first of what would eventually grow to 3 million doses of plague vaccine. Reports of a U.S. entomological Blitzkrieg motivated the citizenry to participate in a massive public health campaign. To deprive the insect and rodent vectors of harborage, the people began clearing away 3 million tons of trash and rubble. By April, some 20,000 medical workers had been organized into 120 "prevention brigades" capable of inoculating more than 100,000 people per day.

Although the crusade was complicated by false alarms from the panicked populace, by May the Chinese and North Koreans were able to claim victory. The authorities pointed out that there had been no major epidemics in North Korea or northeast China—the areas over which American planes had purportedly distributed millions of infected insects. From all appearances, the communists had won the war against disease and gained the upper hand in terms of propaganda. But the U.S. and UN leaders were mounting a political counteroffensive.

The first official response to the North Korean accusations came from the Americans on March 4, 1952. Addressing the U.S. Congress, General Matthew Ridgeway, commander of the UN forces in Korea, flatly denied the allegations and offered a scathing rebuttal: "These charges are evidently designed to conceal the Communists' inability to cope with the spread of epidemics which occur annually throughout China and North Korea and to care properly for the many victims."[21] In his estimation, the whole sordid affair simply revealed the dishonesty and ineptitude of the communist system. Soon, the secretary general of the UN and the chairman of the U.S. Joint Chiefs of Staff were echoing Ridgeway's denial. With this political counterpunch, the entomological warfare charges turned into a diplomat slugfest, and the communists were ready to answer the bell at the next round.

In the spring of 1952, the International Association of Democratic Lawyers, a leftist organization based in New York City, conducted an inquiry into the United States' actions in North Korea. Based on three weeks of interviewing witnesses and reading files, the investigators issued two reports from Beijing. The titles reveal the unequivocal findings: "Report on U.S. Crimes in Korea" and "Report on the Use of Bacterial Weapons in Chinese Territory by the Armed Forces of the United States."

In the midst of the investigation, a well-timed news story from the *Peiping People's Daily* included eyewitness accounts along with a series of grainy photographs showing objects that supposedly had been dropped by American planes. The captions did little to establish the veracity of the report, describing the entomological culprits in such simplistic terms as a "tiny black insect" and "poisonous insects" and the microbiological agents using pseudoscientific terms of "meningitis double globular bacteria" and "consecutive-globular bacteria." Having dropped their scientific guard, the Chinese were sure to take a hard shot from the American experts.

On April 3, the *New York Times* carried a front-page story mocking the laughably naive "evidence" that the communists were touting (see Figure 14.1). The chief curator of insects and spiders at the American Museum of Natural History and a bacteriologist at the Rockefeller Institute for Medical Research were asked to evaluate the photographs and captions. The entomologist identified the "tiny black insect" as a marsh springtail (*Isotomurus palustris*), an utterly benign, flightless insect occurring naturally throughout Europe and Asia and not known to carry any disease. Even more ridiculous were the iden-

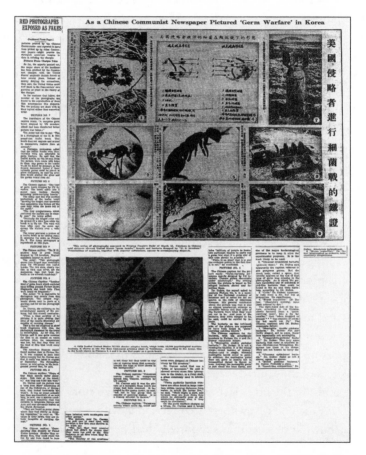

Figure 14.1. The April 3, 1952, front-page article from the *New York Times* refuting the communist charges that the United States was waging biological warfare during the Korean War. The second row of images (left to right) are photographs of a mosquito with its wings having been removed, the formidable silhouette of a springtail, and the remains of a leaflet bomb that the Chinese reported as being loaded with "germ-carrying insects."

tities of the "poisonous insects." The photographs revealed two different creatures: stoneflies (order Plecoptera), which are harmless, weak flying insects devoid of known diseases, and mosquitoes, which for some reason had their wings removed. As for the putative pathogens, an American expert asserted that the supposed meningitis bacteria were misidentified and the "consecutive-globular bacteria" were harmless microbes found commonly in the human throat. Having bloodied the nose of their accusers, the Americans sensed a shift of momentum and tried throwing a diplomatic haymaker.

The United States requested that the United Nations conduct a full inquiry into the North Korean and Chinese charges. The Americans proposed that either the International Committee of the Red Cross (ICRC) or the World Health Organization (WHO) serve as the investigators.[22] Who could object to having a third party serve as the referee?

The communists objected strenuously, claiming that both of these bodies were squarely in the Americans' corner. But because neither the People's Republic of China nor North Korea was a member of the United Nations, it was up to the Soviets to climb into the ring. The Russians first took a swing at the ICRC, arguing that this organization had compromised its integrity by protecting fascist war criminals. (In fact, the Red Cross had known of Hitler's extermination camps and remained deplorably silent.) The Chinese jeered from outside the ring, calling the ICRC a "lackey of American imperialism."[23] As for the WHO, not even the Americans could have believed that this agency was unbiased. After all, the Korean War pitted the North Koreans (with their communist neighbors) against the UN forces (dominated by the Americans). The WHO was a branch of the United Nations, so the Americans were essentially proposing that their personal physician should be the referee.

A ringside announcer might have speculated that the American strategy was to throw out two options in the name of apparent fairness, with one of the possibilities being so absurd that the other would seem to be the pinnacle of propriety by comparison. If so, the U.S. feint-and-jab worked; the United Nations tasked the ICRC with conducting the investigation. But with the referee named, the communists simply refused to fight. The ICRC was denied access to the areas of North Korea and China where the American transgressions were said to have occurred.

Meanwhile, the Soviets landed a solid blow, pointing out that the United States was the only member of the UN Security Council that had failed to ratify the Geneva Protocol prohibiting biological warfare (the Americans finally did so in 1975). The Soviets argued that this was tantamount to the Americans declaring their intent to use such weapons. However, America managed to bob and weave its way out of trouble. The U.S. delegate argued that the accord was an obsolete and impotent paper promise that failed to restrain the Japanese in World War II. The Soviets countered that the U.S. government had granted immunity to the Japanese war criminals of Unit 731, further demonstrating that the Americans could not be trusted. Stung by this shot, the Americans tried to cover up, contending that when the Geneva Protocol was submitted to the U.S. Senate for ratification in the 1920s, the country did not want to risk its neutrality by aligning with the League of Nations. The Soviets scoffed at such a lame excuse, but these rhetorical tactics kept the Americans in the diplomatic fight while the ICRC investigation was ongoing.

After nearly two months of the cold shoulder, the ICRC told the United Nations that the investigation was fruitless and the effort was abandoned. The Americans figured that if their opponents wouldn't come out their corner, then it was time to declare a technical knockout. In July, the U.S. delegation drafted a resolution stating that "the Security Council would conclude, from the refusal of the governments and authorities making the charges to permit impartial investigation, that these charges must be presumed to be without substance and false and would condemn the practice of fabricating and disseminating false charges."[24] Such decisions have to be unanimous in the UN Security Council, and the Soviets were only too pleased to veto the resolution. The diplomatic boxing match of 1952 had ended in a "no decision," with both sides bloodied but neither able to claim victory. But if politics couldn't resolve the issue, then perhaps science could. And when it came to science, the west had the upper hand—at least initially.

The United States asserted that ten Nobel Prize laureates evinced deep reservations concerning the charges. With the scientific heavyweights having backed the Americans, western nations began to fall into line. The Canadians and British expressed their disbelief, citing a lack of scientific evidence. If data were what the world community demanded, then the Chinese and North Koreans would produce records—nearly 700 pages of the stuff.

15

A SWARM OF ACCUSATIONS

The communists realized that the global audience was rolling its eyes during the fight scene between the Soviet and American diplomats. But they also knew that the western nations placed a premium on science. So in 1952, while the United Nations was embroiled in fiery political rhetoric, the Chinese were working in the wings to rewrite the script. By putting scientists in the lead role, they would make sure that the next act would take the world by storm.

To avoid the appearance of gross impropriety, China turned the production over to the World Peace Council.[1] However, the council's bias was very thinly veiled, given that the Soviet-funded organization was founded by Frédéric Joliot-Curie, a winner of the Nobel Prize in physics—and a devout communist. The council drew together an International Scientific Commission for Investigating the Facts Concerning Bacteriological Warfare in Korea and China, thereby hitting all of the right notes ("international," "scientific," and "facts") to create an impression of rigor and objectivity.

The commission was chaired by Joseph Needham, a Cambridge University biochemist. Having been stationed in China in the early 1940s, he was familiar with the Japanese use of insects as weapons. As such, he made an ideal leader for the group: a western scientist experienced in precisely the sort of biological attacks that were under investigation. The cast of supporting characters making up the balance of the commission constituted five other scientists from Brazil, France, Italy, Sweden, and the Soviet Union—can't get much more fair than that, right?

The commissioners arrived in June and conducted a two-month investigation, listening to a slew of witnesses, interviewing captured American pilots, and reviewing reams of documents. However, the investigators did not conduct any field investigations of their own and relied solely on the evidence presented by the Chinese and North Koreans. In August, just weeks after the

United Nations had failed to reach a decision concerning the allegations, the commission called a news conference to issue its verdict: America had waged entomological warfare. The pro-communist French newspaper *Ce Soir* ran a cover story featuring the headline "The Bacteriological War in Korea and China," along with a close-up photograph of flies and a caption noting that the insects were coated with anthrax and cholera.[2]

When the commission's report was released, the findings comprised a concise 60 pages, but the appendices that supported the conclusions ran another 600 pages.[3] The study validated the use of 14 different arthropods, infected with at least eight different pathogens, on 33 separate occasions (along with a weird incident of infected clams, a Hitchcockian tale of diseased voles, and a handful of cases involving fungal pathogens of crops). With science forming the backbone of the report, the authors then added a bit of political meat to the conclusions.

The commission took great pains to draw the link between the U.S. attacks and those of Unit 731. After all, the Americans had sheltered the Japanese culprits. Moreover, according to a Reuters wire report, Ishii and others of his staff were rumored to have been seen on several occasions in South Korea, presumably advising their allies.[4] Three decades after the Korean War, Lieutenant Colonel Murray Sanders—the first of the U.S. government's investigators of Japanese war crimes—claimed that Ishii and one of his associates had been flown to the United States in the early 1950s to collaborate with scientists at Camp Detrick. This assertion has not been independently verified, but the notion is hardly farfetched.

Although the ultimate purpose of the report may have been political, the commission laid out a thorough and intriguing tale of entomological warfare. And the single question in the mind of every diplomat, general, politician, and scientist who read the report was, "Is this horrifying document an accurate account or a fanciful fairytale?"

The first task was simply to figure out what insects and pathogens had been used in the raids.[5] Relying heavily on taxonomic expertise within the Chinese Academy of Science, the investigators derived a comprehensive list of six- (and eight-) legged conscripts and their pathogenic payloads:

- House flies (*Musca vicina*, a cousin of the common house fly, *Musca domestica*, and quite similar to this familiar nuisance) carried anthrax.
- Nonbiting or "false" stable flies (*Muscina stabulans*, a species resembling the blood-feeding stable fly but preferring to dine on excrement and

other nasty stuff) carried typhoid and possibly a disease of pears and apples.

- Anthomyiid flies (*Hylemyia*, a genus similar to that of the house fly but primarily feeding on plants, although some live in dung) carried anthrax, cholera, dysentery, paratyphoid, and typhoid and possibly plant diseases.
- Green bottle flies (*Lucilia sericata*, a species of blow fly named for its coloration; as for its ecology, picture a carcass teeming with maggots); no associated pathogens were detected.
- Sun flies (*Helomyza modesta*, a common, run-of-the-mill fly found in many habitats, where its larvae feed on decaying organic material) carried paratyphoid.
- Midges (*Orthocladius*, a genus of teensy flies that often form swarms in the evening; their larvae are aquatic and the adults, although pesky, do not bite) carried typhoid.
- Culicine mosquitoes (*Culex pipiens* var. *pallens*, a mosquito that may transmit some types of encephalitis); no associated pathogens were detected.
- Aedes mosquitoes (*Aedes koreicus*, a species not known to vector diseases, although its relatives carry yellow fever); no associated pathogens were detected.
- Crane flies (*Trichocera maculipennis*, a species within a family of flies that look like giant mosquitoes, except they neither bite nor transmit disease) carried a neurotropic virus.
- Human fleas (*Pulex irritans*, a species that feeds on humans as well as many other mammals and is known to be a vector of bubonic plague) carried plague.
- Ptinid beetles (*Ptinus fur*, a small, uninspiring brown beetle that feeds on stored grain throughout the world) carried anthrax.
- Grouse locusts (*Acrydium*, a miniature grasshopper, about a half-inch long, with no known or imaginable potential for economic damage or disease transmission); no associated pathogens were detected.
- Migratory locusts (*Locusta migratoria*, a species found in Asia, Australia, and Africa and having an impressive capacity to form migratory swarms and ravage crops but no potential for disease transmission); no associated pathogens were detected.
- Field crickets (*Gryllus testaceus*, a commonplace cricket that some Asian entrepreneurs raise on "cricket farms" to market as fish bait and as pet food for birds and reptiles); no associated pathogens were detected.

- Springtails (*Isotoma negishina*, a minuscule, wingless insect that "hops" via a bizarre spring-loaded, pole-vaulting structure on its abdomen) carried dysentery and an unknown rickettsia.
- Wolf spiders (reported as *Tarentula*, but tarantulas and wolf spiders are in different families; the authors seem to have meant some sort of free-roaming hairy spider) carried anthrax and fowl cholera.
- Lycosid spiders (reported as *Lycosa*, which are the wolf spiders, but presumably this is a different species from the creature noted above) carried anthrax and fowl cholera.
- Stoneflies (Nemouridae, this family of pathogen-free, vegetarian insects spend their larval lives in streams with the clumsily flying adults emerging to find mates); no associated pathogens were detected.

Not satisfied that this was the entire scope of creatures, the commission queried the North Koreans as to accounts found in the earlier report by the International Association of Democratic Lawyers.[6] The Minister of Health confirmed that ants (family Formicidae), bed bugs, and mealworm beetles (*Tenebrio molitor*) had been found, although they lacked any pathogens. The North Koreans corrected an earlier account, noting that a mistranslation had indicated that the Americans had spread ticks when, in fact, the creature was the mite that carries tsutsugamushi fever. The officials stuck with the earlier report of what had to be the strangest entomological weapons: nycteribiid flies, which are rare, wingless, spidery insects that parasitize bats. But merely finding weird flies—or most of the other insects constituting the purported potpourri of living weapons—was hardly grounds for establishing entomological warfare. The commission needed convincing eye-witness testimony and compelling circumstantial evidence to convince the world of American treachery.

The most damning evidence consisted of firsthand accounts provided by villagers within the American drop zones. For example, the commission recounted that

> on the afternoon of March 6, 1952, Shan Wen-Jung and Tu Kung-Chou, inhabitants of Tung-K'an-Tze, Antung, witnessed four American planes passing over, and about ten minutes later discovered objects dropping down like snow-flakes which after reaching the ground were found to be anthomyiid flies, midges, and spiders.

[Inhabitants of K'uan-Tiem] saw eight American fighter planes pass over the city about half-an-hour after noon. . . . From one of them there was distinctly seen to drop a bright cylindrical object. Immediately afterwards, and during the following days, the people of the town including school-boys, organised searches in the region beyond the east gate where the object appeared to have fallen, and collected many anthomyiid flies (*Hylemyia*, sp.) and spiders (*Tarentula*, sp.).[7]

Of course, there were many instances in which the connections between low-flying aircraft and abundant insects were circumstantial, but sufficiently close in time and space. A typical account describes the passage of American planes and the subsequent appearance of insects:[8]

These insects were all discovered in places after American planes had intruded into the areas. For instance, in the morning of March 4th, three planes raided Hung-Shih-La-Tze village, K'uan-Tien hsien. In the same afternoon, large quantities of field-crickets were discovered by the inhabitants on the snow-covered ground outside the village.

Even without any association between a particular flyover and subsequent discovery of insects, the commission found a range of ecological anomalies that pointed to an unnatural source of the creatures. The North Koreans reported finding springtails (order Collembola; see Figure 15.1)—tiny, flightless insects that hang out in damp, shady habitats—"on the cement stadiums about 6 m high in a race course at Fu-Shun and on the top of a neighboring cement silo about 12 m above the ground."[9] On a larger scale, the commission also concluded that the midges and anthomyiid flies were entirely novel to the regions in which they were found. But even insects native to eastern Asia were not above suspicion.

The investigators placed considerable weight on the fact that a dozen of the suspicious endemic species made their appearance months before the normal time of their emergence, sometimes being found in subfreezing conditions. The commission provided mini-tutorials to drive home their point:

For instance, the migratory locust passes winter in the egg stage and the adult dies after laying eggs in the autumn. The eggs hatch out in April and May of the next year. However, at mid-night on March 15 following the

Figure 15.1. The springtail, *Folsomia candida*, is the "white rat" of this insect order, being easily cultured in the laboratory and serving as the standard test organism for the effects of pollutants on soil arthropods. The blind, unpigmented species is 1/20th of an inch in length, and although this insect could be reared in enormous quantities, it seems—as with all collembolans—to have no potential for carrying diseases, despite accusations to the contrary during the Korean War. (Photo by Steve Hopkin)

intrusion by American planes, a large number of locusts were discovered on cement ground still covered with snow inside the city of Shenyang.[10]

The report dismissed the possibilities that the warmth of an early spring or the heat of bomb bursts had accelerated insect maturation. And even if there were some instances in which local factors, such as south-facing slopes, may have hastened the development of insects, another ecological anomaly had no plausible natural explanation.

Insects often amass in impressive numbers, but some of the North Koreans' accounts exceeded the bounds of natural events. While clouds of gnats and swarms of bees are familiar insectan mobs, the investigators came across more startling phenomena:

Besides the anomalies in season and location, the number of insects discovered also shows important abnormalities. . . . For example, the anthomyiid flies discovered at Ku-Chia-Tze, Shenyang, were in tens of thousands, and in Ssu-Ping as many as 6000–7000 houseflies were found in a single group. Even more outstanding was the discovery of tens of thousands of field-crickets at K'uan-Tien on the surface of the snow.[11]

Just as strange in the judgment of the scientific team were the cases of bizarre associations. While some organisms might be expected to hang out together in natural communities, the Chinese and North Koreans reported dense assemblages of springtails and fleas, midges and crickets (family Gryllidae), and flies and locusts. Finding such odd combinations supported the argument that something villainous was afoot. But even more worrisome were the inexplicable pairings of microbes and insects.

Ptinid beetles are brown insects about the size of a typewritten "O." With six gangly legs and two long antennae, they resemble small spiders, hence the common name for this family: spider beetles. Their diet is repulsive, with moldy grain and dried animal excrement being at the top of the list. Despite this disgusting cuisine, these insects weren't known to carry any diseases—until villagers came across infestations in 1952.[12]

According to North Korean sources, American planes passed over several hamlets and dropped some sort of objects. When the curious residents went to see what had fallen, they found no containers but loads of ptinid beetles. In the next few days, several people came down with headaches, body pains, and nausea. When the symptoms escalated to raging fevers and continuous vomiting, the patients attracted the attention of medical authorities. The sputum of the afflicted villagers was rife with anthrax bacilli and at least ten people died.

As to why the Americans would have chosen spider beetles to carry deadly microbes, the Chinese reported that "under the dissecting micro-scope it was clear that the beetle Ptinus would be well adapted for dis-seminating anthrax by this [respiratory] route, for it has an abundance of brittle chitinous spines on its elytra [the hardened hind wings that encase the body] which could be inhaled."[13] Presumably the hairiness of true spiders (order Araneae) explained why these creatures were also used to tote anthrax spores. But not all of the accounts in the commission's report were as incredible as these unprecedented associations of microbes and arthropods.

The centerpiece of the commission's report was use of plague-infected fleas—the modus operandi of Unit 731:

Above all, the fleas appearing were not the rat fleas [the oriental rat flea, *Xenopsylla cheopis*, would be the normal vector in region] which more usually carry plague bacteria in a state of nature, but human fleas (*Pulex irritans*). It was these which were used by the Japanese during the second world war.[14]

According to the commission, Korea had been free of plague for five centuries, with the nearest endemic regions 300 miles to the north. So American planes were the only conceivable source of plague-infected fleas.

In the most remarkable case, after an F-28 fighter flew over the hamlet of Kan-Nan in the middle of the night, the residents awoke to a most alarming sight: "In the morning, the villagers found [many] voles dying or dead in their houses and courtyards, on their roofs, and even on their beds, while others were scattered around the outskirts of the settlements."[15] When the voles (creatures that look like stocky mice with short tails) were rounded up, they were found to be laden with fleas, and subsequent testing revealed that at least one of the rodent raiders was infected with plague. In most cases, the Americans apparently relied on the direct release of fleas from aircraft.

Both military personnel and civilians reported the sudden appearance of masses of fleas in unusual locations, including bare hillsides in subfreezing weather. The most detailed account came from Song Chang-Won, a farmer near the village of Kum-Song Li, who told investigators the following:

> In the morning of March 25, 1952, I went to Pak Yun-Ho's house to consult with him on farming. There I found Pak Yun-Ho returning from the well where he had gone to wash his face. He said there were many fleas floating on the surface of water in a water jar. We went together to the well situated about 80 meters [87 yards] from our houses. I found fleas floating as if dead on the surface of water in the water jar near the well. This reminded me of the fact that at about 4 a.m. this morning an American plane had circled at low altitude without strafing or bombing. I thought these fleas had been dropped by the American plane, and I informed the Village People's Committee of this fact.[16]

North Korean and Chinese scientists confirmed that the fleas were infected with plague, but the news came too late for Pak Yun-Ho, who died of the disease a few days after finding the fleas. Whether such eyewitness statements constitute rigorous evidence is debatable, but testimonials were persuasive, particularly when they came from Americans.

Air Force officers taken as prisoners proved the duplicity of the American military, at least in the minds of those seeking to condemn the west.[17] The communist interrogators provided the commission with page after page of rambling, detailed, hand-written confessions from downed pilots. The writings revealed that the officers had attended secret lectures on the tactics of biological warfare, after which they had flown missions spreading infectious agents over North Korea and China. Based on conversations with the captives, the commissioners "unanimously formed the opinion that no pressure, physical or mental had been brought to bear upon these prisoners of war."[18] For their part, the interrogators noted that the airmen were not themselves monsters but merely good soldiers following the directives of a bestial government. Not only did the communists conclude that the captives had participated in war crimes with the "greatest inner reluctance," but the American officers had even turned the moral corner with the support of their compassionate captors:

> These declarations were made of their own free will, after long experience of the friendliness and kindness of their Chinese and Korean captors had brought to them the realisation that their duty to all races and peoples must outweigh their natural scruples at revealing what might be considered the military secrets of their own government.[19]

These accounts were bolstered by the testimonies of South Korean infiltrators who had been captured while gathering epidemiological data that UN forces hoped to use in assessing the effectiveness of their entomological exploits. The commissioners did not see these confessional statements but accepted the word of the North Koreans that such evidence existed.

The commission was fully aware that such hearsay accounts, along with the incredible scientific evidence, would be pooh-poohed by westerners. The investigators figured that the best way to deflate their detractors was to beat them to the punch. So the report laid out the most problematical elements of the case against the Americans and rebutted these objections before they could be raised.

The anticipated problems centered on the role of insects as weapons of war. One of the most obvious objections would be that the Americans were too smart to drop cold-blooded creatures in the midst of winter. Anticipating this concern, the commission proposed that selective breeding could have "specially endowed [the insects] with cold-resistance."[20] Such genetic manipulations were plausible, but it might have been easier to simply wait a few months before launching an attack with regular insects. And if the Americans employed odd tactics in terms of timing, their choice of targets was no less in need of explanation.

Most of the accounts of entomological assaults came from the front lines, which meant that the Americans were dropping insects near their own forces. The commission observed that such an approach was actually rather clever, given that these were the areas with the highest concentrations of enemy troops and the Americans had means of protecting themselves. According to the report, American scientists had mastered the technology of vector control, with "new and ever more potent insecticides [and] machines of high efficiency for the dissemination of clouds of these substances in large amounts and minimum time."[21] It is certainly the case that when a typhus outbreak threatened in the winter of 1950–51, the UN command mobilized lice-treatment units, dosing their troops with DDT, and when dysentery reared its ugly head, the Americans drenched the afflicted, fly-infested cities with insecticide.[22] And if Americans were experts at killing insects, they were also proficient at producing them.

As to the criticism that entomological warfare on the scale asserted by the commission would have required enormous insect rearing capabilities, the investigators responded with complete faith in American innovation. They noted that "in the scientific literature there are descriptions of methods for the artificial production of insects and arachnids on a large scale."[23] Although instructions for the mass production of springtails, stoneflies, and spiders could not be found in technical journals, the commission seems to have little doubt that Yankee ingenuity would have extended existing methods to these novel creatures. But all of this raises the question of why the Americans would have chosen to use insect vectors in the first place.

The commission had a simple and compelling explanation for the U.S. military's affinity for entomological warfare: Ishii Shiro.[24] Not five years earlier, the Americans had sold out the rest of the world and traded justice for the secrets of the Japanese biological warriors. Given the effectiveness of Unit 731 against the Chinese, and in light of the tremendous head start that the Americans gained from their Japanese tutors, how could the United States have not used insect vectors against their enemies when the military situation became dire? But even allowing that the Americans had the wherewithal to develop and exploit entomological arms, the choice of insects was truly bizarre.

Springtails as weapons? The size of typewritten commas, these primitive insects aren't very hardy, they won't travel more than a few yards in their entire lives, and they don't bite, carry diseases, or destroy crops. However, the commissioners appealed to our ignorance of the enormous diversity of the biological world and argued that we can't be absolutely sure that springtails don't harbor pathogens.[25] From this premise, they raised the possibility that infected springtails could pass along deadly microbes by becoming snacks for domestic animals (hungry ducks were suggested) or feral mammals, by falling into food and water sources, or by chewing on plants. And if one of these pathways worked for springtails, then extrapolating to other farfetched vectors, such as stoneflies, took no more imagination.

But if everything from springtails to fleas had rained down during dozens of American sorties, a skeptic might contend that the communists should have been able to document the consequent suffering and death. The doubter could be expected to demand epidemiological evidence, and the commissioners were ready with their response:

> The Commission is not in a position to give the world concrete figures concerning the total number of Korean and Chinese civilians killed, nor the total morbidity, nor the fatality rate. It is not desirable that this should be done, since it would provide the last essential data for those upon whom the responsibility rests. The information is not necessary for the proof of the case upon which the Commission was invited to express an expert opinion.[26]

The report of the International Scientific Commission may have had a number of unconvincing and mistaken lines of argument, but the commissioners had one thing absolutely right—western politicians, scientists, and militarists were going to launch a withering counterattack.

16

AN IMAGINARY MENAGERIE?

Despite the efforts of the communists to portray the International Scientific Commission as a team of objective scientists, the Americans weren't buying it for a minute.[1] The commission had been formulated by the World Peace Council, an organization that the West saw as a communist front. Its members had made little effort to disguise their anti-Americanism and the chair of the commission, Joseph Needham, was an avowed Marxist (see Figure 16.1).

According to western analysts, the report was simply political tit-for-tat. The communists were getting even for the Americans' having shielded Ishii and his ilk from war-crimes prosecution. Such propaganda had even been worked into putatively scientific outlets, such as the *Chinese Medical Journal.* In an issue of the journal published at the same time as the commission's report, Chen Wen-Kuei reviewed the outbreaks of insect-borne diseases during the Second World War and concluded with a political accusation:

> The fact that the U.S. Government has sheltered and employed Japanese and German "bacteriological warfare experts" should also be mentioned. With regard to this . . . Japanese bacteriological war criminals, such as Shiro Ishii, Jiro Wakamatsu, and Masajo Kitano, are today still at large, and what is more, they are fostered and utilised by the U.S. generals.[2]

Given what they suffered, the Chinese might be forgiven their efforts to paint the Americans as immoral, but the West had no tolerance for what was considered to be a deplorable lack of integrity on the part of the commission. When the press asked what sort of independent corroboration had been applied to the assertions in the report, the Swedish commissioner replied that "delegates implicitly believed the Chinese and North Korean accusations and evidence."[3] And when Needham was asked what evidence he had that the

Figure 16.1. Joseph Needham meeting Zhou Enlai, who was the premier of the People's Republic of China and served as the foreign minister during the existence of the International Scientific Commission for Investigating the Facts Concerning Bacteriological Warfare in Korea and China. Needham, a British scientist and avowed Marxist, chaired the commission, much to the consternation of the United States. (Photo courtesy of The Needham Research Institute)

plague bacilli shown in microphotographs had actually come from the voles scampering around Kan-Nan, he blithely answered, "None. We accepted the word of the Chinese scientists."[4] But what really set off the Americans was not the political bias of the World Peace Council, nor the gullibility of the International Scientific Commission. What really outraged the Americans was the purported testimony of the POWs.

The fiercest political battle concerning the charges of biological warfare was waged over the confessions of the downed airmen. To get a flavor of the acrimony, consider the stinging refutation given by the U.S. representative to the UN: "The so-called 'germ warfare' confessions were not simply a sudden bright idea on the part of the Communists, but were an integral part of a tremendous and calculated campaign of lies."[5] With the diplomatic gloves off, the West started landing some solid blows.

The Americans stipulated that while the written statements included plenty of technical detail that could have come from the POWs, the confessions were rife with communist rhetoric that echoed favored lines from the Chinese press. Consider the wording used in the confession of First Lieutenant John Quinn:

How I was forced to take part in the inhume [*sic*] bacteriological warfare launched by the U.S. Wall Street . . . brought up as I was on the propaganda lies of the Wall Street imperialists. . . . It was a crime against all the peace-loving peoples of the world.[6]

With the other confessions repeatedly alluding to "capitalistic Wall Street war mongers" and the like, the statements appeared to be entirely—and badly—contrived. Moreover, the American airmen all recanted their confessions once they returned to the United States, claiming that the statements had been made under relentless psychological pressure and physical duress. But having mounted a vigorous assault on the veracity of the confessions, the U.S. government overreached, leaving themselves open to an effective counterpunch.

While the Americans argued that the POWs' confessions were coerced, the communists contended that the same could be said of the recantations. The U.S. attorney general sent the returning airmen a blunt warning: "United States prisoners of war who collaborated with their Communist captors in Korea may face charges of treason."[7] And if there was any doubt as to what was expected of the POWs, the secretary of defense told the press, "My views may be extreme, but I believe those who collaborated and the signers of false confessions should be immediately separated from the services under conditions other than honorable."[8] Having been duly warned, the returnees were handed pen and paper and given the opportunity to write retractions. Perhaps they would have done so by their own free will, but the hardball tactics of the U.S. military had some analysts wondering if the Americans hadn't protested a bit too loudly.

In 1998, Stephen Endicott and Edward Hagerman—a pair of Canadian historians with a penchant for unearthing troublesome documents—revealed that the technical substance of the confessions, if not the choice of political phrasing, might have been embarrassingly close to the mark.[9] They came across records from the Office of Special Services of the Inspector General of the U.S. Air Force revealing that lectures on germ warfare had been delivered to the 3rd Bomb Wing at Kunsan air base in 1951—a place and time that coincided with the service of two of the downed officers. To make matters worse, the American records also included enough dots to allow the historians to connect them in some very damning, if somewhat speculative, ways.

The U.S. military's Operations Orders describing the logistics of air raids, along with schedules of attacks on Korean targets in 1952, reveal a curious sequence of events. Some of the Operations Orders called for dropping conventional ordnance, followed by bombs with delayed-action fuses (to dissuade the enemy from attempting to repair the damage), capped off with two M-105 leaflet bombs (see Figure 16.2). According to military sources, these leaflets were intended to "warn non-combatants in the areas adjacent to military targets that those targets were subject to attacks by USAF, thus enabling civilian personnel to avail themselves of an opportunity to escape injury and fatalities."[10]

Giving folks a heads-up sounds compassionate, but one has to wonder about the efficacy of warning folks of an impending attack using leaflet bombs that were scheduled to fall *after* the conventional ordnance had been dropped. Of course, if the leaflet bombs contained something other than thoughtful brochures—say, a load of flies coated with anthrax—the timing made perfect sense. However, circumstantial evidence will get one only so far in making or rebutting an accusation. So American analysts turned their attention to the science at the heart of the commission's report.

The western critics threw a two-punch combination. One line of argument simply maintained that the U.S. military would never have relied on vectors. But the Americans overreached again. Dr. Dale W. Jenkins claimed that the United States had "never investigated the potential of using arthropods for BW [biological warfare]"[11]—a rather remarkable assertion, given that he was the chief of the Entomological Division of the Army Biological Laboratory at

Figure 16.2. A soldier loading an M-16-A1 bomb, capable of holding 22,500 eight-by-five-inch leaflets. During the Korean War, a B-29 would normally carry 32 of these bombs, each of which weighed 170 pounds, and drop them from an altitude of about 20,000 feet. At 1,000 feet above the ground, a fuse opened the device and spread the leaflets. By early December 1950, the Far Eastern Air Force had disseminated 147 million leaflets—or millions of insects—using these devices. (Courtesy of Ed Rouse)

Fort Detrick (formerly Camp Detrick). But Jenkins was supported by Robin Clarke, a science journalist, who maintained that

> certainly they [U.S. forces] would not have relied upon animal carriers or vectors to spread disease for even in the early 1950s it was realized that this was an unreliable method of dissemination and that the spraying of biological aerosols would be a more effective means of waging biological warfare.[12]

This seems to be a plausible argument, except that records from Fort Detrick indicate that aerosol dissemination methods were still under development in 1952.[13] Although the U.S. military was headed in that direction, there is significant doubt that the Americans could have fielded this technology during the Korean War.

The defendants' second line of argument was more potent. Given the absurdity of the entomological weapons, the world would have to conclude either that the Americans were biological buffoons or that the communists were ludicrous liars. No competent entomologist would even consider using ptinid beetles, springtails, stoneflies, and spiders to carry diseases or grouse locusts and crickets to assail crops. The communists were engaged in rank, pseudoscientific propaganda or perhaps a kind of reverse psychology. That is, no real scientist would ever try something as inane as dusting tarantulas with anthrax spores and dropping them by aircraft during the winter, so the accusation had to be true.

As ridiculous as some of the entomological weapons were, the image of parachuting voles was difficult to trump. If the attackers were going to scatter flea-ridden rodents, why would anyone use meadow-loving voles rather than rats that would seek out human habitations? The argument is awfully persuasive, but there is evidence that the Americans had amassed considerable data on rodents other than rats. In early August 1952, the U.S. military field-tested brucellosis bombs at Dugway Proving Grounds by building a mock city and populating it with guinea pigs.[14] The test was repeated three times, with 11,628 of the furry creatures giving their all. The value of the rodent research was not entirely appreciated by the military, as evidenced by an Army Chemical Corps general's wry assessment: "Now we know what to do if we ever go to war against guinea pigs."[15] Such studies notwithstanding, airborne voles seemed rather unlikely. However, refuting the commission's reports of insects not previously known to occur in Asia required a different line of argument.

The Americans maintained that the communists' discovery of new species during the war was the result of two factors. The North Koreans and Chinese had a poor understanding of their countries' insect faunae, and a governmental decree during the war had initiated an unprecedented flurry of entomological collecting. Together, these circumstances resulted in finding various creatures that had been living in obscurity.[16] This rebuttal took care of many suspicious species, but the U.S. government needed to account for the disease-laden native insects appearing in large numbers at unusual times.

The Americans contended that the incidents involving naturally occurring insects and diseases were just that—natural events. Canadian scientists corroborated their allies' alibi, providing ecological explanations for the various accounts in the commission's report. Wartime conditions were well known to foster a panoply of pests and pathogens—and the Korean War was no exception. If every case of pestilence during armed conflict was proof of biological warfare, then every army in history must have used these weapons. And as for the early appearances by the insects, if the communists didn't even know what species lived in their midst, it was unlikely that they had reliable records of insect life histories. Moreover, the claim that the winter and spring of 1952 were normal was belied by the Chinese press. The February 21 edition of the *People's World* newspaper reported epidemics among humans and animals as a consequence of unusually dry, warm weather.

But the entomologists were not the only scientists to take a swing at the commission's report; western epidemiologists also stepped into the ring. The WHO repudiated the charges of biological warfare using seemingly impeccable logic. The medical experts argued that had U.S. planes actually made nearly a thousand airdrops of infected vectors across North Korea and China, the result would have been widespread epidemics with millions of victims. Although the International Scientific Commission was rather circumspect as to the death toll, even the most liberal reading of their report suggests that mortality from disease was probably within the expected norms for wartime conditions.

Washington issued a series of flat denials that the purported devices for disseminating insects would have worked, even if they had been used—which, of course, they hadn't been. Pentagon officials insisted that the bomb casings the communists recovered as evidence of biological warfare were merely those of 500-pound leaflet bombs. And these devices had holes to equalize the pressure, which would have meant sure death for any living organism within. But this neat argument was undermined by Major General E. T. Bullene, who later told a House Appropriations Sub-Committee in unrelated testimony that actually, retaliatory bacteriological warfare does not involve some complicated super weapon. The means of delivering germs to enemy territory are simple and involve equipment of the type with which the services are already well stocked . . . such as the containers used currently for dropping propaganda leaflets.[17]

There were, however, some inexplicable devices described in the commission's report. For example, the North Koreans claimed to have found the remains of a bombshell that was less than 0.04 inches thick; any such casing would have immediately fractured and disintegrated upon release from an aircraft. Although the Americans eagerly attacked every questionable detail in the report, the most compelling rebuttal pertained to military tactics and the essential nature of biological warfare.

The major problem with the report of the International Scientific Commission was not that the evidence was too weak but that it was incomprehensibly strong. The case was lavish beyond belief, with the accusers wallowing in insects, bomb fragments, microbial slides, autopsy reports, eyewitness testimonies, and POW confessions. One of the prime virtues of biological warfare is its covert potential. The Americans claimed that had they waged biological warfare, they would not have engaged in entomological carpet bombing. For years leading up to the Korean War, the U.S. military had treated biological weapons with the utmost security. Was it really plausible that a country developing a weapon under conditions of the highest secrecy would then launch hundreds of daylight attacks in full view of the enemy? The Americans might tolerate being called criminals, but they couldn't abide being called stupid.

With the nations in the left corner and those in the right corner of the political boxing ring having flailed away at one another for months, the political title fight came down to a decision. The international community could believe either that the Americans had launched the most conspicuous and ill-conceived series of insect-vectored biological attacks in the history of the world or that the communists had mounted the most foolishly composed and resource-draining propagandistic conspiracy plot in living memory. The majority of nations favored the latter judgment, unable to swallow the findings of a commission that possessed scientific credibility but lacked political objectivity. However, it is probably fair to say that some, perhaps many, governments harbored a nagging suspicion that where there is smoke there is fire[18]—maybe not a conflagration of overt biological warfare but perhaps the flame of covert field testing.

Back at the UN, the Americans tried to take one final, parting shot.[19] At the 1953 session, the United States submitted a resolution calling for a neutral—as if such existed—commission to render a decision concerning the Soviets' charges. In what western diplomats took as a tacit admission that the case against the U.S. military was a sham, the Russians offered a deal. They would withdraw their allegations if the United States would withdraw its resolution. Three months after the two nations struck the deal, an armistice was signed at Panmunjom. Open hostilities on the Korean peninsula might have ended on July 27, 1953, but the entomological embers continued to smolder.

In the heat of political battle, combatants can become entrenched and unwilling to give ground. Neither side can afford to show weakness in the midst of bombastic tirades and inflammatory accusations. But in the years that follow the conflict, when hindsight provides perspective, the erstwhile enemies often have moderated their positions—and sometimes they have dug in deeper.

Many of the accusers continue to stick to their guns. Joseph Needham, the chairman and lightning rod for the International Scientific Commission, was even more convinced of American duplicity three decades after the investigation. He wrote that based on "everything that has been published in the last few years. . . . I am 100 percent sure."[20] The three surviving members of the commission remain confident of their charges but less certain of the scope of the attacks. In a 1994 interview, one of the representatives maintained that "I am still convinced that the U.S. conducted biological warfare, but not on a massive scale."[21]

The certainty of Americans as to their country's innocence also has diminished with the passage of time. A decade after the Korean War, Dale Jenkins, chief of the Entomological Division at Fort Detrick, acknowledged that the U.S. military had the capacity to use insect vectors.[22] He further admitted that entomological weapons would likely have been a part of a biological offensive, but did not go so far as to state that his country had engaged in such attacks. And in 1979, Professor George Wald, a Nobel Prize winner in Physiology or Medicine, said that "as for the allegation that the U.S. used germ warfare in the Korean War, I can only say with dismay and some shame that what I dismissed as incredible then seems altogether credible to me now."[23]

Others remain steadfastly undecided. The prestigious Stockholm International Peace Research Institute waffled in the early 1970s, deeming some of the charges to be plausible, ascribing others to natural phenomena, and considering the rest as fabrications.[24] Western scientists seem to harbor similar ambivalence, typically concluding along the lines expressed by Seán Murphy, Alastair Hay, and Steve Rose, a trio of British biologists who concluded that

> although the Americans suffered some political damage over the tribunal's conclusions the case still remains one which is essentially unproven. It must be said, however, that there was a good deal of circumstantial evidence to support the tribunal's findings.[25]

Not surprisingly, U.S. military historians tend to stand by their nation's denial. Moreover, contemporary analyses also offer sound reasons for why the communists might have implemented a systematic program of deceit.[26] One theory proposes that accusing the Americans of biological warfare was a savvy tactic to motivate and mobilize the Chinese and North Korean people during a genuine public health crisis brought on by the war. Another theory suggests that the communists, knowing that the U.S. military had biological weapons, made the accusations to attract the world's attention to an impending crime—something like screaming "Help!" before the mugger takes the gun out of his pocket.

These efforts to explain away the communists' charges have not convinced two of the nation's military experts with the greatest knowledge and experience of biological warfare. When asked whether the United States waged some sort of entomological warfare in Korea, Lieutenant Colonel Robert Kadlec—a medical doctor who was a member of the Homeland Security Council and serves as the staff director for the U.S. Senate Subcommittee on Bioterrorism and Public Health—was cautious but forthright: "I would say more likely than not, particularly in the context of that conflict and where we were in the evolution our offensive capabilities."[27] His sentiments are echoed by Colonel Charles Bailey—who has a Ph.D. in medical entomology and is now director of research at the National Center for Biodefense at George Mason University. Concerning the claims of entomological attacks, Bailey asserts that

> it's not outside the realm of possibility that something was done. During that time there was a very active offensive program ongoing at Fort Detrick [Bailey was commander of Fort Detrick in the early 1990s]. . . . The Americans had a big vector program, so they obviously must have tested it somehow or another. What would have stopped them? They tested all the

others, including simulants offshore. I doubt that they would have used a live pathogen because they never did so to my knowledge in the other tests. I can certainly envision them dropping vectors to test their distribution and survival. But actually infecting them? I doubt that.[28]

To many, the Americans were vindicated in January 1998, when a Moscow-based Japanese newspaper reporter acquired documents from the Russian Presidential Archives that strongly suggested that the North Koreans, Chinese, and Russians conspired to falsely accuse their common enemy of waging biological warfare.[29] Although the plot is not entirely clear, glimpses into the workings of the respective governments provide a series of damning snapshots. Among the communiqués was a memo from the Soviet chief of the secret police, Lavrenti P. Beria, to Soviet Premier Georgi Malenkov and the Presidium of the Central Committee stating that

> two false regions of infection were simulated for the purpose of accusing the Americans of using bacteriological weapons in Korea and China. Two Koreans who had been sentenced to death and were being held in a hut were infected. One of them was later poisoned.[30]

Another memo sent from the Soviet leadership to the Chinese was remarkably—perhaps suspiciously—explicit in exonerating the Americans:

> For Mao Zedong
> The Soviet Government and the Central Committee of the CPSU were misled. The spread in the press of information about the use by the Americans of bacteriological weapons in Korea was based on false information. The accusations against the Americans were fictitious.[31]

One might think that the coconspirators would be repentant, but being caught red-handed is a matter of perspective. Nothing about the Korean War was simple, including the interpretation of these documents. Skeptics point out that all but one of the documents date from 1953, a full year after charges were brought against the Americans. Thus, it could be that the communiqués refer to events after the first round of (actual) attacks. While the Soviets and North Koreans evidently collaborated to fake a crime scene, this only shows that there was a single act of deceit, not that every incident was a sham. And so the two sides continue the 50-year political slugfest.

The old wounds linger, in large part because the Cold War relied on creating psychological distance between "us" and "them." During the Korean War, the communists vilified the West and the Americans dehumanized the "Reds." The U.S. soldiers referred to the enemy as "gooks," while General Ridgeway's famous message entitled "Why We Are Here" established the moral certitude of the American position:

> To me the issues are clear. It is not a question of this or that Korean town or village. Real estate is, here, incidental. It is not restricted to the issue of freedom for our South Korean Allies, whose fidelity and valor under the severest stresses of battle we recognize; though that freedom is a symbol of the wider issues, and included among them.
>
> The real issues are whether the power of Western civilization, as God has permitted it to flower in our beloved lands, shall defy and defeat Communism; whether the rule of men who shoot their prisoners, enslave their citizens, and deride the dignity of man, shall displace the rule of those to whom the individual and his individual rights are sacred; whether we are to survive with God's hand to guide and lead us, or to perish in the dead existence of a Godless world.[32]

The Cold War was about the very survival of western culture, indeed of God. When the stakes are cosmic, the development and use of almost any weapon could be rationalized. And cold-blooded tactics called for cold-blooded warriors.

17

THE BIG ITCH

Better dead than Red. The western mantra of the 1950s. And people meant it. With this sort of sociopolitical view, it is little wonder that the U.S. military pursued every conceivable means of defeating the soulless communists—even conscripting insects as unwitting patriots. After all, if entomological weapons based on bungled Japanese science had killed nearly half a million people, just think what an advanced nation might be able to do with these creatures. Or imagine what a technologically sophisticated foe might accomplish.

With the Soviet Union detonating an atomic bomb in 1949, Cold War anxiety paved the way for a series of incredible experiments in the name of protecting democracy. The official rationale for these studies was to prepare a civil defense plan, but of course any data concerning how the American populace might be put at risk by biological warfare were eagerly converted into the development of tactics for attacking the Soviets and their allies.

Field studies of biological warfare most infamously involved the secret releases of "simulants" to mimic clandestine attacks with deadly microbes.[1] In collaboration with the CIA, Camp Detrick scientists began by dispersing theoretically harmless and easily traced bacteria in the Pentagon's ventilation system, then moved on to contaminating subway systems and office buildings. Still larger targets were fogged with zinc cadmium sulfide, a florescent compound that could be formulated to drift like airborne bacteria. In 1950, whole cities were sprayed from naval ships, with 800,000 clueless people inhaling some 4 billion particles in the course of an experiment off the coast of San Francisco. Over the next few years, clouds of bacterial simulants and chemical markers were released in more than 200 unsuspecting communities across the country. There were no reports of serious infections or adverse reactions among the citizenry, but the military had little incentive to go looking for problems (see Figure 17.1).

Figure 17.1. Camp Detrick's "Black Maria" (ca. 1945) was a wooden building covered in tar paper; it served as the shell housing the nation's first laboratory for offensive research on biological weapons. The smokestack was part of the boiler system used for incinerating experimental animals. Within the compound, each scientist was armed with a .45 caliber pistol. A soldier in the guard tower stood ready with a Thompson submachine gun. (Courtesy of the Department of the Army)

Despite reams of data from experiments on the American public and a doubling of the research program at Camp Detrick, the future of germ warfare did not look promising.[2] Technological sophistication had been the driving principle behind aerosol delivery. The bacteriologists' dream was to isolate pathogens, which could be cultured in staggering quantities, and directly infect the enemy. While insect-borne diseases were some of the best candidates for weaponization, the vectors were difficult to produce, handle, store, and disperse. However, the American scientists ran into the same problems that vexed the early work of Unit 731: pathogens are immobile wimps. So the microbiologists reluctantly but ineluctably turned to the entomologists to provide the means for delivering the virulent but vulnerable pathogens.

The researchers soon began to see the advantages of using insects as miniature missiles for delivering pathogenic warheads. While microbes are aerial plankton, biting insects are consummate hunters, able to track their victims using an unparalleled sense of smell. As an added bonus, while gas masks excluded airborne germs, insects could circumvent this defense by squirming beneath clothing or finding patches of exposed skin. Moreover, the vectors could survive for days in the environment, their hunger mounting until an unwary enemy removed their protective gear or moved into the area.

The U.S. military used a bit of discretion when it began field-testing entomological weapons. Named with a touch of gallows humor, Operation Big Itch took place on the bleak landscape of Dugway Proving Grounds.[3] The goal was to determine if fleas could be reared, transported, loaded into munitions, and then delivered to a target in sufficient numbers to transmit disease to an enemy. The insects were not infected with plague, but the ultimate endpoint of this venture was to improve upon the success of the Japanese.

So it was that in 1954, from the skies over the Utah desert, fleas rained down on cages of guinea pigs. The most uncertain aspect of the operation was whether the vectors would survive and disperse after being launched from rather novel devices. The E-14 and E-23 munitions worked along similar lines: a cardboard cylinder about the size of a container of oatmeal was equipped with a mechanism to expel a burst of carbon dioxide from a pressurized cartridge. The force of the gas would rupture a bag of fleas within the cylinder, expelling them like shotgun pellets as the device tumbled earthward from a height of 1,000 to 2,000 feet.

Except for a couple of glitches, Operation Big Itch was a success. Very few of the fleas died during transport, the munitions worked brilliantly, the insects descended without incident, and the guinea pigs became infested. The only biological drawback was that the fleas gave up on finding hosts within 24 hours, so it was evident that this entomological weapon would need to be used in close proximity to the enemy—at least if we were waging war on furry rodents in a desert. The other operational drawback was fixable but a bit more worrisome.

In one of the trials, the E-23 components "malfunctioned," a euphemism for what might have been a disaster if the fleas had been carrying plague.[4] The munition was supposed to become armed when a strip of sealing tape was removed, but one of the devices discharged while still in the plane. The hungry insects demonstrated their host-seeking capacities, biting the pilot, bombardier, and a military observer. But the Americans had an even bigger problem than insubordinate fleas.

While the entomologists were planning a facility to produce 50 million fleas per week, the microbiologists were struggling to mass-produce plague bacteria. During the Cold War, the Soviets managed to culture and weaponize *Y. pestis*, but U.S. researchers never cracked the problem. So, the military turned its attention to another insect-borne disease with perhaps even greater potential.

The golden child of the American entomological warfare program was the yellow fever mosquito, *Aedes aegypti*. With the medical community pursuing the eradication of this insect from the United States, the possibility of mass-producing the vector put military interests squarely at odds with public health ideals. This was a battle that the generals would not lose. The real challenge would be finding a way to produce enormous quantities of infected mosquitoes. Feeding millions of the adult insects on sick animals seemed to be impossibly complicated. Although this was the natural means of infection, the scientists at Camp Detrick were not constrained by such limitations.

The breakthrough came when researchers attempted a seemingly absurd experiment, adding the virus to the watery medium in which mosquito larvae squirmed and fed.[5] No such route of infection was possible in nature, so nobody held much hope that the wrigglers would uptake the pathogen from their aquatic surroundings. But when adult mosquitoes emerged a few days later, they were fed on mice that—to the sheer delight of the scientists— contracted yellow fever. With the newfound capacity to efficiently mass-produce infected mosquitoes, the next step was to determine if the insects could be weaponized. And this meant testing the vectors in the real world.

Operation Big Buzz was a simulated, mosquito-based attack.[6] More than a million uninfected *Aedes* mosquitoes were reared and stored for two weeks to simulate operational conditions. In May 1955, about a third of the insects (the others were used in loading and storage tests) were packed into E-14 munitions and dropped on rural Georgia because the southern United States was a hospitable environment for mosquitoes. Human volunteers (and guinea pigs) were placed at regular intervals from the target. *Aedes aegypti* spread into the countryside and managed to find hosts nearly a half-mile downwind from the release site. The first field test of vectors against human targets had been a rip-roaring success.

Although fragments of declassified military records reveal the workings of Operation Big Buzz, only the general nature of the next two projects can be inferred. Both Operation Drop Kick and Operation Grid Iron likely involved releases of mosquitoes, but the details of the experiments are not publicly available.[7] The shift to a football theme in the naming of these operations is curious, but perhaps the military did not want to provide the enemy with clues as to the essence of the projects. Or perhaps cute names no longer had a place in what was becoming a deadly serious military program. For the United States had new reasons to worry about a biological Armageddon.

In 1956, Soviet Defense Minister Georgy Zhukov announced that biological and chemical warfare would be used by their armed forces in future wars.[8] In response, the United States frantically reassessed the nation's vulnerability to these weapons—and the military's capacity to retaliate in kind. If the policy of "Mutually Assured Destruction" was viable for nuclear arms, then extending this strategy to other weapons of mass destruction seemed logically consistent, although it is difficult for any government to be rational when planning to kill millions of people.

The United States had long maintained strategic ambiguity regarding biological warfare, but the communist threat provided the perfect opportunity to make explicit the American policy. The National Security Council bluntly expressed the country's willingness to retaliate in kind:

> To the extent that the military effectiveness of the armed forces will be enhanced by their use, the United States will be prepared to use chemical and bacteriological weapons in general war. The decision as to their use will be made by the President.[9]

As for the matter of international law, the army's position was that "the United States is not party to any treaty now in force that prohibits or restricts the use [in] warfare of toxic or nontoxic gases, of smoke or incendiary materials or of bacteriological warfare."[10] With the moral and political obstacles out of the way, the development of biological weapons was limited only by science's capacity to conscript and coerce living organisms.

By this time, the scientists at Camp Detrick had a growing inventory of arthropod vectors available for further testing in terms of defense—and continued development with respect to offense. There were colonies of mosquitoes infected with yellow fever, malaria, and dengue; flies harboring dysentery, cholera, and anthrax; fleas carrying plague; and ticks loaded with tularemia, relapsing fever, and Colorado fever. The thriving entomological warfare division was attracting a cadre of outstanding researchers, drawn by the military's advertisements that sidestepped the ultimate goal of the research while promising an unparalleled opportunity to carry out "basic studies of effects of rearing procedures for various insects on longevity and fecundity [and] the effects of different environmental factors on infection of insects and on virulence of microorganisms."[11] But with the heightened international tensions, the scientists were expected to focus on more pragmatic goals. Like killing the enemy.

After investigating other insect-borne diseases, the U.S. Army Medical Command concluded that the yellow fever virus was the best pathogen to use against—and the most likely agent to be used by—the communists.[12] Military analysts had coldly calculated that virus-infected mosquitoes were more cost-effective than germ-laden aerosols if the goal of an attack was to maximize mortality (a bacterial fog was more economical in terms of generating casualties). The southern states were viewed as a prime target—the backwater of the nation, with poor medical services and crude pest-management practices. DDT was a viable defense, but only if an afflicted region had modern equipment to apply the insecticide. Nor were these analyses merely mental exercises. The seriousness with which the U.S. military took the Soviet threat was exemplified in the elevated status granted to Camp Detrick, which became Fort Detrick in February 1956 (see Figure 17.2). Along with this came a new round of secret studies of mosquitoes—and the human test subjects were no longer volunteers.

From April to November 1956, within residential areas of Savannah, Georgia, the U.S. military conducted simulated attacks using uninfected mosquitoes. Although many of the details are lacking, the first experiment—

Figure 17.2. Camp Detrick's infamous Building 470 was built in 1952 as a pilot plant for the production of pathogens for use in biological weapons. The seven-story facility contained large tanks for culturing anthrax. Decontaminating the building proved extremely difficult and costly, so it sat for years as a symbol of the dangers lurking within. Not until 1988 was it finally turned over to the National Cancer Institute for renovation. (Courtesy of the Department of the Army)

code-named Operation May Day—was intended to determine whether mosquitoes released as if they'd dispersed from bombs and warheads (the Sergeant missile was being considered as a potential delivery system) would find and bite people.[13] The insects reliably fed on the unsuspecting public, and these encouraging results provided the foundation for the next study.

In the Avon Park Experiment, 200,000 mosquitoes were released from aircraft over a Florida bombing range.[14] The dispersion system is not specified in the available records, but the army likely used its new XM28, a "bagged-agent dispenser" with a 700-pound payload consisting of 2,090 paper bags loaded with hungry insects. Although the target was likewise unnamed, it apparently extended into the communities surrounding the bombing range. The inclusion of the public as nonconsenting experimental subjects can be inferred from Chemical Corps report: "Within a day, the mosquitoes had spread a distance of between one and two miles, and bitten many people."[15] We know even less about Operation Quickhenry, except that whatever was done must have worked extraordinarily well. For in the late 1950s, the U.S. military undertook a remarkable series of experiments that culminated in a chilling analysis of the potential of entomological warfare.

The Bellwether tests were the most extensive experiments on the use of mosquitoes as weapons, and many of the relevant details can be gleaned from declassified records.[16] Bellwether One consisted of 52 field experiments in the fall of 1959. The research was designed to assess the role of environmental factors on the capacity of mosquitoes to find and feed on hosts. Wind speed was found to be crucial in determining whether the vectors could locate a host, although temperature, solar radiation, and relative humidity also mattered. A typical scenario consisted of seating human volunteers around the perimeter of a 30-foot-diameter circle, with hungry mosquitoes released from the center point. Under presumably optimal conditions, for every 100 mosquitoes released, 40 blood meals were taken—a very acceptable return on investment. These small-scale investigations were encouraging, but in the course of an actual attack, the enemy would not be in such close proximity to the release point. The next step was to create more realistic conditions to challenge the mosquitoes.

In Bellwether Two, the mosquitoes had to find their human targets under a wide range of possible scenarios. Using a series of 14 tests, the researchers discovered that the frequency of biting was nearly constant out to 100 feet from the release point, but began to drop off at about 200 feet. In other words, a single release would effectively saturate an area equivalent to three football fields with bloodthirsty insects. The detailed analyses provided a vivid picture of what made for an ideal target. The mosquitoes were particularly attuned to irregular motion; people who alternately moved and rested had significantly more bites than those who continuously walked or sat. Somewhat surprisingly, individuals near buildings were bitten more often than people in open areas. Such results could not have been better in terms of an urban target, in which people typically engage in sporadic movements while surrounded by buildings.

Although there is no declassified information concerning the nature of Bellwether Three, the next experiment left no doubt that the U.S. military was developing an offensive capability. Bellwether Four was explicitly designed to determine the biting activity of mosquito strains being produced at Fort Detrick. This was the tryout to see which player would make it to the major leagues of biological warfare. Unfortunately, the competition between the Detrick and Rockefeller strains of *A. aegypti* was inconclusive because the researchers flubbed the experiment.

The study was intended to simulate an urban attack, with volunteers playing the role of targets for the invading insects. The problem came when the people were told that they could swat the mosquitoes, which meant that the scientists couldn't figure out whether a strain was uniformly aggressive or it had a few persistent individuals. There was, of course, the classic conclusion drawn from a badly implemented experiment—funding should be provided for further studies to resolve the issue. But even without knowing which strain constituted the optimal weapon, the military had gathered enough evidence from the Bellwether tests to decide the place of entomological weapons in the U.S. arsenal.

In 1960, the U.S. Army Chemical Corps underwent a subtle but profound shift in policy, reflecting the acceptance of entomological warfare. The military was no longer asking *if* insect vectors could be used against the enemy but *how* they could be most effectively deployed. The Corps produced an "Entomological Warfare Target Analysis," the purpose of which was to identify the qualities of susceptible targets.[17] Much of the report has been excised for security reasons, but what remains paints a chilling picture. Although there were estimated to be 75 arthropod-borne diseases with military potential, the study concentrated on a subset of 18—with the Americans' golden child analyzed in grim detail.

In terms of yellow fever, a viable urban target needed to have mosquitoes penetrate just 3 to 5 percent of the houses and buildings. Because *A. aegypti* typically feeds on humans while they are indoors, the exposure of a population to the vector declines with the use of air conditioning and the consequent closing

of windows. But the study revealed that if even a small portion of a city's structures were accessible to mosquitoes, the disease could gain a foothold. With
respect to ecological conditions, if ambient temperatures were between 61 and
101°F, the insects could readily find their hosts. Once a locus of infection was
established, female mosquitoes had only to feed on a diseased host within three
days of the onset of symptoms and the insect would become a carrier for life.
This multiplier effect would create an escalating cycle of infection and transmission, leading to a full-blown epidemic within weeks. The primary limitation
would be the immunological condition of the target population; if people had
previously survived yellow fever, they would not become reinfected.

Much to the delight of the American military planners, the Soviet Union
was loaded with susceptible targets—cities within the proper temperature
range and packed with bodies that had never been wracked with yellow fever.
Even better, some prospective targets already had *A. aegypti*, so it would be
nearly impossible to detect the added mosquitoes or to trace their source.
Although the Soviets had a yellow fever vaccine, in the judgment of the analysts
it "would be impossible for a nation such as the USSR to quickly undertake a
mass-immunization program to protect millions of people."[18] The American
planners were clearly thinking of initiating an epidemic of enormous proportions. Potential targets can be inferred from the third appendix of the report,
including Moscow, Stalingrad, and Vladivostok, along with the Soviet-allied
cities of Basra and Cairo. The Chinese urban centers of Canton, Shanghai,
and Tsingtao also made the list. But spreading millions of infected mosquitoes
over communist metropolises would require not only a sophisticated delivery
system and finely honed logistics but also an incredible stockpile of insects.

By 1960, the entomologists at Fort Detrick could produce half a million
infected mosquitoes per month. As impressive as this was, it fell far short of
the number of insects needed to attack a major city. So the military drafted
plans to increase production by nearly a thousandfold.[19] Pine Bluff, a small
cotton-producing town in southeast Arkansas, would house the largest insect-
rearing facility in the world, a mosquito mill with the capacity to produce 100
million infected vectors per week. While some historians put the production
figure at a more modest 130 million mosquitoes per month, even this figures
leaves no doubt that the Americans considered entomological warfare to be a
deadly serious enterprise.

Producing enormous quantities of living organisms had been the forte of
Pine Bluff Arsenal for a decade. In the fall of 1950, the U.S. Congress secretly

appropriated $90 million to renovate the aging installation. The new plant boasted a ten-story building that housed enormous fermentation tanks for the production of pathogenic organisms. When the military decided that brucellosis would make a keen weapon in 1953, the plant cranked enough bacteria to fill more than 2 million bombs a month.[20] Just two years later, Pine Bluff took on the large-scale production of tularemia bacteria. When the order came to rear insects, the arsenal had plenty of "can do" experience.

With an annual operating budget of more than $5 million and nearly 2,000 workers swarming over a site that had grown to cover more than two square miles, there would be no problem retooling the factory. If the military wanted to grow mosquitoes, then a few 45,000-gallon stainless steel vats would be the ticket.[21] But just as the Pine Bluff Arsenal was cranking up to mass-produce insects, the microbiologists were preparing to pull the rug out from under entomological warfare.

The 1960s saw scientists begin making rapid progress in the safe, large-scale production of pathogenic microbes.[22] They were mastering techniques for purifying and stabilizing formulations of bacteria, viruses, and rickettsiae. Weaponization was just around the corner, with the engineers developing munitions that would release aerosols of optimally sized particles. In these heady days of biological warfare, the operational problems of microbial aerosols were profoundly underestimated. Today's experts realize the difficulty of formulating a pathogen so that it both stays airborne long enough to be inhaled and survives desiccation and ultraviolet radiation while aloft. We now appreciate that even with such a weapon, the aerosol must be released at a time and place in which wind speed and direction ensure effective dispersal.

Logistical challenges notwithstanding, advances in germ warfare meant that insect vectors began to fall out of favor. A political superpower should base its biological warfare program on stainless steel vats and sophisticated spray apparatus, not pools of squirming larvae and Rube Goldbergesque dispensers of insect-filled paper bags. The insects represented everything that a high-tech army found undesirable: a swarm of mosquitoes was disobedient, inefficient, and unreliable. Insects might have been entirely drummed out of the service had not geopolitics interfered with the microbiological hegemony.

The Cold War was heating up in an area of the world in which insects called the shots: the steamy tropics of Southeast Asia. In August 1964, the Gulf of Tonkin Resolution converted a clandestine conflict into an open war. In Vietnam, the American military was about to learn some very difficult lessons about guerrilla tactics—and entomological weapons.

18

YANKEE (AND VIETNAMESE) INGENUITY

Entomologists working to convert insects into weapons for the U.S. Department of Defense could not have asked for a more propitious conflict than a war in Southeast Asia. While Korea was at the latitude of northern Colorado, Vietnam was 2,000 miles closer to the equator—and insects flourish in tropical climes. In 1965, the American military conducted Operation Magic Sword to assess the biting habits of the yellow fever mosquito after being released from a ship anchored off the warm, humid shores of the southeastern United States.[1] This was the best approximation, without mounting an insect invasion of another sovereign nation, for determining if vectors released from naval vessels off the coast of Vietnam would make landfall and attack the enemy.

The scientists ascertained that the insects, assisted by sea breezes, could cross up to three and a half miles of ocean and establish a beachhead. Operation Magic Sword also revealed that the entomological-warfare folks could keep their charges in battle readiness during a transpacific journey. By cooling batches of mosquitoes to 64°F to reduce their metabolic rates and maintaining them at 80 percent humidity to prevent dehydration, the insects remained viable for 52 days. Despite their promise, mosquitoes were probably not loosed on Vietnam, although other unconventional weapons were used with abandon.

As encouraging as biological warfare seemed to be, the chemical arsenal was even more promising. While the insecticide industry had spawned the German nerve gases, herbicide research provided the U.S. military with plant killers—and in a battleground choked with vegetation that concealed a cunning opponent, eradicating the plants became tantamount to annihilating the enemy.[2]

Operation Ranch Hand began in November 1961, and over the next six years sprayed more than 17 million gallons of herbicide to denude 2,000 square miles of jungle and 420 square miles of crops (see Figure 18.1). The North Vietnamese were outraged by what they considered to be an overt act of chemical warfare. The Americans argued that because the targets were plants, not people, there was no violation of international laws or treaties.

With chemical weapons scorching the countryside, the communists accused the Americans of releasing the larvae of "killer insects" to ravage Vietnamese agriculture. In October 1966, the North Vietnamese News Agency reported:

> These larvae were let loose on 30th September 1966 on the Cham Thanh district of In Tan province. Route 21 from Duong Zian Hoi to Vinh Cong was affected. All rice, plants, fruit trees in a band of 2 kilometers wide either side of the road have been destroyed.[3]

The North Vietnamese did not file official charges against the Americans, perhaps because an inquiry would likely have revealed that the infestations consisted of local pests rather than exotic creatures dropped by the U.S. Air Force. However, it is conceivable that the Americans played an indirect role in the outbreaks; sublethal doses of herbicides can weaken plants and make them vulnerable to insect attack. But there was an even more nefarious environmental effect of Operation Ranch Hand.

A 1969 United Nations study found that human disease can follow on the heels of large-scale deforestation, as occurred in Vietnam.[4] The process is rather simple, if not easily foreseen. After the original trees are killed,

Figure 18.1. Operation Ranch Hand waged war on Vietnamese plant life in a controversial effort to deprive the enemy of cover and food. However, nobody anticipated that herbicides (17 million gallons applied from 1962 to 1971) would set the stage for arthropod-borne disease outbreaks. The secondary growth that followed chemical deforestation fostered the deadly trio of rickettsiae, rats, and mites that conspire to inflict scrub typhus on the people.

secondary forest or grassland develops, and these habitats are often condu-
cive to blood-feeding arthropods and their hosts. In Southeast Asia, both the
mites that transmit scrub typhus and the rats that harbor the pathogen flourish in secondary forests.[5] Although the U.S. military did not plan to induce
scrub typhus outbreaks among the enemy, the Americans were fully aware of
another, indirect form of entomological warfare.

Savvy commanders know that whichever side better protects its troops from
the diseases that flourish in the detritus of war gains a strategic advantage. But
despite extensive medical efforts, at least 2.5 million U.S. troops contracted
arthropod-borne illnesses in Southeast Asia between 1962 and 1973.[6] At the
height of the conflict, disease accounted for three-quarters of the army hospital admissions, and malaria was the primary culprit. With nearly 40,000 cases
in the latter half of the 1960s, the U.S. military imposed "malaria discipline."
Soldiers who were not taking chloroquine to prevent the disease—and this
was evident from a simple colorimetric test of their urine in the field—were
punished. Although the malaria rate dropped precipitously, the medical corps
continued to battle other diseases, including Chikungunya and dengue carried
by mosquitoes, scrub typhus transmitted by mites, and various enteric diseases
spread by flies. But perhaps the most frightening pathogen was the military's
oldest foe and ally—bubonic plague.

Rats, fleas, and bacilli lie in wait throughout much of Southeast Asia, with
a few people killed by plague every year. But with the devastation of war,
hundreds of civilian cases began to swamp medical clinics. In short order,
U.S. troops were vaccinated and the military undertook a large-scale hygiene
campaign that involved applying insecticides to infested sites and reducing
filth in friendly villages and army camps. Such measures would seem to fall
into the realm of good soldiering, but cleanliness can be next to wickedness.
The key to turning the rubble of war into a strategic asset lay in ensuring that
the enemy was more vulnerable to disease.

While only a dozen American troops contracted plague during the war,
the situation was dire in areas controlled by the Viet Cong, where vermin
thrived. While plague was reported in just a single province of South Vietnam
in 1961, 22 of the 29 provinces north of Saigon were afflicted in 1966. The flea-
bitten communists were surely suffering from a lack of medical supplies and
infrastructure, but whether the Americans exploited the logistical difference to
wage passive entomological warfare can be debated.[7] What can't be doubted,
however, is that the North Vietnamese were ready, willing—and able—to use
entomological weapons.

The communist forces lacked technological sophistication, but they possessed a sort of anti-Yankee ingenuity. Large sectors of Vietnam were riddled with an underground network of first-aid posts, armories, kitchens, dormitories, classrooms, and even small theaters. From the miles of tunnels, the Viet Cong could decide when and where to fight—sometimes lobbing wasp and hornet nests into U.S. positions to disrupt defenses before launching an attack.[8]

After unwittingly building a camp on top of the tunnels near Cu Chi, the Americans realized that their subterranean enemy had to be defeated if there was any chance of controlling the region. Special volunteer commandoes, armed only with pistols and knives, descended into the narrow passages. These "tunnel rats" encountered a nightmarish assortment of booby traps.[9] Feeling his way through a dank tunnel, the lone commando might overlook a thin trip wire. Suddenly a load of poisonous arthropods would rain down from a hidden cavity in the roof. While sharing a two-foot-wide, three-and-a-half-foot-high burrow with a few dozen angry scorpions was not as bad as stepping into a pit of punji sticks, the scuttling creatures provided a horrifying ambience. Indeed, at least in peacetime, psychologists have found that people rank spiders and insects on a par with snakes, bats, terrorists, and—according to one survey—death. The Viet Cong, however, did not limit their entomological weaponry to subterranean venues.

In Rudyard Kipling's *Second Jungle Book*, Mowgli sees his forest threatened by a pack of ferocious red dogs. The hero enlists the aid of the Little People of the Rocks—a colony of bellicose bees. The clever boy lures the vicious dogs into the domain of the bees, who kill the invaders. Although Kipling did not provide a scientific name for Mowgli's entomological weapons, the creature that he had in mind might well have been *Apis dorsata*.[10]

The giant honeybee of Asia, occasionally called the "rock bee," has been described by tropical entomologists as "the most ferocious stinging insect on earth." This insect is far more aggressive than its cousin, the European honeybee, which accounts both for the vast majority of honey produced in the United States and for more deaths than any other poisonous animal. Not only is the giant honeybee 50 percent larger than its placid cousin, but the Asian species attacks in huge numbers (colonies build a single, open comb that can be as much as ten feet across) and pursues an intruder for 100 yards or more (see Figure 18.2). The distribution of this belligerent bee stretches from Mowgli's forests of India to the jungles of Vietnam. And real guerrillas are just as cunning as fictional boys.

Figure 18.2. Giant honey bees cover a single, exposed comb up to ten feet across. Colonies high in the forest canopy served as "tree mines" for the Viet Cong, who would set off a small charge near the bees and convert several thousand enraged one-inch workers into living shrapnel. Realizing the potential of conscripting these fierce insects, the Americans attempted to develop chemicals that could be used to direct the bees to attack the enemy. (Photo by Stephen L. Buchmann)

The Viet Cong carefully relocated colonies of these bees to trails used by the enemy and then attached a firecracker to the comb.[11] When a patrol passed within striking range, a patiently waiting VC would set off the charge. The infuriated insects delivered painful stings and drove the soldiers into dangerous disarray. There were also intriguing but unconfirmed reports that the North Vietnamese trained their insect conscripts to attack anyone wearing an American uniform. Such a tactic is not implausible, given that bees are capable of associative learning (for example, relating particular colors and shapes with rich sources of nectar). While communist forces were running training camps for bees, the Americans were using their scientific acumen to turn the tables on the enemy.

Like many insects, bees use an elaborate system of chemical messages, or pheromones, to coordinate their behavior. Odors are used to attract mates, identify colony members, and mark food sources. But perhaps the most spectacular olfactory signal is the alarm pheromone. This chemical cocktail is contained in tissues surrounding the sting, although a worried bee can send an alarm without actually stinging; just opening the sting chamber in preparation for unsheathing the barbed lance will emit the odor. The primary component of the honeybee alarm pheromone was discovered in 1962. Isopentyl acetate, a chemical with the odor of bananas, functions like a cavalry bugle drawing bees into the battle. And the Pentagon was also enticed.

The U.S. military funded a top-secret research program to devise an apparatus to spray the enemy with the alarm pheromone of bees, thereby converting the local insects into fierce allies.[12] Such a tactic might seem a bit absurd, but this is the same military that was pursuing Project Aquadog, an aborted attempt to train dogs to swim underwater on seek-and-destroy missions against enemy frogmen. The thought of turning the giant honeybee against the communists was tempting, but the idea of stinging swarms chasing black-clad, banana-scented guerrillas never came to fruition (so to speak). However, bee pheromones may still be part of the American arsenal. In 2003, *Harper's* magazine published a portion of the glossary from a report on "Nonlethal Weapons: Terms and References" by the U.S. Air Force's Institute for National Security Studies that included this entry:

> Pheromones: The chemical substances released by animals to influence physiology or behavior of other members of the same species. One use of pheromones, at the most elemental level, could be to mark target individuals and then release bees to attack them.[13]

Although bees were hardly decisive weapons, these insects played one of the strangest roles in the history of unconventional weaponry in the days following the ignominious end of the Vietnam War. In 1981, U.S. Secretary of State Alexander Haig announced that "for some time now the international community has been alarmed by continuing reports that the Soviet Union and its allies have been using lethal chemical weapons in Laos, Kampuchea [Cambodia], and Afghanistan."[14] This stunning accusation was based on accounts of "yellow rain" falling on the Hmong people, who suffered tortuous deaths.[15] American government experts isolated a fungal toxin from field samples and claimed that this was the basis of the chemical attacks. But the case fell apart when independent scientists identified the yellow residue as pollen-laden insect feces. Bee poop, to be precise. It seems that the giant honeybee spatters the forest during a colony's morning constitutional—and fungi can grow in these insect latrines, although not at dangerous levels.[16] Perhaps the allegations by the United States were an expression of a frustrated superpower that had, at least officially, abandoned its biological weapons.

As the Vietnam War was reaching a low point for the United States, President Nixon announced the unilateral cessation of his country's biological warfare program. Historians continue to debate Nixon's political rationale (nobody seems to suspect a moral motive), but whatever the reason, the U.S. disarmament stimulated a series of international discussions.[17] These culminated in the 1972 Biological and Toxin Weapons Convention, which banned the development, production, stockpiling, transfer, and acquisition of biological arms. In 1975, the United States ratified this treaty along with the 50-year-old Geneva Protocols. But by this time, flaws in the BTW Convention were becoming widely recognized.

In making compromises to get the accords passed, the signatories relented to Soviet pressure and weakened the verification provisions. The ink had not even dried on the signatures before the Russians were violating the treaty.[18] For the next 20 years, the Soviet Ministry of Medical and Biological Industry produced weapons-grade anthrax and smallpox by the ton. When a 1991 British-American inspection team asked about the fermentation tanks at one site, the Soviets claimed that the facility produced insect pathogens for pest management. So even when insects weren't being used in weapon systems, they were providing alibis.

With regard to entomological weapons, the treaty seemed clear, but no wording can prevent bureaucrats from muddying the waters. Although insects would seem to be "biological agents," the United Nations, along with counter-proliferation agencies and the U.S. military, interpreted the ban to include only microbial agents and biological toxins. On the other hand, the United Nations also expressed concern that mosquitoes and ticks could be used as vectors, so perhaps these creatures would fall under the convoluted provisions that prohibited biological "weapons systems."

The bottom line is that a nation could pursue entomological warfare and contend that insects fell through the cracks of international law. And if the Cubans are to be believed, this is precisely what the Americans were doing.

19

CUBAN MISSILES VS. AMERICAN ARTHROPODS

On October 22, 1962, President John F. Kennedy addressed the nation with this chilling announcement:

> Good evening my fellow citizens. This Government, as promised, has maintained the closest surveillance of the Soviet Military buildup on the island of Cuba. Within the past week, unmistakable evidence has established the fact that a series of offensive missile sites is now in preparation on that imprisoned island. The purpose of these bases can be none other than to provide a nuclear strike capability against the Western Hemisphere.

So began a game of brinkmanship the likes of which the world had not seen before. Tensions escalated when Kennedy ordered a naval blockade and the Russian Premier, Nikita S. Khrushchev, responded by authorizing his field commanders in Cuba to launch the nuclear missiles if U.S. forces attacked the island. All the while, Kennedy's advisers played out a range of scenarios, including armed invasions, tactical air strikes—and unleashing crop-destroying insects (see Figure 19.1).[1]

Entomological warfare drew rather less attention than more conventional responses, but herbivorous insects had their moment in history. An individual planthopper or leafhopper (superfamily Fulgoroidea and family Cicadellidae, both in the suborder Auchenorrhyncha) is hardly an imposing weapon, being a green bullet-shaped creature about the size of a grain of rice. But females can lay 300 eggs in their month-long adult lives. Like an entomological shotgun blast, these insects inflict their damage by virtue of overwhelming numbers rather than individual potency. Planthoppers feed by piercing the plant tissue

Figure 19.1. During the Cuban Missile Crisis, the American government considered unleashing planthoppers against Castro's crops. Entomologists at Fort Detrick had been working with the U.S. Department of Agriculture to weaponize planthoppers and leafhoppers as vectors of plant diseases to destroy sugarcane and rice. This photo shows the black-faced leafhopper (*Graminella nigrifons*), which transmits diseases of corn. (Photo by Stephen Ausmus, USDA/ARS)

and sucking the sap through their straw-like mouthparts; a heavy infestation can wilt a crop in a matter of days. But the military had far greater plans for the planthoppers.

Because these insects tap directly into a plant's fluids, they are extremely effective in transmitting microbes. An insect-vectored plant pathogen gains the same advantages as an animal pathogen: protection against adverse environmental conditions and direct transport to a susceptible host. With these advantages, an outbreak of a plant disease can devastate an agrarian nation, such as Cuba.

Available evidence suggests that the Army Chemical Corps was in cahoots with the U.S. Department of Agriculture in weaponizing planthoppers.[2] The goal was to destroy Cuba's most important economic asset—sugarcane. The sugarcane leafhopper (*Parkinsiella vitiensis*) transmits the virus that causes Fiji disease, which stunts plant growth and induces tumorlike deformities. Although the details of the proposed entomological assault remain murky, it seems that substantial progress was made.

Not long after the Cuban Missile Crisis, Dr. Charles L. Graham, an entomologist with the Crops Division at Fort Detrick, received a special award for his work on virus transmission by leafhoppers. And declassified documents show that the U.S. military had also considered the *Hoja blanca* virus, a planthopper-borne pathogen of rice (another of Cuba's major commodities), to be among the highest priority biological warfare agents.[3] The importance of insect-borne plant diseases might be debated, but there's no doubt that the Americans were pursuing the weaponization of pathogens capable of destroying crops. According to William Patrick, Fort Detrick's chief of product development, the U.S. military had programs that covered the waterfront of rice and wheat diseases and had a stockpile of 40 tons of wheat rust spores—a pathogenic arsenal that was replenished every three years.[4]

In the end, Khrushchev backed down. On October 28, the Soviet premier ordered the removal of the missiles. Although the superpowers avoided launching nuclear warheads, a long series of suspicious insect invasions convinced the Cubans that the Americans had no reservations about launching insects. Throughout the early years of the Cold War, Fidel Castro often sprinkled his epic speeches with bombastic accusations that the United States was waging entomological warfare against his nation. His claims seemed to be little more than propagandistic rants and drew little international attention.

It took nearly 20 years after the Cuban Missile Crisis before Castro finally had what he wanted: credible evidence to support a plausible allegation. And in this instance people, rather than plants, were the supposed targets.

In May 1981, the citizens of Havana began to come down with a series of alarming symptoms: raging fevers, crippling muscle and joint pains, searing headaches with severe pain behind the eyes, and bright red rashes.[5] The terrible pains in people's joints tipped off the doctors seeking a diagnosis, for the common name used to describe the disease is breakbone fever. An outbreak of dengue fever (the technically accepted name) would have been bad enough had the disease followed its typical course of merely debilitating victims, but in some cases the illness turned lethal. After four or five days of agony, patients began to develop bruises and started bleeding from the nose, mouth, and gums. Children died of internal bleeding, while adults succumbed to shock as their circulatory systems collapsed. Hemorrhagic dengue was sweeping through Cuba's capital—and new cases were cropping up in two other cities.

At the peak of the epidemic in early July, more than 10,000 patients per day were being reported by an overwhelmed medical system. The Cuban health officials knew that to control the disease they had to suppress its vector. Dengue is spread by mosquitoes, and heavy spring rains had fostered enormous numbers of *Aedes aegypti*; the infamous yellow fever mosquito turns out to be a versatile vector. A massive campaign was waged to suppress the biting insects, and this program quashed the disease before it spread to the entire island. During the five-month outbreak, the Ministry of Public Health recorded 344,203 cases, with 116,151 people hospitalized and 158 deaths.[6]

The episode was traumatic, but at first glance nothing appeared particularly unusual. Dengue is a tropical disease and Cuba was a poor nation—an epidemiological slam dunk. However, to the keen investigatory eye, the outbreak was not so easily dismissed as a natural event. William H. Schaap, a man possessing a healthy—his critics would say paranoid—distrust of the U.S. government, argued that the epidemic was artificially induced.

As a codirector for the Institute for Media Analysis (an organization devoted to "uncovering the deceptions of the mass media, which functions as a mouthpiece for the military-industrial complex") and founding editor of *Covert Action Information Bulletin*, Schaap may not be the most objective analyst in the world, but his case against the American government is worth considering.[7] Perhaps this is a classic conspiracy theory in which genuine coincidences are framed so as to form a pattern. On the other hand, any government that seriously considers assassinating the leader of another nation via exploding cigars would seem capable of using infected mosquitoes as weapons.

The initial anomaly that set the tone for the more speculative elements of the case was the timing of the outbreak. The 1981 outbreak in Cuba was the first occurrence of dengue in nearly four decades and included the first large-scale irruption of hemorrhagic cases in the Caribbean since the turn of the century. (The hemorrhagic form of dengue can develop when a person is reinfected with the virus after an earlier bout of the disease.) Mexico and parts of Central America were also stricken by the disease in the early 1980s, so the outbreak was not localized to Cuba. However, the geographic pattern of the outbreak within Cuba was peculiar.

Dengue emerged contemporaneously in three widely separated cities. While disease outbreaks may have multiple loci, Cienfuegos is 150 miles southeast of Havana and Camagüey is another 180 miles southeast of Cienfuegos. Somehow, infected mosquitoes had simultaneously arrived in cities more than 300 miles apart. The widely spaced epicenters would have been less surprising if the initial victims had visited hot spots of the disease elsewhere in the world and brought the virus back to their communities. However, Cuban medical authorities asserted that patients had not engaged in international travels—a plausible claim given the economy and politics of Cuba. Officials reported that in May, when dengue first appeared, only a dozen people had entered the country from endemic regions (Vietnam and Laos), and the Institute of Tropical Medicine had certified them as being free of the disease.

Based largely on these epidemiological aberrations, Schaap concluded, "There appears to be no other explanation but the artificial introduction of infected mosquitos."[8] The scientific community was intrigued, but unusual patterns of disease do not necessitate intentional human agency. For example, medical entomologists pointed out that at the time of the epidemic, Cuba had extensive military operations in Africa, a continent rife with hot spots of type-2 dengue. A garrison of soldiers returning to their homes in cities around Cuba could have triggered the outbreak. Facing such counter-explanations, the conspiracy theorists tapped into other lines of evidence.

The Biological and Toxin Weapons Convention Treaty of the Nixon administration notwithstanding, the U.S. military had developed the means of mass-producing *Aedes aegypti,* and the army's inventory of pathogens had included dengue, so the essential ingredients and technical know-how for entomological warfare were on the shelf. And in 1981, the Americans would only have needed enough infected insects to seed an epidemic and then let Cuba's abundant mosquito population do the rest. Skeptics retorted that the United States, not known for its patience, would not have kept a colony of dengue-infected mosquitoes going for years while waiting for the optimal conditions to develop naturally in Cuba.

Confronted with this argument, America's accusers proposed that the U.S. military had altered the weather, so as to foster the conditions necessary for mosquito-borne diseases to flourish. Nobody contested the facts that in the three afflicted cities rainfall totals were 42 to 146 percent above normal or the observation that mosquito populations exploded as a result. And it was true that since the 1940s, the U.S. military had been experimenting with cloud seeding—using silver iodide or dry-ice pellets to wring rainfall from clouds. By connecting rather distant dots, Schaap contended that the Americans first seeded the clouds to induce torrential rains, which fostered an outbreak of native mosquitoes, and then U.S. agents seeded the urban centers with dengue-infected mosquitoes to trigger the disease cycle.[9] Aside from such fanciful conjectures, one thing is known: Americans were still deadly serious about entomological warfare.

In 1981, the U.S. Army Test and Evaluation Command at Dugway Proving Grounds completed "An Evaluation of Entomological Warfare as a Potential Danger to the United States and European NATO Nations."[10] Although most of the report is hidden from public view, the declassified portions reveal the ruthless calculus of military strategists. As the title suggests, much of the analysis is expressed in terms of defending America and its allies, but there was no difficulty substituting a communist target for a western city.

The document includes an economic comparison of two biological weapons based on a pair of warfare scenarios. Assuming that a battalion was attacked with a cloud of yellow fever–infected mosquitoes or an aerosol of the bacteria responsible for tularemia, the entomological weapon proved to be substantially less costly. But battlefield releases were considered largely impractical owing to the unpredictability of the vectors' movements. There would be no such problem when attacking a city, however. The army's report laid out the scenario:

> The cost of attacking an urban area covertly with yellow fever–infected mosquitos was estimated. It was assumed the cost of planning a city attack with yellow fever-infected mosquitos is comparable with the cost of planning an aerosol attack on Washington, DC (scenario 7 of reference 10). In the present hypothetical attack, 16 simultaneous attacks were planned.[11]

The detailed cost accounting reveals that the analysts had extremely precise knowledge of every phase of entomological warfare, suggesting that economics and logistics were derived from the perspective of an attack by the U.S. military. A table listed the costs of planning ($547), agent production ($9,066), munition acquisition ($500), and weapon deployment ($360).[12] So as not to miss any facet of an operation, the report included travel and per diem costs of covert agents. The total budget worked out to $10,473—less than a tenth of the cost for an attack with a bacterial aerosol (tularemia was used for comparison). But what really mattered was not total cost but dollars per death, and the military bookkeepers had this carefully figured. From the data in a table titled "Resource Cost Summary for Yellow Fever–infected Mosquito Attack on a City" along with a bit of back-calculation from other information in the document it appears that an attack using mosquitoes would set the military back only 3 cents per corpse.

Much of the rest of the report remains classified and therefore hidden from public scrutiny. However, the conclusions pertaining to the communist threat survived the censors and demonstrate how earnestly the U.S. military considered the possibility of entomological warfare:

> Intelligence information gathered about the Warsaw Pact countries indicates that in the past, they have attempted development of an EW [entomological warfare] capability. Indirect evidence, e.g., mass rearing of potential insect vectors and working with microbiological agents compatible with EW that are not a problem in these countries, comprises the evidence available to

indicate present activity in this area. The Warsaw Pact nations certainly have the capability to conduct EW. . . .

EW systems are not likely to be employed on military units because the agent vectors must be released too close to the target area. This would make a covert attack on a military unit very difficult to achieve. EW could be very effectively used against civilian urban populations or it could be used to cause great economic losses in the cattle and livestock industry.[13]

In fact, no evidence has ever emerged that the Soviets or their allies were poised to use yellow fever–infected mosquitoes as a weapon.[14] The only nation with such a program was the United States, so one must wonder to what extent the purported analysis of the Warsaw Pact was a diversion, with the primary purpose of the report being an assessment of the Americans' capacity to wage entomological warfare.

Based on the declassified segments of the report and historical snapshots of the work conducted by military scientists, there can be little doubt that the Americans had the motive, means, and opportunity to conduct an entomological attack on Cuba. And as this tale was unfolding, another remarkable coincidence emerged on the other side of the globe.

Less than a year after the dengue epidemic in Cuba, a similar story emerged in Afghanistan. In February 1982, the Soviet weekly *Literaturnaya Gazeta* reported that CIA operatives posing as malaria-eradication workers in Lahore, Pakistan, were experimenting with the spread of dengue and yellow fever. The Afghan and Soviet governments accused the United States of waging biological warfare via mosquito-borne disease.[15] Nothing much came of these claims, other than adding fuel to the political fire of the Cold War.

Throughout the 1980s, the Cubans kept the pot simmering by making the occasional accusation that the Americans were engaging in clandestine entomological warfare. Then came the collapse of the Soviet Union in 1991. With his greatest ally in disarray and the Cold War ending, Castro turned up the political heat.

In the years following the dissolution of the USSR, Cuba and its sympathizers peppered the international community with a series of accusations that the United States was launching a virtually continuous series of entomological assaults on the island nation's agriculture. The claims came fast and furious as politicians and their advisers tried to sort the plausible from the absurd. The diplomatic food fight began in 1992, when the *Green Left Weekly* reported:

In December 1992, citrus fruits—in great export demand—were affected by biological warfare; the black plant louse, the most efficient transmitter of the disease known as tristeza de citrico (citrus sadness), was identified. The insect vector was traced to the Caimanera municipality, where the U.S. naval base in Guantanamo is located.[16]

The accused was, strictly speaking, not a louse but an aphid—most probably the brown citrus aphid, *Toxoptera citricida*. The Americans did not contest that this tiny creature, resembling a matte black pinhead, had been discovered in Cuba, but teensy insects constantly arrive on various shores courtesy of natural processes and inadvertent human activity. The decline and death of citrus orchards was certainly lamentable—at least for the Cubans—but if every new pest outbreak in a country hostile to the United States was going to be blamed on the superpower, then the Americans would need a government department dedicated to rebutting these accusations.

The international community required more than the appearance of a new, crop-eating insect somewhere in the world to trigger an investigation. The Cubans later claimed to have caught a U.S. scientist with four test tubes of citrus tristeza virus in his camera case, but without physical evidence to support the accusation, other governments would not be drawn into the spat.[17] Undeterred, Castro continued his program of wearing down international reticence.

A year after enemy aphids had infiltrated the island, a Cuban official reported the arrival of a new pest—the citrus leafminer, *Phyllocnistis citrella*.[18] This tiny moth first appeared in La Habana province, just 100 miles from the Florida Keys, leaving no doubt (to the Cubans) that the Americans were behind the invasion. Although its wings would not quite span a hole punched into a sheet of paper, the insect spread like wildfire on the tropical winds. The adults lay eggs on citrus leaves, and the hatching larvae destroy the foliage by burrowing between the layers of leaf tissue. The little caterpillars can also tunnel beneath the citrus rinds, ruining what few fruits the trees manage to produce. With a new generation every month, the insect outpaced control efforts on the part of the Cuban farmers. Moreover, by hiding within the plant, the larvae were protected from many insecticides. The strategy of boring into plant tissue also paid off handsomely for another insect that was purportedly introduced by the Americans.

When the coffee crop in the province of Granma Santiago de Cuba began to fail, the culprit turned out to be the coffee berry borer, *Hypothenemus hampei*. Like a caffeine addict's ultimate reincarnation, the larvae of this beetle bore directly into the maturing beans. Although this pest is found in most of the coffee-growing regions of the world, Cuba was spared its ravages until 1994. With the beetle running amok, coffee production in infested districts dropped by more than 80 percent, and even where harvests could be salvaged the market value of the damaged beans was abysmal. According to Cuban officials, "The coffee borer is exotic in our country and there is no plausible explanation for its natural appearance in the island. On the contrary, there is ample evidence indicating that it was intentionally introduced and showing how this was done."[19] No evidence was actually presented, but perhaps the Cubans were too busy figuring out how to repulse a battalion of eight-legged invaders that had come ashore just a few miles from the infamous Bay of Pigs.

A clear understanding of how the Cubans perceived the mite outbreak in their rice crop can be garnered from transcripts of a Cuban television program, broadcast on May 24, 2002:

RANDY ALONSO: Good afternoon, viewers and listeners. For more than four decades our people have been exposed to horrendous acts of sabotage, underhanded attacks against economic and social facilities, banditry, mercenary invasions, biological warfare, military threats and hundreds of other terrorist acts organized and financed by successive U.S. governments. This afternoon we will continue with our round table meeting "Who are the real terrorists?" Joining me on the panel are [seven other participants and] Jorge Ovies, Director of the Plant Health Research Institute . . .

JORGE OVIES: The last plague in the plant kingdom, the rice mite, appeared in September 1997, in a seed farm in the Nueva Paz municipality; a crucial center for the production of rice seeds at precisely the same time as widespread rice production was being promoted. We all know how important this has been in guaranteeing self-sufficiency in many enterprises, families, cooperatives.

This mite causes blemishes and empty grains, and it facilitates the entrance of a fungus, *Sarocladium oryzae*, that already existed in Cuba, but that was not widespread. The symbiosis with this mite, which causes rotting in the husk, led to greater penetration by the fungus with the resultant damage caused; this is why the vulgar name "mite-fungus complex" is applied to this phenomenon in the case of rice.

It is interesting to note, Randy, that this plague did not exist on the American continent. The only incident was a plague reported 20 years before, in Taiwan and the People's Republic of China. We did not even have any monitoring programs in place. . . .

No model could ever predict that this plague would arrive directly from Asia to our rice farms, and to a seed farm nonetheless. What is more, seeds do not transmit the mite that was introduced. When we import, we always import seeds. . . . What cannot be scientifically explained is how this plague arrived in Cuba and how exactly it coincided with the other factors; when the rice popularization program was at its peak.[20]

Of course, the conclusion was that the outbreak of panicle rice mite (*Steneotarsonemus spinki*), which cost Cuban agriculture $44.3 million in annual losses and control measures,[21] was the work of the Americans. And not only were the capitalists using mites to devastate crops, but in a strange twist, insects themselves also became targets of eight-legged agents.

In 1991, just as honey was becoming a major export for Cuba, their bee industry was crippled by an epidemic of "acariasis disease."[22] Although this was not a true disease in the sense that a pathogen infected the insects, the size of the culprit was very nearly microbial. This malady is due to the infestation of a tiny mite (*acariasis* is derived from *Acari*, the order to which mites belong). The culprit, *Acarapis woodi*, sets up house within the host's trachea, the network of breathing tubes that carry oxygen directly to the tissues. The mites crawl into the bee's spiracles, the openings of the tracheal system along the sides of the insect (which would be the equivalent of a half-inch creature crawling up your nose), and suck their meals through the thin walls of the moist, oxygenated tubes. Between draining the bee's vital fluids and stuffing its respiratory system with offspring, the mites grievously weaken their host. Infested bees are lethargic, unable to fly or gather nectar, and suffer a lingering death. Honey production plummets as the colony collapses.

The Cubans were well aware that acariasis spreads when the mites crawl out of a dead host and into healthy individuals. But a microscopic mite isn't going to get very far by walking, so the only way that a nation's hives can become infested is by the introduction of infested bees. And while mite-laden honeybees can't fly across the Gulf of Mexico, an American plane laden with sick bees could make the crossing in less than an hour. As Cuban beekeepers struggled with this new affliction, an even more devastating mite found its way to into their hives.

In 1996, varroasis was found in three apiaries in Matanzas.[23] Previously unknown in Cuba, this "disease"—which is caused by another mite, *Varroa*

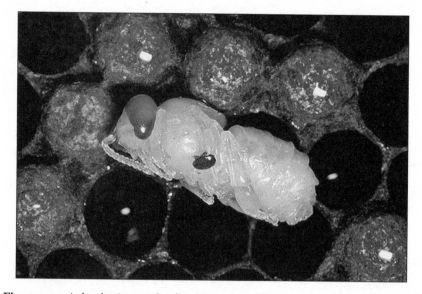

Figure 19.2. A developing worker bee is parasitized by a Varroa mite (the dark, oval object on the midsection of the host). The mite feeds on the bee's hemolymph, or "blood." In addition, these parasites can transmit deformed wing virus, a pathogen that causes lethal deformities, including stunted wings, distorted appendages, and paralysis. Castro's government accused the Americans of introducing the Varroa mite to Cuba in 1996. (Photo by Scott Bauer, USDA/ARS)

jacobsoni—is perhaps the most feared malady of bees (see Figure 19.2). Resembling a minuscule crab, the mite clings to the outside of the bee, feeding through the body wall of adult and larval bees (by comparison, imagine a five-inch tick latched onto your back). In short order, the honey industry wracked up $2 million in losses, along with the destruction of 16,000 beehives in a desperate effort to contain the epidemic.[24] The Cubans were quick to note that the epicenter of the outbreak was on that part of the island nearest to southern Florida. Knowing that the United States had been battling the Varroa mite since its arrival in Wisconsin a decade earlier, Cuban officials drew what had become the obvious conclusion for a besieged island nation—imperialist mischief.

The accumulation of cases didn't seem to be headed anywhere, but the camel's back was about to break. For years Castro and his officials had been griping through unofficial channels about America's assault on Cuban agriculture. But never had the communist regime taken its complaints to an international forum. For that matter, no government had ever formally invoked the provisions of the Biological and Toxin Weapons Convention. That is, until the final straw descended on Cuba in 1996.

20

A TINY TERRORIST IN CASTRO'S CROPS

The Cuban Ministry of Foreign Affairs presented the U.S. government with a written complaint of entomological warfare the day after Christmas in 1996. The allegation concerned the release of insects by an American plane that passed through the Giron corridor, a designated flight path over Cuba. Per the diplomatic drill, the U.S. State Department explained away the charge in early February, maintaining that the incident was merely the release of warning smoke—not a cloud of crop-eating insects—meant to ensure visual contact with a nearby aircraft.[1] Such a cursory denial would have completed a typical tête-à-tête between the two countries, but the Cubans had other plans.

Perhaps the U.S. government should have suspected that something more than another spat was in the offing when the Cubans took their case to the United Nations, issuing a strangely worded communiqué to the secretary-general on April 28:

> The Permanent Mission of Cuba to the United Nations presents its compliments to the Secretary-General and, with regard to item 80 of the preliminary list, has the honour to convey hereby a report on the appearance in Cuba of the *Thrips palmi* plague.[2]

Thrips are obscure little insects that resemble animated hyphens on a page. Only under a microscope can one see that the clear yellow creatures are bedecked with a pair of feathery wings, formed from extraordinarily fine hairlike structures (see Figure 20.1). Although these delicate insects might not be the sort of organisms that one associates with a plague, what thrips lack in terms of body size they make up for with respect to their eclectic feeding and wanton reproduction. The species that was plaguing the Cubans was known to consume citrus, cotton, cucumbers, mango, melons, peas, potatoes, soybeans,

Figure 20.1. *Thrips palmi* is about 1/15th of an inch in length and bears frail, featherlike wings. Despite its feeble appearance, this species is phenomenally fecund and feeds on a spectacular range of crops, including beans, cantaloupe, melons, peppers, potatoes, soybeans, and sugarcane. The Cuban government formally charged the United States with releasing these pests over agricultural lands east of Havana in October 1996. (Photo by J. Marie Metz, USDA/ARS)

sunflowers, tobacco, tomatoes, and watermelons, along with various flowers.[3] The insects suck fluids from the host's tissues, yielding stunted and deformed plants. And as if to mock the unfortunate farmer, thrips add an insectan Midas touch, causing damaged leaves to appear silvered or bronzed.

With females producing as many as 200 offspring during their two- to three-week lifetimes, and the larvae maturing into adult egg-laying machines in about a week, a population of thrips can reach staggering numbers in a very short time. If humans reproduced and matured as quickly, Adam and Eve would have populated the earth with 6 billion offspring in just 110 days. The Cubans may not have calculated the biotic potential of thrips, but they had enough information to inform the UN Secretary-General that "there is reliable evidence that Cuba has once again been the target of biological aggression."[4]

By involving the United Nations, Cuba forced the United States to dignify the accusation with a reply in front of a global audience. In response, the Americans issued a terse dismissal on May 6: "The United States categorically denies the outrageous charges made by the Cuban Government regarding the alleged discharge of the *Thrips palmi* insect over Cuba to damage agriculture there."[5] The U.S. government figured that their response had cued the curtain fall in the diplomatic theater of the absurd. But after a two-month intermission, Castro raised the curtain for a spectacular encore on the world's stage.

On July 7, the Cubans filed an official request through the Russian government, acting in its role as one of the original treaty depositories of the Biological and Toxin Weapons Convention.[6] Article V of the Convention stipulated that its signatories could initiate "consultation procedures"—a formal hearing concerning violations of the treaty. In a quarter century, no government had ever pursued such an action. But now the United States was officially charged with having engaged in biological warfare. With the British ambassador, Ian Soutar, serving as the chair, a Formal Consultative Meeting convened later that year in Geneva, Switzerland, to hear the case.

Both sides agreed on the essential facts.[7] On the morning of October 21, 1996, an American plane operated by the U.S. State Department had taken off from Patrick Air Force Base in Cocoa Beach, Florida. The aircraft was initially bound for Grand Cayman, with authorization to fly over Cuba via the Giron corridor. The ultimate objective of the mission was to participate in a drug-eradication program in Colombia, via the spraying of coca fields with herbicide. While flying through the Giron corridor, the pilot made visual contact with a Cuban aircraft and released several puffs of a smoky substance (see Figure 20.2). When the Cuban pilot reported the gray mist, the air traffic controller radioed the American plane to determine if there was a mechanical problem. The pilot replied that he was having no difficulties and continued on his way. The nature of the material released from the American spray plane was the crux of the case.

The Americans explained that their pilot, having observed the Cuban plane in adjacent airspace, released a smoke signal to ensure visual contact and avoid a midair collision. Officials noted that such a warning was standard operating procedure: "Aircraft used for crop eradication by the International Narcotics and Law Enforcement Affairs Bureau of the Department of State are equipped to generate smoke and with an aerosol sprinkling system."[8] The Cubans countered that such equipment and procedures were anything but ordinary. Rather than a prudent puff of smoke, the Cubans contended that the cloud was a mass release of thrips. And their case was based on more than rank suspicion—they had some damning circumstantial evidence.

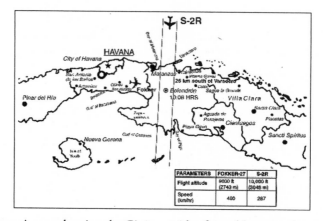

Figure 20.2. A map showing the Girón corridor from the report given to the UN Secretary-General, which asserted that "on 21 October 1996, at 10:08 hours, crew members of scheduled flight CU-170 of Cubana de Aviación (Cubana Airlines) . . . noticed a single-engine airplane flying from north to south, at about 1000 feet (300 meters) above them, apparently spraying or sprinkling unknown substances—some seven times—in an intermittent manner." The Cubans claimed the substance was a mass release of thrips from a U.S. plane. (Courtesy of Granma International)

Two months after the American flyover, *Thrips palmi* irrupted with a vengeance. This period corresponded with the time necessary for the insects to complete three generations, an entirely plausible lag from introduction to outbreak. Moreover, the pattern of the infestation was extraordinary. The pests were decimating potato fields within the Giron corridor of western Cuba, while farms to the east were unscathed.[9] The nearest natural sources of the thrips were island nations to the east: Haiti, Jamaica, and the Dominican Republic. A natural invasion of thrips from one of these islands would not have skipped 400 miles of vulnerable farms to infest fields that just happened to lie in the flight path of the Americans' "smoking gun."

In rebuttal, the U.S. representatives argued that there were many cases in which airborne organisms had invaded Caribbean islands—without invariably infesting the fields closest to their point of arrival. Furthermore, various pests had long made their way into new lands via agricultural trade, so any port of entry could serve as a locus of infestation. But the Cubans weren't done yet.

A thorough analysis by Cuban scientists suggested that *T. palmi* was an ideal agent for waging covert biological warfare.[10] The evaluation pointed out that this species could be mass-produced in the laboratory. Thrips would disperse like aerial plankton on the winds after release from a plane, and their rapid development and fantastic fecundity meant that even a couple of dozen adults could seed an outbreak. Moreover, the tiny invaders were difficult to detect until their numbers and damage were beyond control. And finally, control with insecticides was notoriously difficult given the insects' tendency to shelter themselves within the plants.

The Cubans' analysis might have been brushed aside as so much speculation were it not for the Federation of American Scientists. This scientific association seemed beyond repute: an independent, nonprofit organization founded by members of the Manhattan Project in 1945, directed by prestigious researchers and dedicated to ending the worldwide arms race. Eight months before the purported entomological attack on Cuba, the federation issued their "Report of the Subgroup for Investigation of Claims of Use or Escape of Agents Which Constitute Biological or Toxin Weapons." The scientists included *T. palmi* among the organisms of concern to the signatories of Biological and Toxin Weapons Convention.[11]

The American response to the Cubans' so-called evidence was an exasperated plea to use reason and science, albeit not that of the Federation of American Scientists. The U.S. government argued that there would always be ecological coincidences to provide rich fodder for those bent on creating conspiracy theories. But no rational collection of jurists could conclude that clandestine military operations were the most plausible explanations for such unusual occurrences. Surely the United States could not be blamed for having released every creature that had the qualities of a biological weapon and managed to find its way to the shore of a backward nation seeking to vilify the West while lacking the infrastructure to protect itself from pest outbreaks.

In closing, the Cubans argued that the timing and location of the pest outbreak were so perfectly matched to the mysterious emission from the American plane that the United States simply had to be the culprit. The Americans scoffed at such a farfetched interpretation, contending that normal flight operations and natural events readily accounted for all of the supposed evidence. Moreover, western observers noted that Cuba had taken its case to the international community via Article V of the Convention, which meant that the decision would rest on arguments provided by the adversarial governments. Had the Cubans invoked Article VI, the United Nations Security Council would have been compelled to undertake its own investigation. The implication was that Castro's regime sought to avoid a credible, independent inquiry.[12]

The participants in the Formal Consultative Meeting were given until late September 1997 to ponder the allegations and render their verdicts. Hungary and the Netherlands concurred with Denmark's assessment:

> [The United States] convincingly demonstrated that the occurrence of *Thrips palmi* in the Matanzas province of Cuba . . . could have resulted [from] a number of causes, including natural phenomena as well as the normal movement of trade and goods.[13]

Germany agreed and further asserted that "insects such as *Thrips palmi* couldn't be dispersed from an aircraft as a dry substance," presumably because the tiny soft-bodied creatures were thought to be too fragile to survive such treatment. In the end, all but two of the nations found the Cuban charges to be without substance.

Cuba's communist allies could not bring themselves to side with the Americans. The Chinese chose to simply prevaricate, stating that their experts had found it "hard to draw conclusions." The North Koreans were more direct, finding it "regrettable that the incident of spraying of biological substances by the United States against Cuba has taken place."[14]

With these verdicts by the representatives, a dismissal of the charges from the Formal Consultative Meeting might have been expected. However, when the summary judgment was issued by the chairman, Ambassador Soutar of the United Kingdom, the wording could not have been more equivocal. In mid-December 1997, he reported, "Due *inter alia* to the technical complexity of the subject and to the passage of time, it has not proved possible to reach a definitive conclusion with regard to the concerns raised by the Government of Cuba."[15]

Use of the British legal term *inter alia*, meaning "among other things," only obfuscated an already enigmatic statement. In the minds of some critics, this unwillingness to render a clear decision signaled political weakness among the parties to the Biological and Toxin Weapons Convention and encouraged the Cubans to persist in their accusations. According to Milton Leitenberg (a researcher from the University of Maryland who also energetically sought to discredit the charges that the United States had employed entomological weapons during the Korean War):

> If there had been a quite aggressive investigation of the 1997 Cuban charges, and a definitive and noticeable report at the end stating that the charges

were unsupportable and concocted, that would most certainly have served to impede further politicized allegations every time a natural disease outbreak occurs in a country in which there is an ongoing conflict, or in which local political elites think that there is some domestic political gain to be obtained by making charges of biological weapons use by an external actor.[16]

Based on subsequent charges by America's adversaries, Leitenberg's analysis was on the mark. In a 1999 interview, Lieutenant General Valentin Yevstigneyev, deputy director of the Russian Defense Ministry's Office of Nuclear, Biological and Chemical Weapons, alluded to the Cuban case in insinuating that America was behind an insect outbreak along the Volga:

Last year in the Saratovskaya region, we fought against locusts and managed to save the harvest, nearly destroyed by these insects. When we started to determine the type of these insects, it turned out that they originated on the Apennine peninsula [a very distant place from the Volga]. So, it's up to you to decide whether it was a *gift of nature* or a secret form of diversion, especially with regard to recent developments in Cuba.[17]

Seizing the opportunity to piggyback on the Russians' claim that locusts could be used as covert weapons, Iraq launched its own diplomatic offensive. In July 1999, Iraqi officials accused a United Nations mine-removal specialist of planting locust "land mines." According to a letter sent by the Iraqi foreign minister to the UN's humanitarian relief representative, a New Zealand specialist was seen burying several boxes filled with locust eggs near the village of Khanaqeen. The day after the complaint from the ministry, Saddam Hussein ranted in a radio address about American duplicity:

They buried the locusts' eggs so that they will later become fully-grown locusts that would eat people's food. People would then be deprived of food and consequently die in this beleaguered state. . . . This is not the work of the individual, but the work of a state and its intelligence services. . . . The entire world said that there are spies who worked under the cover of the United Nations for the United States.[18]

Every entomologist knows that locusts are endemic to the Middle East, so finding their eggs in the soil is hardly surprising, assuming that any biological samples were actually a part of the Iraqi story. In a move of remarkable diplomacy—or weakness, depending on one's perspective—the United Nations actually launched an investigation of the charges "out of respect for the Iraqi government." In short order the case was determined to be preposterous and, to nobody's surprise, there was no subsequent locust outbreak. The Russians and Iraqis were, in fact, plowing some well-tilled political ground, given that the potential of locusts as entomological weapons had been discussed 30 years earlier. The notion was assessed and dismissed by a UK delegate during a 1969 disarmament conference in Geneva.

With the Russians and Iraqis suffering no adverse consequences for their mudslinging, Castro's government continued to accuse the United States of entomological skullduggery. Castro may have truly believed that the Americans were out to get his people by using insects as mercenaries on the Caribbean front. Or he may have been borrowing a page from the North Koreans. That is, public allegations could preempt the imminent use of entomological arms if one does not cry wolf too often. Occasionally a complaint by Castro's regime has been forwarded through recognized channels, but most of the allegations have been issued during speeches by high-ranking officials and targeted for local audiences and sympathetic media outlets. While their recent political tactics have been lacking in originality, the Cubans have added a novel, legalistic maneuver to their efforts.

Cuban organizations have twice brought suit against the U.S. government, seeking compensation for damages inflicted by biological weapons. In the summer of 1999, a suit was filed in a Havana civil court seeking $181.1 billion for the loss of life and suffering inflicted by the United States since the onset of biological attacks in 1959.[19] Most recently, an 18-count indictment filed in January 2000 charged that American biological attacks had involved both the release of plant-feeding insects and the spread of plant, livestock, and human diseases, many of which were vectored by insects (including the dengue epidemic of 1981).[20] As in the earlier case, the Americans did not bother to defend themselves in what they presumed would be a kangaroo court.

Political observers suggested that the Cuban suits were legal tit-for-tat in response to a previous U.S. District Court case won by the relatives of four Americans who were killed when Cuban MiG fighters shot down two private planes involved in a "Brothers to the Rescue" operation.[21] Despite American and Cuban judges' rulings in favor of their respective plaintiffs, there appears to be no chance that anyone will collect on the awards. Such legal maneuvers are intended to make points of principle rather than yield payments of court-assessed damages.

If Leitenberg is right in claiming that international ambivalence fosters politically motivated accusations of biological warfare, then perhaps his concern also applies to the American government. Various government agencies and private interests have contended that Castro's scientists and biotechnology industries have the capacity to develop biological armaments—and some have insinuated that the Cubans might have moved into the production stage. In 1998, a workshop on agricultural warfare hosted by the U.S. Departments of Agriculture and Defense was introduced in these terms:

> What would it take to induce Castro to undertake covert biological and chemical attacks on Florida, New York City, or other U.S. locations? Or could Cuba currently be undertaking biowarfare experiments in the Caribbean Basin or in the U.S.? There are a lot of animal and plant diseases popping up in unexpected places around the world and in the U.S.[22]

For the most part, however, the United States had the luxury of believing that it was safe from outside attack. Although a civil war once wracked the nation and assassins had taken the lives of public figures, 20th-century Americans had not contemplated the possibility that a homegrown insurgency could inflict horrific suffering. The country's tranquility was shattered on Wednesday, April 19, 1995, at 9:02 A.M. when a truck packed with 5,000 pounds of explosives detonated in front of the Alfred P. Murrah Federal Building in Oklahoma City. And the reality of domestic entomological terrorism was not far behind.

FIVE

THE FUTURE OF ENTOMOLOGICAL WARFARE

MAYOR BRADLEY:

WHEN COMMUNITY AND POLITICAL LEADERS FAIL TO LISTEN TO REASON, OR PUT PROFIT ABOVE COMMON SENSE, IT BECOMES NECESSARY FOR THE PEOPLE TO CREATE BY FORCE OR SUBVERSION THE MEANS TO ENLIGHTEN THEM.

. . . THE INVOLUNTARY AERIAL SPRAYING OF CARCINOGENIC POISONS UPON OUR HOMES IS THE LAST STRAW. WE DECIDED TO DO SOMETHING. WE COULD PROBABLY MAKE OUR POINT, ALBEIT A MINOR ONE, BY ASSASSINATING A FEW SELECT COUNTY OFFICIALS, OR DO THE SAME TO UNIVERSITY ENTOMOLOGISTS WHO SUPPORT THE PRO-PESTICIDE POSTION [*sic*] . . . WE HAVE A BETTER IDEA.

WE DECIDED TO MAKE THE MEDFLY "PROBLEM" UNMANAGEABLE AND AERIAL SPRAYING POLITICALLY AND FINANCIALLY INTOLERABLE.

EIGHTEEN MONTHS AGO WE IMPORTED EGGS AND LARVAE OF FERTILE MEDFLIES AND RELATED GENERA INTO SOUTHERN CALIFORNIA, AND STARTED BREEDING THEM HERE . . .

THIS IS NO JOKE. WE STRONGLY URGE YOU AND OTHER OFFICIALS TO TAKE US VERY SERIOUSLY, OR THE MEDFLY "PROBLEM" OF 1990 WILL BECOME A NIGHTMARE. WE ESTIMATE THAT FOR EVERY DOLLAR WE SPEND, IT WILL COST THE CITY AND COUNTY OF LOS ANGELES A MILLION OR MORE TO THWART OUR EFFORTS.

(signed)

BREEDERS

—Excerpt of the letter sent to the mayor of Los Angeles and other officials by a domestic terrorist group threatening to use insects to wreak havoc.

21

MEDFLIES, FRUITS, AND NUTS

The Mediterranean fruit fly (*Ceratitis capitata*) is the world's most destructive pest of fruits, capable of causing staggering agricultural losses and triggering massive control programs. It is also one of the prettiest pests, with the house fly–size adults sporting wings that look like stained glass in earth tones (see Figure 21.1). The larvae, on the other hand, are voracious little maggots that gorge themselves on everything from avocados, coffee, olives, and tomatoes to bananas, citrus, mangos, and peaches. Although the species is found in Hawaii, it has not become established on the mainland—and the agricultural industry desperately wants to keep it that way.

The reproductive potential of the Medfly means that even a few stowaways on a fruit shipment could spell disaster. During warm weather, the insect can produce a new generation every two weeks—and a female can crank out 800 eggs in her lifetime. Assuming that just half of each generation survived, in just two months a founding population of 100 individuals could balloon to 80 billion flies. But the direct economic losses would only be a drop in the bucket.

Countries without Medflies understandably impose strict quarantines. If just one major importer of American produce, Japan, imposed an embargo on California fruits and vegetables due to Medflies, within a year the industry would lose a $600 million market. To scale up the potential disaster, an international embargo on Californian fruits would cost 35,000 jobs and at least $3.6 billion. And in a near worst-case scenario—just to make clear the economic scale of agriculture—with a total quarantine of California fruits, both nationally (after all, Florida and other fruit-growing states don't want to host Medflies, either) and internationally, the loss would be 132,000 jobs and $13.4 billion.[1] By way of comparison, the Loma Prieta earthquake that

Figure 21.1. The Mediterranean fruit fly, if permanently established, would cost California agriculture as much as $1.9 billion annually—more than the gross domestic product of 30 of the world's poorest countries. The pest can devastate both farmers and politicians. In 1982, then-Governor Jerry Brown's mishandling of the Medfly outbreak sent his approval ratings plummeting and derailed his bid for election to the U.S. Senate. (Photo by Scott Bauer, USDA/ARS)

struck the greater San Francisco area in 1989 was the costliest natural disaster in California's history. The total property damage was $6 billion.

It's no wonder that when localized Medfly infestations are detected, colossal eradication programs ensue. In 1980–82, an incipient outbreak in the San Francisco Bay area triggered the aerial application of malathion-laced bait over an area of 1,400 square miles at a cost of $100 million. As a nerve poison, malathion is toxic to a broad spectrum of insects and vertebrates, including people. Although it is one of the safer organophosphate insecticides, malathion had more than sufficient hazards to raise the ire of environmental activists.[2] So when pockets of Medflies were found again in 1988, radical environmentalists were primed to sabotage the insecticide program.

By 1989, the state's aerial campaign against the Medfly was in full swing. The entomologists were armed with tons of chemical weapons, and the ecoterrorists were armed with biological weapons. In early December, the environmentalists' bizarre counteroffensive was announced to the public. The headline in the *Los Angeles Times* announced: MYSTERY LETTER PUTS A STRANGE TWIST

ON LATEST MEDFLY CRISIS. The upshot of the story was captured in the first few lines:

> The Mediterranean fruit fly crisis has taken an odd turn with law enforcement officials tentatively investigating a mysterious letter from a group that claims to be breeding and spreading the pest throughout Southern California. The group, calling itself the "Breeders," said in an unsigned letter to Mayor Tom Bradley, agricultural officials and media that it was angered by repeated aerial spraying of pesticide to eradicate the fly and "decided to make the Medfly 'problem' unmanageable and aerial spraying politically and financially intolerable."[3]

The two-page typewritten letter opened with a diatribe concerning "aerial spraying of carcinogenic pesticides over populated urban areas" and then asserted that "the biosphere of the planet is on the verge of collapse . . . pesticides are in the food, the water, the soil and the air. All this because some of us suffer rapacious and narrowminded greed."[4] According to the letter, the Breeders had been clandestinely producing Medflies and releasing them to continuously expand the spray areas in Los Angeles and Orange counties. Having already forced the treatment program to encompass 232 square miles, the Breeders were prepared to up the ante. If the spray program was not terminated, they threatened to spread the infestation into the enormously valuable and productive San Joaquin Valley.

As evidence of their handiwork, the Breeders pointed out, "State officials have probably noticed an increase as well as an unusual distribution of Medfly infestations in Los Angeles County since March of 1989." The allusion to an "unusual distribution" of Medflies described the patterns that were being seen by field workers. Rex Magee, associate director of the California Department of Food and Agriculture, noted that "this particular infestation has had some characteristics that we have not seen in the past." Although officials couldn't be certain that ecoterrorists were to blame (some attributed the expanding infestation to private citizens smuggling infested fruit out of quarantined areas), there was no doubt that someone was desperate to stop the spray program, and ordinary Californians were being drawn into the fray. Pat Minyard, the director of Plant Health and Pest Prevention Services for the California Department of Agriculture, was sent a package labeled "Danger" and "Toxins." When the police bomb squad opened the box, they found oranges apparently sent by a citizen from within a treated neighborhood.[5]

Many officials suspected that the Breeders were a hoax, but those who understood the ecology of the Medfly and the operations of the control program were alarmed.[6] An internal report by the USDA strongly favored the possibility that the Breeders were for real. According to this document (the author was anonymous, although clearly an expert in the field), four lines of evidence gave credence to the ecoterrorists' claims. First, the timing of the fruit flies' appearance was unusual: "We have never had these types of population buildups. . . . Programs in the past have had detection in late August and September, but [populations] rapidly dropped off once cooler fall and winter temperatures took hold."

Next, the distribution of the insects was worrisome: "This spread from one location to another, 15 to 20 miles apart is not normal. In addition, in my recollection, we have not had flies show up in an area repeatedly without our having found a substantial larval infestation." Third, a piece of evidence never revealed to the press pointed to an unnatural situation. Both the Medfly and its cousin, the Oriental fruit fly (*Bactrocera dorsalis*), were found in the same areas, a situation apparently without precedent (neither insect is native to the United States, so the chances of both arriving at the same time is rather remote). Finally, the author noted that although in earlier infestations mated females had occasionally been detected prior to finding males, this odd sequence was "an area of major concern" in 1989. The bottom line was that the circumstantial evidence pointed to a human agency in the ongoing outbreak.

Roy Cunningham, a USDA entomologist and dean of the state's Medfly scientific advisory panel, asserted that the Breeders' logic was terribly flawed, as any expansion of the infestation would simply result in more spraying—precisely the opposite outcome than the terrorists sought. But he also understood that terrorism can proceed along utterly unexpected and seemingly unreasonable paths. Cunningham launched his own investigation to determine the veracity of the Breeders' claims.[7]

On December 6, he convened a panel of experts in Los Angeles to evaluate the possible reasons for the peculiar features of the Medfly infestation. The meeting was considered a private gathering, so no public announcement of the findings was made. However, the *Los Angeles Times* managed to glean a sense of the proceedings. It seems that the panel was deeply divided, with some members interpreting the pattern of trap catches to be the result of natural processes and others concluding that someone must have been purposely breeding and releasing Medflies (see Figure 21.2). Those who suspected human duplicity would soon have further evidence of ecoterrorism.

Figure 21.2. Medfly traps are essential for survey and monitoring, and during the early stages of an infestation these devices can play a role in reducing pest densities. This trap uses a blend of chemical attractants. In addition, the cylindrical shape mimics the three-dimensionality of host fruit, and the clear panels exploit the flies' tendency to move toward light, where sweet and lethal bait awaits them. (Photo by Peggy Greb, USDA/ARS)

The January 4, 1990, edition of the *Los Angeles Times* reported not only that scientists were trapping flies in odd locations but also that the biology of the insects raised suspicion:

> The latest find of a female fly just outside a previous spray zone continues a pattern that has baffled scientists associated with the state and federal Medfly project. The scientists believe they should be trapping more males and that it is curious that so many recent fly finds have occurred close to the borders of previous spray zones. . . . Scientists also are unable to fully explain why they have not found larvae in large quantities in the infestation areas.[8]

Whatever the origin of these pests, the immediate result was precisely what the Breeders had claimed to be their short-term objective. Officials were forced to expand the spray program to encompass an additional 15 square miles, on top of the already treated 300 square miles. And public opinion was also beginning to shift toward the ultimate goal of the ecoterrorists. People in the treated areas were not at all pleased to find themselves within an entomological war zone. With the economic and political costs of the eradication program mounting, law enforcement stepped up its efforts to find the perpetrators.

The Los Angeles Police Department (LAPD) pursued its investigation in parallel with the USDA's Office of Inspector General (OIG). The Breeders toyed with law enforcement, sending copies of their letter to various newspapers and mailing postcards with messages such as "For Real. 'The Breeders.'" In February, LAPD detectives and OIG special agents repeated their earlier efforts to contact the terrorists by placing an advertisement in the "Personal Messages" section of the classified advertisements of the *Los Angeles Times*, per the instructions of the Breeders. The notice stated:

Breeders
If you're for real send one of your little friends to: P.O. Box 1549, LA CA 90053.
We want to talk. Call John at USDA 213/894–5828 bet. 9 & 10 a.m.[9]

This time, an unidentified male called the listed number and said, "John, this is the Breeders, a Medfly is on its way to you in the mail. We are definitely going to release again if any more spraying is done. Bye."[10] No package arrived, but the bizarre pattern of pest dispersion continued to befuddle and worry agriculture officials.

Meanwhile, investigators pursued new leads but they all led to dead ends.[11] Federal agents looked into the actions of several disgruntled state and federal employees who might have had reason to undermine the Medfly eradication program, but investigators found no hard evidence of duplicity. The LAPD received a tip on the California Crime Hotline alleging that a science teacher had been raising and releasing Medflies, but the fellow had only been turning loose common house flies as part of an experiment with his students. Although law enforcement could not find the bioterrorists, agricultural agents had no difficulty finding troubling evidence.

A February 10 story in the *Los Angeles Times* reported that

county fly trappers have found far fewer larvae than expected in an infestation this large, and the discovery of new adult flies often have [*sic*] occurred just outside of the infestation boundaries, forcing a steady expansion of the spraying zone.[12]

Many found the Breeders' plot just too convoluted to believe, but a state official noted that "you got 8½ million people out there. You never know, there's always the chance that someone out there has a loose screw." The most reasonable tactic may not occur to the terrorist, who is motivated by passion. And the Breeders' tactic—whether rational or insane—seemed to be working. Not only was the public increasingly upset with the health, environmental, and economic costs of the program, but pest managers were starting to second-guess their operation.

The *Los Angeles Times* article noted that "the mere possibility [of reared and released insects] has forced officials to become a bit more cautious in their campaign against the Medfly." The California Department of Food and Agriculture announced a two-week delay after finding new flies in their traps on the Cal Poly Pomona campus. "We're always cognizant of the Breeders thing," said the county agricultural commissioner, "We wanted to make sure these were really wild flies. We can't afford to start a new treatment area until we know what we've got."

Over the next few months, the infestation in southern California declined. With fewer and fewer Medflies, the controversial insecticide spray program wound down. Agricultural officials, law enforcement agencies, and policy analysts were left to debate whether they'd been duped or whether they'd dodged a bioterrorist's bullet.

An independent analysis by the Belfer Center for Science and International Affairs at the Kennedy School of Government in 2001 concluded that the extortionists had been for real: "The Breeders used a biological means, Medfly, to attack crops in California. By contrast, most attacks have either been hoaxes or relied on chemical agents to attack agriculture."[13] People closer to the case seem to waffle, unable to fully embrace the bioterrorist explanation. James Reynolds, the Western Regional Director for USDA at the time of the incident, maintains that

> the Breeders claimed to be raising flies in a garage somewhere in the LA basin. It seemed possibly credible at the time since we were chasing Medfly throughout the LA basin and we had several situations where we found flies just outside the treatment area requiring us to increase treatment and regulatory boundaries. While there was a biological explanation, sabotage also could have been an explanation . . . I am not aware that OIG ever came close to any credible lead which would suggest the letter was a hoax.[14]

Other officials are less convinced, but unwilling to dismiss the possibility. Pat Minyard, the California official who had been sent the box of fruit labeled

"Toxins," now believes that the Breeders were probably a hoax, based on the screwy logic of the purported terrorists.[15] To really bring California agriculture to its knees, the Medflies should have been released in the state's agricultural Mecca, the Central Valley, rather than the Los Angeles Basin. Moreover, Minyard asserts that because these insects are difficult to rear, the enormous numbers of flies necessary for an effective attack would challenge the capacity of a highly skilled and well-organized terrorist cell—and the Breeders did not appear to be such an organization.

Minyard finds some relief in the realization that even a rather weak system of survey and inspection might be adequate to protect California farms from Medflies, which is fortunate in today's world, for he also contends that agriculture is more vulnerable now than it was before 9/11. According to Minyard, Homeland Security officials have much less interest in protecting crops and livestock than did the USDA's former Plant Protection and Quarantine program. The fact that we've not experienced an attack on agriculture by foreign terrorists is, in his assessment, a matter of either the terrorists being inept or the nation's defense being competent. He suspects the former.[16]

Perhaps Minyard's most penetrating insight regarding the case of the Breeders was in recognizing that even a hoax is a wickedly effective weapon for bioterrorists when the target is greenbacks rather than green plants. Economic damage can be inflicted without actually harming crops or livestock. His concerns are echoed by James Carey, a professor of entomology at the University of California, Davis, who has made the following argument:

> If you find what they call a "Class-A pest" like the Mediterranean fruit fly and you have someone with a bottle full of Medflies deliberately planting them in traps, what happens is that it sets in motion the eradication campaign programs, the quarantines, because it's very difficult to distinguish between a real outbreak and one that's deliberately planted. So I can see that as the worst nightmare situation.[17]

While hoaxes are problematical, the focus of national defense is on actual releases of organisms that would directly damage agriculture or spread disease among people, livestock, or crops. And in such cases, the country's counterattack would depend heavily on its stockpile of chemical weapons—insecticides. But insecticides are a double-edged sword.

In 1996, a disgruntled supplier of National By-Products dumped chlordane—a banned insecticide, related to DDT—into a load of animal carcasses destined for conversion into animal feed for Purina Mills.[18] His note to officials in Wisconsin said to expect widespread animal mortality. The authorities realized that even if dairy cattle didn't die in droves, their milk would be rendered worthless. As little as 300 parts per billion (about 3 cups of liquid in an Olympic swimming pool) of chlordane would render milk unfit for human consumption. Remarkably rapid action on the part of the company minimized the damage to their business and reputation by ensuring that the contaminated feed never entered the human food chain.

The adulteration resembled a case 15 years earlier, in which a herd of beef cattle was poisoned with an organophosphate insecticide that was dumped into a farm silo. Although the culprit was never found, police suspected that the attack was meant to settle a grudge against the farmer.

The cases of angry environmentalists releasing insects and aggrieved farmers dumping insecticides illustrate several important aspects of agricultural bioterrorism. First, relatively localized acts of sabotage (or even the threat of damage) have the potential to generate enormous economic and social costs. Next, the perpetrators are extremely difficult to apprehend. And finally, government agencies struggle to neutralize even middling acts of entomological terrorism.

Such attacks on agriculture were the work of a few angry citizens lacking the funding, fervor, and facility to hijack airplanes and turn them into missiles. But on September 11, 2001, the people of the United States realized the terrible potential of a well-organized and utterly ruthless enemy. And Americans must now ask (and their government ought to answer): What could happen if such an opponent targeted the nation's heartland?

22

FEAR ON THE FARM

The role of entomological weapons in the modern world is changing as rapidly as the nature of human conflict. Conventional military engagements between uniformed troops equipped with planes and tanks battling to seize control of land have given way to insurgent forces using improvised weapons to attain cultural and political victories. Stealth, sabotage, and subterfuge even the odds. And insects can be an ideal means of waging an "asymmetrical" war.

For decades, military planners assumed that humans were the most likely targets. But 21st-century conflicts with unconventional enemies create different scenarios for security and defense planners. From a terrorist's perspective, American agriculture has the 3 Vs of a good target: valuable, vital, and vulnerable. Food and fiber production accounts for 13 percent of the gross domestic product, a trillion dollars in economic activity, and one in every six jobs in the United States. Without the export of farm products, the nation's trade deficit—which is already dangerously out of kilter—would slide toward catastrophic imbalance. But agriculture means more than material wealth.

Farms are the cultural lifeblood of America. Fields, silos, fences, and barns affirm the nation's most cherished mythology. As is so eloquently expressed by Floyd Horn and Roger Breeze, top administrators in the USDA,

> Agriculture in America is a lot more than these statistics. It is cowboys and vintners, and the runners in the pits at the Chicago Board of Trade. It is biotechnologists, florists, forest rangers, tenant farmers, rural cooperatives and the county fair. It is FFA and 4-H. It is human nutritionists giving us sound advice that could save us billions in health care costs and prevent suffering on a massive scale. Agriculture is a huge part of the American investment portfolio. It is an unequaled "jewel in the crown" of this great nation. It is our great concern that the U.S. agricultural production, processing, and

marketing system are more vulnerable than ever to deliberate assault by a wide range of biological warfare agents.[1]

The bureaucrats and politicians have good reason to worry. For history leaves no doubt that invasive species—whether introduced by accident, ignorance, or malice—have the potential to cause enormous losses to the economic and social well-being of Americans.

From 1906 to 1991, an estimated 553 nonnative organisms successfully settled in the United States, and two-thirds were insects. A complete assessment of the damage done by the invaders has not been attempted, but the 43 insect species for which careful analyses have been conducted account for $93 billion in losses.[2] This cost is thought to be more than 20 times that of all other exotic organisms combined. Of course, the insects were—for the most part—attacking crops and livestock rather than humans. But both cunning bioterrorists and foresighted leaders understand that western societies can be grievously damaged without killing large numbers of people, or perhaps any at all. As Shintaro Ishihara, the governor of Tokyo and one of the most influential politicians in Japan, predicted, "The twenty-first century will be a century of economic warfare."

American agriculture is an ideal objective for bioterrorists. The farms and ranches of the United States are virtually unprotected. Some analysts contend that because agriculture is spread across so vast an area, it would be difficult to mount a damaging attack. And this is true if an enemy used nuclear or chemical weapons. But while radioisotopes and nerve gases have half-lives, organisms have doubling times. Simply put, insects disperse and reproduce—and they do so very well.

To make matters still worse, agriculture is a potential target for a frightening array of enemies, both foreign and domestic. Along with the usual suspects—state-sponsored terrorists, rogue states, and religious zealots—comes a gang of other assailants including national and international competitors of industrial agriculture and profiteers hoping to manipulate futures markets. Then we have groups driven by moral righteousness, such as advocates of animal rights; opponents of genetically modified foods; and ecoterrorists concerned with the chemical sins of modern agriculture. And we can't overlook the possibility of organized crime, militias, and copycats getting in on the act.

If it sounds like the list of evildoers is overreaching, consider *The Turner Diaries*, a source of inspiration for Timothy McVeigh and others of his ilk. This piece of bizarre fiction was written in 1978 by William Pierce, founder of the white supremacist organization known as the National Alliance. The diaries describe a right-wing revolution from within the United States, starting with an effort to shock Americans out their complacency: "[The militia] began appealing to things they [Americans] can understand, fear and hunger. We will take food off their tables and empty their refrigerators." Although this was fiction, there are plenty of actual events to encourage the hopeful terrorist. The real world provides vivid cases in which localized, accidental (as far as we know) releases of foreign insects exploded into full-blown disasters.

The nursing staff was horrified. On the bed lay an elderly patient covered in ants, which had crept through a break in the wall of the nursing home. Most Floridians have heard stories of fire ants attacking newborn animals, but few ever witness these insects killing a fellow human. By the time the staff came to the aid of Mary Gay, the 87-year-old woman had been stung over 1,600 times. The pain must have been excruciating, as fire ants owe their name to the intense burning sensation caused by the venom that they inject. One sting feels like being pricked by a hot needle; hundreds of stings can induce shock. Mary Gay died two days later.[3]

In the southeastern United States, few creatures are more loathed than the red imported fire ant.[4] Ninety percent of residents report that they or someone in their family has been stung by these insects, causing 80,000 people a year to seek medical attention. Nearly a quarter of those stung manifest an allergic reaction. In severe cases, a victim experiences anaphylactic shock, a condition that accounts for the majority of 100 annual deaths from fire ants.

This insect's reign of terror began in 1933, with the arrival of a ship from South America in Mobile, Alabama. At that time, vessels often filled their empty holds with soil from home, and this ballast was emptied along the waterfront as goods were taken aboard. A load of soil from Brazil was apparently seeded with fire ants, which spread into the port city. For many years, the citizens and officials in Alabama were not alarmed. After all, the black fire ant (then thought to be a color variant) had been accidentally brought to the United States in 1918 without causing havoc. Not until nearly 40 years later did entomologists recognize that the black and red imported fire ants were distinct species—and the latter was far more aggressive and invasive. In 1953, the USDA conducted its first survey for this pest and found it had a foothold in ten states. William Buren, the entomologist who named the creature, called this red form *Solenopsis invicta*, meaning "invincible"—a name intended to reflect his realization that this species would be extraordinarily difficult to defeat. He was right.

By 1997, red imported fire ants had spread throughout the Southeast, infesting 430,000 square miles of land. With each colony capable of producing 6,000 winged reproductive ants each year, there seems to be no stopping the invasion until the species has occupied every potential habitat. Only when the insects encounter conditions that are too cold or dry does the invasion grind to a halt.

A new colony develops rapidly, with the queen laying several hundred eggs each day until the growth of the loamy mound stabilizes after about six months. By this time, there are between a quarter and half a million workers—an area the size of a football field may be crawling with more than 10 million of these insects.

The red imported fire ant is a voracious feeder on both plant and animal tissues. Not only do the foragers consume seeds and fruits of valuable agricultural commodities, but the ant mounds block irrigation systems and make harvesting a nightmare. Wildlife species—particularly amphibians, reptiles, and ground-nesting birds—are ravaged by the fire ants. Livestock are blinded and killed by suffocation when they are stung around the face. In Texas alone, the cost of protecting plants and animals from fire ants exceeds $1.2 billion each year. And various attempts at eradicating the pest with large-scale insecticide applications and cleverly designed biological control programs have only demonstrated that Buren was right—the ant is invincible. But there is another insect pest that makes the fire ant seem like a welcomed visitor at a picnic.

In August 1996, a beetle was discovered in Brooklyn—and this insect's damage potential is estimated at $669 billion.[5] For comparison, the terrorist attacks on September 11, 2001, resulted in direct economic losses of $27.2 billion. Millions of urban trees are in the path of the Asian longhorned beetle (*Anoplophora glabripennis*), with nearly half of New York's trees and two-thirds of Chicago's at risk from this wood-boring insect.[6] In these two cities alone, tens of millions of dollars have been spent removing lethally infested trees, and there is no end in sight. When urban foresters find the beetles—often in maples, although ash, birch, boxelder, buckeye, chestnut, poplar, sycamore, and willow are vulnerable—the afflicted trees are summarily cut down. Trees have no natural defense against an infestation, and no insecticide can reach the pest once it has tunneled into the wood.

The newly hatched larvae bore deep galleries, where they spend the winter. In late spring, the insect pupates and matures into the adult stage, an inch-long insect with long, graceful antennae. Its jet-black body with white mottling resembles the night sky, giving rise to the alternative common name, the "starry-sky" beetle. The adults emerge from the galleries, feed on the bark of twigs, and reproduce in what is, for insects, a rather modest manner. The female lays only about 80 eggs in her lifetime, but this is plenty to fuel the invasion.

The beetle most likely arrived as a stowaway in wooden pallets or packing crates from China. Although the insect had attacked North American sugar maples in Chinese plantations long before making its way across the ocean, nothing was done to enhance vigilance among inspectors. Once the insect was discovered in the United States, a nationwide quarantine program was established to contain its spread. However, this effort may only delay the inevitable. The beetle larvae are readily transported in firewood and lumber—and with the insect having been found in warehouses in 14 states, it seems only a matter of time before enough beetles slip through our defenses to establish infestations across the country. To make matters worse, unlike many other forest pests, the Asian longhorned beetle attacks healthy young trees as well as old stressed trees.

The wholesale destruction of urban forests to stem the tide of the infestation is a staggeringly expensive proposition. To some, the summary execution of infested trees seems extreme. But the chainsaw is botanical euthanasia, for even if the tree hangs on for a few years it is condemned to a lingering death. And the dying plant is a serious risk to people and property as hollowed-out branches snap off in the wind.

The prognosis for natural forests is similarly discouraging, particularly with respect to economics. The beetles' tunnels degrade the suitability of the wood for lumber or veneer. Even small amounts of damage can halve the value of a tree. The ecological irony of cutting trees to save forests (or at least profits) is a bitter pill to swallow. All of this would be bad enough, but in recent years the longhorned beetle has been joined by an Asian ally—the emerald ash borer (*Agrilus planipennis*).[7]

This gorgeous iridescent green insect was first discovered in June 2002 in Michigan, and two years later 15 million ash trees were dead or dying. The initial quarantine seems to be working, but with no means of controlling the pest, foresters are deeply concerned. The cost of removing—let alone replacing—the urban ash trees in the United States is pegged at $40 billion. People don't eat trees, so the damage by wood-boring beetles doesn't raise the specter of crop failures and agricultural disasters. However, other invasive insects ably fill this role.

Few people are in a better position to develop realistic scenarios of bioterrorism than Lieutenant Colonel Robert Kadlec, who in the 1990s was a member of the U.S. delegation to the Biological Weapons Convention, worked as an inspector with the United Nations Special Commission in Iraq, and taught courses on biosecurity as a professor in the Department of Military Strategy and Operations at the National Defense University. He also is a medical doctor with master's degrees in both national security studies and tropical medicine and hygiene. As such, the detailed, hypothetical cases involving entomological attacks in his analysis *Battlefield of the Future* make one wonder if Kadlec knows something—maybe a whole lot—more than he can openly share.[8]

One of Kadlec's intriguing scenarios describes a group of European vintners who are angry over the economic juggernaut of the American wine industry. Posing as tourists, the saboteurs travel to the Napa and Sonoma Valleys. They carry unremarkable canvas bags, filled with maps, sweaters, cameras, and cans labeled as pâté, a most sensible food to complement their numerous tastings. However, rather than containing foie gras, the tins hold thousands of grape lice (*Phylloxera vitifoliae*)—tiny aphid-like insects that kill vines by infesting and destroying the roots. Moving through the heart of California's wine country, the disgruntled Europeans spread their deadly cargo.

Within a few months, the vineyards begin to show signs of stress, but it is too late to stop the invasion. Carried by wind and water, the insects spread to thousands of acres by the time anyone identifies the problem. Kadlec estimated that such an attack could kill two-thirds of the infested plants and cost the industry at least $1 billion.[9]

The wine industry is fortunate that the grape louse actually arrived in the United States more than a century ago, when there were few vineyards to fuel a major outbreak. Although the insect has been kept in check, Kadlec's central concern—the vulnerability of American agriculture to foreign insects—is indisputably valid. And while direct insect damage to crops is a serious concern, there is an even greater worry in terms of bioterrorism: the use of insects to vector plant diseases. As the wine industry is now realizing, all of the tactical advantages found with human disease vectors apply to carriers of viruses and bacteria that infect plants.

In 1989, a bizarre insect, native to the southeastern United States, was found in California.[10] How it crossed the country is anyone's guess; how it threatens one of the nation's most profitable agricultural industries is increasingly evident. The glassy-winged sharpshooter (*Homalodisca coagulata*) looks like a bug-eyed, mottled-brown, half-inch toad with clear wings (see Figure 22.1). But its clownish appearance obscures its deadly potential. The sharpshooter sucks up tremendous quantities of plant juices through its strawlike

Figure 22.1. The glassy-winged sharpshooter is the vector of Pierce's disease, a lethal bacterial infection of grapevines. The pest invaded California in 1989; if uncontrolled, the insect-pathogen duo could inflict $20 billion in losses—and with the possibility of the insect-borne disease moving into the state's almond, stone fruit, and citrus orchards, the potential loss to agriculture exceeds the gross domestic product of Costa Rica. (Photo by Reyes Garcia III, USDA/ARS)

mouthparts. If scaled up to the size of a human, this insect would drink 4,300 gallons of liquid a day, but its voracious feeding is not what terrifies agriculturalists. Rather, this species is the vector of Pierce's disease—a deadly and incurable bacterial ailment of grapevines.

Pierce's disease was first found in California in 1884. However, the local insects were not very effective vectors and growers limited the pathogen to local hot spots for a century. But the glassy-winged sharpshooter dramatically shifted the balance of power in favor of the disease. Within a year of the insect's arrival, thousands of grapevines looked like they'd been scorched. In effect, they were drought stricken, for the bacteria block the water-conducting system within the plant, causing it to die of thirst even if the roots are well irrigated. While the bacteria reproduce at an alarming rate, the same cannot be said for the vector.

The glassy-winged sharpshooter produces only two generations per year, and a female lays perhaps a hundred eggs—a nearly puritanical sex life by entomological standards. Rather than reproduction, this insect's success as a disease vector lies in its catholic diet (more than a hundred different plants), impressive longevity (six-month-old individuals are common), and extensive movement (typically wandering three miles from its birthplace). While Kadlec forecasted a billion dollars in damage in his entomological sabotage scenario, California growers might be delighted if losses from Pierce's disease could be held to this level. Just the rearguard action of protecting vineyards with insecticides amounts to $35 million per year. And if this line of defense fails, infected vines must destroyed and replanted—a process that could cost $20 billion. But there is cause for hope.

The state's quarantine seems, for the moment, to have curtailed the spread of the insect and disease. Meanwhile, releases of tiny wasps (*Gonatocerus triguttatus*) that parasitize sharpshooters' eggs have yielded promising results. In times of war, the enemy of my enemy is my friend—and the stakes are even higher than initially believed. Recent analyses suggest that grapes are just the tip of the agricultural iceberg. If the wine industry's pest-management program fails, this insect-pathogen alliance is likely to attack other agricultural fronts, including California's almond, stone fruit, and citrus orchards—which are worth tens of billions of dollars, not counting the annual value of the harvest.

Being cognizant of the historically devastating consequences of accidental insect invasions and of the prospective struggle against terrorism in the years to come, western nations have begun to seriously consider how they will protect themselves from entomological attack. The driving question for defense departments, agricultural offices, public health services, and intelligence agencies becomes, "What are the most likely targets for a bioterrorist who is armed with a suitcase-full of insects?" Although many agricultural systems would seem to make excellent targets, two examples are sufficient to illustrate the potential of an entomological attack.

Orange juice—it's not just for breakfast anymore. But suppose orange juice wasn't on the table at any meal, or to be more realistic, what if agricultural sabotage reduced retail sales by 50 percent for a period of five years? The economic damage would be staggering: $9.5 billion—or the cost of rebuilding the World Trade Center towers from scratch. And several biological agents have the potential to inflict such losses.

There are half a dozen exotic diseases of citrus that keep orchardists on edge, and all but one are transmitted by insects. A couple of these exemplify the growers' anxieties.[11] The bacterial ailment called citrus variegated chlorosis was first reported in Brazil in 1987, where one-third of the 150 million trees in the state of Sao Paulo were infected. Consider that the state of Florida is about the same size, with nearly as many citrus trees. But there is another, more worrisome similarity. Florida already hosts the leafhopper that transmits the disease. The only missing ingredient is an infusion of the pathogen to get the ball rolling.

Florida also plays host to the Asian citrus psyllid (*Diaphorina citri*), a relative of aphids that, like its cousins, possesses syringe-like mouthparts perfect for injecting microbes into plants. The psyllid arrived in the Sunshine State in the 1990s and has since flourished. While it is a pest in its own right, the deeper concern comes with the recent arrival of citrus greening disease to Florida. Infected trees become yellowed and produce small, discolored, bitter fruits.

As devastating as the losses would be from one of the insect-borne citrus diseases, vast sectors of the agricultural system would be untouched. However, there is another highly vulnerable target whose role in food production is exceptionally pervasive. In this case, rather than insects being the terrorist's weapons, they could become the bull's-eye.

In 1998, a workshop was held in Washington, D.C., to assess food and agricultural security in the United States. Various scenarios were explored by the foremost analysts in the field, and the most twisted but potentially potent tactic was not a frontal assault on fields, orchards, or pastures. Rather, the experts set aside particular crops and asked, "What if a bioterrorist took aim at a fundamental ecological process upon which much of agriculture depends?"

Without insect pollination, 90 different U.S. crops would not yield fruit or seed. Eliminate bees and billions of dollars disappear too. And America's honeybees are sitting ducks.[12] Moving easily through unguarded apiaries, an enemy could sprinkle the hives with spores of fungi, including those causing an incurable disease called chalkbrood, which transforms the larvae into mummified lumps of white mold. Bacteria also would make fine weapons, such as the malady called foulbrood, which digests larvae and pupae within their wax cells, leaving a sticky black crust in their place. And the viruses have considerable potential, including Kashmir bee virus, which causes adults to stagger about as if intoxicated as the colony goes into decline. Or a terrorist might introduce a protozoan such as *Nosema*, which leads to debilitated workers fouling the hive with excrement as the bees suffer from an insect version of dysentery. And finally, a saboteur could easily infiltrate commercial beeyards and release mites, such as those which crawl into the respiratory system of bees and suffocate the insects.

Some of these maladies are already present within our borders and could be spread from infected colonies, but the greater concern is for those ailments that have not been introduced to the United States. Given the vital importance of bees to food production, the mobility of these insects (which would facilitate the spread of disease), the prevalence of commercial and hobby apiaries, and the ease with which one could access the hives, it is no wonder that bees were deemed by the experts to constitute "an inviting and largely unappreciated target."

Farms and ranches across the industrialized world are easy marks for the wannabe terrorist. This situation would be bad enough, but we've managed to paint a bull's-eye on agricultural commodities through our own cleverness. The concern began in the context of human disease. As early as 1969, the United Nations warned governments that diseases that had been eliminated from a region, such as yellow fever from tropical countries or typhus from developed nations, could be reintroduced as biological weapons.[13]

More recently, public health officials have become concerned that through eradicating smallpox and then eliminating the vaccination program, the human population has become profoundly vulnerable to bioterrorists armed with the virus.[14] Of course, obtaining cultures of eradicated diseases is not a trivial matter. Opportunities are far more auspicious when it comes to analogous scenarios with pests of crops and livestock.

The United States is aggressively pursuing a program to rid the nation of the cotton boll weevil (*Anthonomus grandis*), a pest with an annual price tag of $780 million. But nobody appears to have considered whether boll weevils could become to agriculture what smallpox is to humans—an ideal weapon. This possibility is well beyond mere academic speculation. A parallel case of pest-management success breeding agricultural vulnerability has worried those charged with protecting the U.S. livestock industry.

Even for entomologists who have an affinity for six-legged animals, the New World screwworm is a loathsome creature.[15] The species name (*hominivorax*) alludes to its appalling diet—*hominis* means "man" and *vorax* pertains to "voracious." The screwworm is a man-eater (see Figure 22.2). The insect was named in 1858 by the French physician Charles Coquerel, who noted that this fly was responsible for hundreds of gruesome deaths within the Devil's Island penal colony.

An infestation is seeded by a female fly, who lays a mass of 200 to 300 eggs around a wound. In just 12 hours, the eggs hatch and the larvae begin tearing at the injured flesh with their mouth hooks. The gore of this feeding frenzy

Figure 22.2. The business end of a screw-worm larva, showing the creature's fanglike mandibles that are used to tear the flesh of living, warm-blooded animals. Though it was eradicated from the United States in 1966, a study by the National Research Council revealed that reintroduction of this pest by a bioterrorist would not be difficult and would cost the country tens of millions of dollars. (Photo by John Kucharski, USDA/ARS)

attracts more females as the maggots expand the open sore by the hour. Death almost inevitably results unless the wound is treated—a most unlikely event for a convict on Devil's Island. After five days of feeding, the larvae (by this time, about the size of macaroni noodles) drop to the ground, burrow into the soil, and form pupae. A week later, bluish-green metallic adults emerge, mate, and seek fresh meat. While most people in the 19th century had access to medical interventions that would keep screwworms from eating them alive, livestock were not so fortunate.

As early as 1825, ranchers in the western United States were reporting heavy losses from screwworm infestations. When the insect made its way to the southeastern states in the 1930s, producer profits and animal health went into a precipitous decline. Cattle, horses, sheep, goats, pigs, and dogs were fair game. The navels of newborn animals were often packed with writhing larvae, but any sort of wound rapidly became infested with maggots. Within a week, a pinprick from barbed wire could expand to a couple of inches in diameter, with larvae boring six inches into the animal's body. After another week, the oozing mass of raw tissue might support as many as 3,000 maggots. At this point, the animal was doomed and, if it was fortunate, secondary infection and accumulation of metabolic toxins ended its suffering within another few days. Large animals could be tormented for weeks, but their fate was sealed.

With livestock producers facing annual losses of more than $400 million, entomologists came under tremendous pressure to solve the problem. Edward

Knipling, an influential scientist in the U.S. Department of Agriculture, speculated that screwworms might be controlled through the mass release of sterile males, which would saturate the reproductive capacity of females with fruitless matings. The development of an artificial diet and the discovery that low doses of radiation caused sterility in the flies provided federal scientists with the essential elements to manufacture enormous quantities of infertile males.

A production facility was built in Florida to produce 20 million sterile males a week for a program of aerial dispersal. Beginning in 1958, planes worked their way westward from Florida, systematically distributing 800 flies per square mile to inundate the females with behaviorally lustful, but reproductively worthless, mates. By 1966, the country was declared free of this insect menace and attention turned southward.

To keep the screwworm from returning, the United States joined forces with Mexico. A plant built in Chiapas cranked out half a billion sterile insects a week until screwworms were eradicated from Mexico in 1990. Since then, the United States has extended the no-fly zone progressively southward. A fly factory in Panama ensures that the pest will not make it across the bottleneck between Central and South America. The eradication of the screwworm fly is a spectacular success for the livestock industry—and a golden opportunity for bioterrorists.

According to the USDA's Veterinary Services, a reestablishment of the screwworm to the United States would result in a $750 million annual loss to the livestock industry—not counting the $400 million accrued investment in the current eradication program.[16] And American agriculture's vulnerability to this insect was laid bare in a stunning analysis commissioned by the federal government.

A panel of distinguished scientists was asked by the U.S. Department of Agriculture to assess the United States' preparedness for biological attacks. The National Research Council (a division of the National Academy of Sciences) issued a summary of their alarming findings in 2003, but the most worrisome elements were hidden from public scrutiny.[17] Prior to releasing the report, the USDA requested that Appendix E be withheld under an exemption of the Freedom of Information Act.[18] Despite the government's concern, the concepts in this restricted document have been discussed openly in scientific and policy literature, although some of the details concerning terrorist logistics and targeting criteria of hypothetical attacks are unique.

The text of Appendix E consists of seven scenarios describing attacks on U.S. agriculture, including a case in which bioterrorists release screwworm flies. The experts imagine an incident beginning with the discovery of screwworm infestations from Florida to California. Such a large-scale synchronized outbreak points to an intentional introduction, a conclusion that the media are quick to grasp and exploit. Newspaper headlines trumpeting the arrival of flesh-eating flies and television footage of maggot-ridden animals are a terrorist's dream come true.

State and federal officials realize that a rapid, coordinated response is essential. With a fly being able to travel 180 miles and deposit 2,500 eggs in its lifetime, the government must immediately quash the hot spots. While entomologists rush to implement control measures, intelligence and law enforcement agencies reconstruct the origin of the attack.

Investigators discover that the perpetrator bribed a worker at a foreign facility that mass-produces sterile male flies. It was a simple matter to sneak screwworm pupae out of the factory before the insects had been sterilized. The terrorist then flew to Miami, smuggling a few quart jars of pupae in his baggage. Driving across the southern United States in a rental car, he seeded feedlots and other aggregations of livestock with his bloodthirsty conscripts.

Meanwhile, according to the hopeful folks constructing the scenario, the government agencies marshal their expertise and suppress the outbreak within a year. Even so, the price tag of the emergency control program is in the tens of millions of dollars, representing an enormous return on the terrorist's investment. Such an attack is entirely plausible, given the state of readiness in the United States.

The National Research Council asserted that the current inspection system makes it unlikely that an intentional introduction would be caught at the port of entry. Despite the council's hopeful claim that sophisticated scientific understanding would be required to pull off an attack, the scenario demonstrates that a saboteur would need only elementary knowledge of the insect's biology and the country's agriculture. As for the government's capacity to effectively suppress a screwworm outbreak, the scientists harbored no illusions that the imagined response was typical of national preparedness.

A rosy ending to the hypothetical attack is plausible only for a well-known insect for which there is already a pest-management program. From this realization, the experts warned that the nation lacks the infrastructure, methods, and knowledge to rapidly detect or suppress intentional releases of most insects. In effect, the screwworm scenario was the exception that proved the rule: the United States is ill-prepared for entomological terrorism.

The situation would be cause for serious concern if the only targets were crops and livestock. But let's face it, a farmer watching his fields wither under an onslaught of insects or a rancher seeing his cattle infested with maggot-ridden sores doesn't evoke near the alarm of a human suffering from a deadly vector-borne disease. And it appears that those responsible for protecting the health of the American people may be no better prepared than the agencies charged with safeguarding the country's food and fiber from terrorists armed with insects.

23

WIMPY WARMUPS AND REAL DEALS

A critical lesson of the last few years is that terrorists can rely on simple weapons: box cutters and car bombs. Whether simplicity is a matter of choice or necessity is difficult to know, but the latter is certainly relevant with respect to biological attacks. Without the scientific and technical support of a military-industrial complex, terrorists may be unable to culture and formulate pathogenic organisms into effective weapon systems. Insects, however, offer a low-tech, "safe and effective" alternative: they are easily collected or reared, robust to environmental adversity, and able to disperse on their own. Until recent years, entomological weapons were discounted by some military analysts because of a pair of perceived weaknesses: insect invasions were deemed too slow and too imprecise to alter the course of a modern war. But terrorists are engaged in what we might call postmodern warfare.

For today's radicals, slow-acting agents are not necessarily a problem. Although military planners in industrial nations are not enamored of tactics that take months to play out, terrorists appear willing to engage in interminably protracted conflicts. Blowing up buildings and buses can be a powerful tactic, but there's also a place for low-cost, high-impact operations that take time to unfold. A 1986 report from the Stockholm International Peace Research Institute (SIPRI) provided early glimmers of this realization in noting that, while biological agents had not be considered as effective battlefield weapons because they act more slowly than chemical agents, microbes were judged by military strategists to be suitable weapons for damaging whole populations and agricultural assets.[1]

As for imprecision, what is a problem for modern militaries may well be a virtue for postmodern terrorists. The SIPRI analysis dismissed conventional military uses of living organisms by reasoning that pathogens were generally uncontrollable and particularly unpredictable when carried by insect vectors.

These limitations echoed the conclusion drawn by Fred C. Iklé, director of the U.S. Arms Control and Disarmament Agency, in the mid-1970s: "The military utility of these [biological] weapons is dubious at best, the effects are unpredictable and potentially uncontrollable."[2] Earlier analysts even argued that nations were *unlikely* to employ insects or other organisms because such weapons would yield high rates of civilian casualties, disrupt vital ecological processes on a large scale, and cause the greatest harm to children, sick people, and older adults. What were once moral disadvantages in conventional warfare have become the very qualities that are sought in today's asymmetrical conflicts. But even the notion that rational analysis can reveal the tactics most likely to be used by an enemy has become problematic in the modern world.

The military historian Alastair Hay has warned that a credible assessment of the vulnerability of the United States to attack cannot rely on our past experience with military tactics or geopolitics.[3] Some modern insurgents use unconventional tactics to seek the conventional goal of land, as in Chechnya, Northern Ireland, the West Bank, or the Basque homeland. However, Hay also contends that contemporary enemies may have unusual objectives that preclude western nations from setting aside possibilities, "even if the reasoning behind the attack appears irrational." An act of dubious military value might be favored by the jihadist who seeks to conquer cultures, control minds, and capture souls rather than possess land. And what better weapon to attract attention, disrupt society, or instill fear than insects, the organisms that are consistently one of the frontrunners when pollsters ask what generates the greatest anxiety in people (ironically, according to some studies, more than terrorism itself)?

The vulnerability of the United States and other western nations to terrorist attacks using insects is evident from several incidents in recent years. The most compelling cases concern newly arrived organisms that were not—at least insofar as government officials either know or are willing to admit—introduced by enemy agents. But there is no reason that a moderately educated, reasonably motivated, minimally funded terrorist could not have initiated these outbreaks.

If terrorists were to conduct an entomological attack on an industrialized country, perhaps the greatest impact could be had by using insects to spread disease. Consider what Eric Croddy, one of foremost experts in chemical and biological warfare in the United States, recently listed as the qualities of diseases suitable for weaponization.[4] The pathogen should infect the victim in small doses (as with a single bite of a vector); cause acute and severe illness soon after infection (as in many insect-borne diseases); remain potent during production, storage, and handling (as with infected vectors); and survive environmental stresses during dissemination (as within a vector). So what might a terrorist attack with an insect-vectored disease look like? In 1999, the answer was provided in vivid detail.

On August 23, New York City health officials received a call that set into motion a series of events that graphically demonstrated the incapacity of the wealthiest nation on earth to stop an insect-borne disease.[5] Deborah Asnis, an infectious disease specialist at Flushing Hospital Medical Center in Queens, reported that she was attempting to diagnose and treat two elderly patients with a mysterious neurological illness. Their fever, confusion, and case histories pointed to mosquito-borne encephalitis, although the victims' severe muscle weakness was rather unusual in such illnesses. Two more patients would exhibit similar symptoms by the end of the week.

Asnis sent tissue samples from her patients to the State Department of Health and called the Centers for Disease Control (CDC, now the Centers for Disease Control and Prevention)—America's frontline defense against new and exotic illnesses. The experts suspected St. Louis encephalitis, and sophisticated tests at the CDC laboratory in Ft. Collins, Colorado, yielded positive results for this disease. Delighted to have an answer, the New York City Department of Health and the CDC announced their finding on September 3. And the Big Apple responded with a flourish.

New York was the best prepared city in the nation to deal with a disease outbreak.[6] After the 1993 bombing of the World Trade Center, Mayor Rudolph Giuliani put counterterrorism at the top of his agenda. New York had a cadre of emergency-response officials poised to handle attacks involving unconventional weapons—skills that readily transferred to an outbreak of encephalitis. Jerome Hauer, New York City's head of emergency management, initiated a $6 million insecticide-spraying campaign to quash the mosquitoes and cornered the national market on insect repellent. An army of 500 city employees delivered nearly a half-million cans of repellent to neighborhood firehouses and police precincts for distribution to the citizenry. And a quarter-million brochures printed in eight languages provided residents with vital information about St. Louis encephalitis. There was only one problem—nobody actually had this disease.

While public health officials were worrying about sick people, Tracey McNamara was fretting about dead birds. As the head of the Bronx Zoo's pathology department, she was trying desperately to figure out what was killing her feathered friends. McNamara suspected that eastern equine encephalitis was to blame, but this disease also should have wiped out the emus, which were doing just fine. The perplexed pathologist sent tissues samples to the USDA National Veterinary Services Laboratory in Ames, Iowa. Then she contacted the CDC to call their attention to the possible link between dying birds and sick people.

The CDC laboratory in Ft. Collins, Colorado, was buried under an avalanche of samples from New York City hospitals, and the scientists had no interest in complicating their tidy explanation of the emerging epidemic. So the bird and human investigations proceeded independently. When McNamara learned that the tissue samples she'd sent to the USDA failed to match any pathogen in that laboratory's expansive database, she bypassed the recalcitrant CDC.

The U.S. Army Medical Research Institute of Infectious Diseases (USAMRIID)—the legacy of the military's biological warfare program at Fort Detrick—did not normally involve itself in civilian health matters. But McNamara had a personal connection to a pathologist working at the military facility, who convinced his superiors to make an exception. Within two days, USAMRIID scientists had ruled out St. Louis encephalitis as the culprit.

With this news, the CDC had no choice but to join the army back at the drawing board. And on September 27, the two laboratories announced that a "West Nile–like virus" had been found in the tissue samples. Birds and people were dying from a disease that had never been seen in the Western Hemisphere. Those responsible for biodefense were stunned.

The nation's dress rehearsal for a terrorist attack had been woefully inadequate.[7] The public health system had taken 35 days to identify a new disease, and only a stroke of good luck allowed the initial misdiagnosis to trigger an appropriate response—insecticides against the mosquitoes and repellents for the people were the right tools for either St. Louis encephalitis or West Nile virus. One might contend that had New York maintained a more aggressive mosquito-abatement program as part of normal operations, the outbreak might have been avoided, but the city's capacity to respond to a medical (and entomological) emergency arguably made up for the deficiency. The CDC had failed to live up to its motto of "expect the unexpected." The initial screening involved only a half-dozen possible viruses, and there was no consideration that wildlife disease specialists could provide vital clues in solving a human disease mystery.

Nor was the bureaucracy up to the task. Daily conference calls among officials dragged on for hours, so that agencies spent precious time confabulating rather than acting. Biodefense was an "orphan mission" in which dozens of city, state, and federal agencies had say-so but nobody had responsibility. The problem was exacerbated by jurisdictional disputes, which often had to be resolved before anything could be done. But there was little time for agency mea culpas—a city's outbreak was developing into a nation's epidemic.

Despite public health officials having taken impressively, if somewhat fortuitously, swift action to suppress a mosquito-borne disease, West Nile virus moved rapidly beyond the city limits of New York.[8] By the end of 1999, the disease had been spread, primarily by infected birds, to four neighboring states. In 2000, cases were reported in 12 states, then 27 states, and by 2002 the disease afflicted people in 44 states. Although fewer than 100 human cases and ten deaths were reported in each of the first three years of the epidemic, this number jumped to 4,156 cases and 284 deaths in 2002. The next year, there were 9,682 cases while the death rate was unchanged. Since that time, morbidity and mortality have declined as people have acquired immunity via low-grade infections, animals capable of serving as reservoirs for the disease have been vaccinated or wiped out, and dry summers in the West have put a damper on mosquitoes. There is precious little evidence that much, if any, credit for this reprieve can be attributed to public health or pest-management programs.

The view among experts in biological warfare that insect vectors would not be effective in disseminating disease was dramatically dispelled. For years, conventional wisdom maintained that biting insects would be too unreliable in locating human hosts, that industrial nations had pest-management systems capable of exterminating any such threat, and that insect-borne diseases were restricted to tropical countries with poor public hygiene. A new disease, carried by indigenous insects, had spread like wildfire across the United States. The public has to wonder whether the government would have been any more capable of suppressing a bioterrorist attack. For that matter, can we be sure that West Nile virus arrived accidentally?

The U.S. intelligence community seriously considered the possibility that this mosquito-borne disease had been loosed upon the American public by the Iraqis.[9] And there were reasons for suspicion. In April 1998, the British tabloid *Daily Mail* published a chilling excerpt from a book entitled *In the Shadow of Saddam*. The author, Mikhael Ramadan (likely a pseudonym, as nobody has been able to contact him), purportedly served as one of Hussein's doubles. His story is a bizarre account of a madman's plan:

In 1997, on almost the last occasion we met, Saddam summoned me to his study. Seldom had I seen him so elated. Unlocking the top right-hand drawer of his desk, he produced a bulky, leather-bound dossier and read extracts from it. . . . The dossier holds details of his ultimate weapon, developed in secret laboratories outside Iraq. . . . Free of UN inspection, the laboratories would develop the SV1417 strain of the West Nile virus—capable of destroying 97 pc [per cent] of all life in an urban environment. . . . He said SV1417 was to be "operationally tested" on a Third World population centre. . . . The target had been selected, Saddam said, "but that is not for your innocent ears."[10]

This sounds rather batty, but not completely absurd. Skeptics point out that no strain of West Nile virus has been designated SV1417, but this may well have referred to a code particular to the Iraqi biological weapons program.[11] Furthermore, New York City is not a Third World center, but Hussein was surely capable of changing his mind. And, of course, the fatality rate of West Nile virus is not 97 percent, but Iraqi scientists would have wanted to please their irascible leader.

Various experts, including Ely Karmon, senior researcher at the International Policy Institute for Counter-Terrorism in Herzliya, Israel, have pooh-poohed the notion that Hussein would have used West Nile virus: "There are victims but West Nile isn't one of the biological agents considered to be a warfare or terrorist agent. . . . I don't know what Iraq would achieve from spreading a relatively mild sickness that can be treated. . . . [Hussein] would want to do something more major."[12] If Ramadan's account is accurate, then Hussein believed that he was going to accomplish something spectacular. Moreover, Karmon is mistaken about West Nile virus not having been considered as a biological warfare agent.

Ken Alibek, the former deputy chief of research for the USSR's Biopreparat, reported that the Soviet biological warfare program had evaluated West Nile virus because of its potential for mosquito transmission in cities—and the Soviets shared their research with the Iraqis. And members of the U.S. military took seriously the possibility that West Nile virus was used as a simulant by one of America's enemies to assess the nation's vulnerability to insect-borne diseases in preparation for a far more devastating attack.[13] But not only are the skeptics' concerns less than reassuring, there are several pieces of circumstantial evidence that make Ramadan's extraordinary claims eerily plausible.

Iraqi scientists had the capability to produce West Nile virus.[14] Isolating a virus prior to production can be difficult, but the United States made it easy for foreign researchers by providing pathogens for medical studies. And in 1985, the CDC received a request from Iraq and dutifully supplied samples of West Nile virus. Likewise, building the research facilities to culture viruses is a pretty sophisticated project. However, the French stepped in to help in the 1980s. The pharmaceutical giant Rhône-Poulenc constructed a foot-and-mouth-disease vaccine plant at Al Manal and trained the locals to operate the facility. During the Gulf War in 1990–1991, the Iraqis converted a section of the factory into a production facility for botulism toxin as part of their biological and chemical warfare program. In 1992, the United Nations razed that portion of the plant but left intact the laboratories used for virus research.

A spate of West Nile virus outbreaks in various nations furthered suspicions that something unnatural was afoot.[15] While New York was battling its outbreak, the disease struck Volgograd and Rostov, where 600 Russians were sickened and at least 32 died. The next year, simultaneous outbreaks of West Nile virus irrupted in Israel and Saudi Arabia, two of Hussein's top enemies after their duplicity in the Gulf War. The strain of West Nile virus in Israel was apparently the same as that which was sweeping across the northeastern United States. But even if we allow that Hussein's military had the pathogen, did the Iraqis have the capacity to trigger outbreaks in New York City and these other locales?

Experts have been unable to trace how West Nile virus made its way to the United States.[16] Some scientists speculate that an infected mosquito or bird might have carried the disease. These creatures could have been stowaways on a ship or plane, or they could have been planted—and nobody would ever know the difference. The CDC has argued that it would require a large number of mosquitoes to trigger an outbreak, which seems unlikely to happen by accident. Other scientists contend that a sick person could have arrived at JFK Airport and then been bitten by a mosquito who spread the virus to local birds. But, again, whether such an individual arrived by accident or was sent by Hussein can't be determined. George W. Bush's secretary of the navy, Richard Danzig, who played a leading role in encouraging the government to prepare for bioterrorism, summarized the problem: "Even if you suspect biological terrorism, it's hard to prove. It's equally hard to disprove."[17] But many officials have attempted to deny this essential uncertainty.

In October 1999, the CIA rushed to dismiss the possibility that West Nile virus had arrived via an act of bioterrorism. The agency assured a worried public that "a thorough review of the Iraqi biological weapons program found no evidence that the Iraqis had experimented with West Nile virus at any of the laboratories investigated."[18] Of course, absence of evidence is not evidence of absence, and Hussein might have been telling Ramadan the truth when he said that the virus was being produced in laboratories outside of Iraq.

The CDC echoed the intelligence community's claim that the disease was a natural event, and they likewise presented no data in support of their contention. For that matter, neither agency even indicated what direct, empirical evidence *could* be used in this regard. But the difficulty in sorting out the facts may run deeper than the existence of scientific information, for some bioterrorism analysts contend that the government has good reasons for providing the American public with less than a full account of disease outbreaks. Jason Pate and Gavin Cameron argued in a study published by the prestigious Belfer Center for Science and International Affairs at Harvard's John F. Kennedy School of Government:

> Differentiating between naturally occurring outbreaks of disease and those caused purposefully by subnational entities is extremely difficult and may be impossible if no group or individual comes forward to claim responsibility for the outbreak. . . . If it were discovered that a particular outbreak had been intentionally caused, would it be in the public's best interests to make that information widely available? Doing so could create panic and incidentally assist the goals of the perpetrator.[19]

Based on what we know, it is extremely improbable that West Nile virus was clandestinely loosed on the United States. However, the phenomenal rate at which this disease blanketed the country surely drew the attention of those seeking to harm western nations. Given this lesson, the question becomes, "What insect-borne diseases might yet be used for bioterrorism?"

In 1983, SIPRI published a meticulous analysis of the most likely pathogens to be developed as biological weapons.[20] Of the 22 prime candidates, half were arthropod-borne viruses. A similar study in 2000 by the World Organization for Animal Health generated a watch-and-worry roster of livestock diseases, and 6 of the 15 A-list diseases were carried by insects. One disease appearing on both agencies' databases is currently sending chills up the collective spine of U.S. government agencies tasked with protecting humans and agriculture. The virus causing Rift Valley fever might spread as readily as West Nile virus—and the former pathogen makes its African cousin look like a head cold.

After unusually heavy rains in 1930 and 1931, veterinary officers in Kenya began to notice extremely troubling symptoms among sheep in the Rift Valley.[21] Within months, thousands of animals were aborting their fetuses, and many of the newborns were dying. The malady started with fever and malaise in the lambs, quickly followed by horrendous bouts of bloody diarrhea, and death within a day or two. These were the first cases of what is now known as Rift Valley fever, a disease that soon afflicted other livestock as well. In cattle, the virus almost invariably caused abortions, and calves suffered 10 to 70 percent mortality; goats, camels, and buffalo were also susceptible.

The pathogen spread within a herd by aerosolized saliva and body fluids, but scientists realized that these forms of transmission could not explain the fast-moving epidemic. The virus' expansion across the countryside would be a slow ramble, rather than a mad dash, if it relied solely on direct transmission. Soon after the disease ravaged Kenya, veterinary scientists grasped the importance of rainfall as a precursor to the epidemic. For a continent in which insect-borne diseases had long been a source of misery, the connections among moisture, mosquitoes, and mortality were all too familiar.

During dry periods, mosquito eggs can survive in a state of suspended animation for months or years, with the Rift Valley fever virus biding its time along with the insect embryo. When the rains return, the infected eggs hatch, the larvae and pupae rapidly mature, and within days a swarm of hungry, disease-carrying adults begins to search for blood. Once the virus infects a mammal, the pathogen replicates at a phenomenal rate. Other mosquitoes then feed on the sick animal—indeed, some species preferentially attack feverish hosts. The infected females pass the virus through their eggs into the next generation and the cycle is complete. With the enormous reproductive potential of the virus and its vectors, Rift Valley fever spreads like living wildfire.

Thanks to a relatively dry period, Kenya was spared another major outbreak until wet conditions prevailed in 1950. Within two years, half a million ewes aborted their fetuses and 100,000 lambs and sheep succumbed to the disease.[22] Over the next quarter-century, the disease spread throughout sub-Saharan Africa. The toll on the people of this protein-deficient continent was terrible, but at least the disease was limited to livestock. Or so the experts believed.

In 1977, Rift Valley fever crossed the Sahara, perhaps transported by mosquitoes northward through the irrigated farmlands of Sudan. Whatever the mode of arrival, the disease hammered Egyptian agriculture, sickening a quarter to a half of all sheep and cattle. If the expansion of the disease into northern

Africa weren't bad enough, next came reports of people falling ill with terrible symptoms.

Before the 1977 outbreak, there had been reports of humans suffering from a flulike condition in areas hit by Rift Valley fever. The sick typically recovered within a week. But what the Egyptian medical authorities saw was beyond anyone's experience. Patients were complaining of weakness, back pain, and dizziness. Some victims became anorexic, while others exhibited extreme sensitivity to light, as the virus invaded the retina. Physicians worried that a lethal progression was underway. The retina connects to the optic nerve, meaning that the pathogen had found a bridge into the brain. The doctors' worst fears were realized as patients began to exhibit the classic symptoms of encephalitis: headache, seizure, and coma as their brains became inflamed and then shut down. Others suffered a more grisly fate as the virus infected the liver and other organs, followed by internal hemorrhaging, shock, and death. By the end of the epidemic, 200,000 people had fallen ill, with perhaps 1 percent of the victims becoming blind. In some communities, a third of the populace was afflicted. A virus that previously had not been considered lethal to humans had killed 598 people.[23]

The outbreak might have been worse, but Fort Detrick's USAMRIID just happened to have 300,000 doses of vaccine for Rift Valley fever on hand. Although the vaccine provided only partial protection, it played a significant role in suppressing the disease among livestock. The stockpile of vaccine left no doubt that the U.S. Army worried about Rift Valley fever being used in a biological attack, at least against the nation's livestock. Whether they knew of its potential to kill people is not entirely clear, but medical experts now peg the expected human mortality during an outbreak at 1 percent (about ten times that of West Nile virus), with 10 percent being possible under some conditions. Among the survivors, as many as one in ten will suffer some permanent loss of vision. The military's concern was validated by subsequent outbreaks in Africa.

The first epidemic of Rift Valley fever in western African was reported in 1987, after construction of the Diama Dam at the mouth of the Senegal River unwittingly created the ideal habitat for mosquitoes. The outbreak wiped out entire herds and killed 200 people. This story was repeated when the Egyptians provided breeding grounds for mosquitoes with the opening of the Aswan Dam. Then, rather than a dam project, an El Niño weather pattern in 1997 created immense tracts of flooded land across Kenya, Somalia, and Tanzania. And this time, along with the tremendous losses of domestic animals and 300 human deaths, devastating economic sanctions were imposed. An embargo by the countries of the Middle East banned exports of meat from eastern Africa for a year and a half. But this effort to keep the disease restricted to the African continent soon failed.

In September 2000, Rift Valley fever struck the Tihama plain of Saudi Arabia and Yemen.[24] Tens of thousands of sheep and goats aborted their fetuses, and 855 people suffered from what medical authorities termed "severe cases" of the disease, with one in four dying (see Figure 23.1). A year after this outbreak of Rift Valley fever, exploding planes and collapsing buildings changed how Americans perceived the world. As the pathology of terrorism spread across the world, the potential of insect-borne disease to devastate agricultural production and human health could not be ignored by national security agencies—or, presumably, terrorists.

How difficult would it be to introduce Rift Valley fever to the United States or other western nations? The man with the answer to this question is Colonel Charles Bailey, the director of the U.S. National Center for Biodefense. Nobody in the world is better prepared to assess the matter, given that Bailey earned a Ph.D. in medical entomology, rose from research scientist to commander of U.S. Army Medical Research Institute of Infectious Diseases

Figure 23.1. A pen of goats in Saudi Arabia during the 2000 outbreak of Rift Valley fever. The disease caused tens of thousands of sheep and goats to abort their fetuses, adding the hunger and poverty to villages stricken by the illness. More than 200 people contracted lethal cases of the disease, either via mosquito bites or through contact with the blood or other bodily fluids of infected animals. (Photo by Abigail Tumpey, courtesy of CDC)

at Fort Detrick, served with the Defense Intelligence Agency, and published more than two dozen papers on Rift Valley fever in scientific journals.[25]

Bailey contends that there are many ways that a would-be terrorist could smuggle the pathogen into a country. The virus is transmissible by aerosols, so a feverish passenger could serve as the carrier, but this tactic would not be terribly efficient. This might be how West Nile virus arrived, but Rift Valley fever has a biological feature that would make it far simpler to introduce: The virus can survive in the dried-out eggs of mosquitoes. According to Bailey:

> The easiest [method of introduction] by far—and there's no way authorities would ever detect it—is to simply go to an endemic area during an outbreak, collect floodwater *Aedes* [mosquitoes] from prime habitats, let them feed on a viremic [infected] animal, collect the eggs, put them on filter paper, let them dry, put them in your shirt pocket, come into the country, go to a suitable habitat, drop the filter paper into the water, and walk away.[26]

Bailey is not alone in his assessment of the ease with which a bioterrorist could introduce the virus.[27]

According to Geoff Letchworth—who is a doctor of veterinary medicine and has a Ph.D. from Cornell University, researched insect-borne diseases at Plum Island (the nation's highest biosecurity laboratory, just a mile and a half northeast of Long Island), and recently retired as the research leader of the USDA's Arthropod Borne Animal Diseases Research Laboratory—the challenge for the virus under natural conditions is to overcome several unlikely steps that put the pathogen into intimate contact with its vector and hosts. But Letchworth contends that humans can readily circumvent these limitations:

> The terrorist effectively skips the early steps and goes right to the feed lots in the southern U.S. where exposure to mosquitoes is very high, directly inoculates animals and gets it going. I can't believe that you'd be unsuccessful. When Rift Valley fever starts spreading around an area, the attack rates are amazing—20 to 50% of people and animals become infected. The amount of virus that comes out of animals is just amazing. Once it gets cooking it blasts through the whole population.[28]

Assuming that a terrorist succeeded in initiating a localized outbreak, could the emerging epidemic be contained? Like West Nile virus, the pathogen responsible for Rift Valley fever would find a welcoming committee of native vectors. Nearly every corner of North America harbors a species capable of transmitting the disease. Bailey contends that the staggering reproductive capacity of the pathogen allows it to hijack almost any kind of mosquito. Reaching concentrations in the host's blood that are typically 10,000 times greater than other viruses, the Rift Valley fever virus is ingested by the vector, overwhelms the natural barriers that many mosquitoes have to infection, and then sets up shop in the insect's salivary glands.

Bailey, dismayed with the Department of Agriculture's lethargic and underfunded biodefense program, predicts that the country would be unable to keep the disease from becoming established. Extrapolating from the case of West Nile virus, Rift Valley fever would likely race from coast to coast in five years. Comparing the diseases can be useful, but the tendency to equate West Nile virus, a relatively mild disease, with Rift Valley fever could well cause government agencies to egregiously underestimate the risk.

According to Corrie Brown, who headed the pathology section of the biocontainment facility on Plum Island, Rift Valley fever "would make West Nile look like a hiccup."[29] She contends that the disease would shut down U.S. beef exports—a $3 billion loss to the economy. And its impact on people would be painfully evident.

"To say that Rift Valley fever makes West Nile look like a hiccup is an understatement," according to Mike Turell, a USAMRIID specialist.[30] He notes that while most people infected with West Nile virus don't even know they have it, 90 percent of those contracting Rift Valley fever are stopped in their tracks by debilitating symptoms. Most survive, but the effect on the nation's health and sense of security would be devastating. Nevertheless, little is being done to prevent or prepare for the arrival of this disease.

According to C. J. Peters, director of the Center for Biodefense at the University of Texas Medical Branch in Galveston, the United States should be aggressively pursuing three lines of defense, none of which is adequately funded or fully under way.[31] He argues that the government should be funding the development of an animal vaccine, a human vaccine, and reliable diagnostic tools. What Peters does not include others put front and center: mosquito control.

Based on the fragmented, uncoordinated, and mistimed mosquito programs that typified responses to the outbreak of West Nile virus, the prognosis for curtailing Rift Valley fever by suppressing its vectors is poor. While many communities improved their pest-management systems, the current infrastructure is far from sufficient. And with the decline of West Nile virus, control programs are already being dismantled and cannibalized as people, equipment, and funds are directed toward more pressing matters. But this appears to be a very dangerous gamble. Letchworth put the prognosis in starkly unambiguous terms: "If we get Rift Valley fever, we'll forget that West Nile virus ever happened. And taking the long-term view, getting Rift Valley fever in the United States is a matter of when, not if."

Given the expertise that was brought to bear during the National Research Council's study of agricultural terrorism, it is no surprise to find that Appendix E included a scenario exploring the possibility of using Rift Valley fever as a weapon. After conducting a chillingly detailed, hypothetical case study, the scientists concluded that a terrorist would need only a modicum of scientific understanding to select viable times and places for a release of the Rift Valley fever virus. Although Hurricane Katrina struck Louisiana and Mississippi in 2005, which was after the National Research Council's report was published, we can readily apply their analysis to this real-world situation. After the natural disaster, a prospective terrorist would have had no difficulty finding promising mosquito habitat, a collapsed medical system, a shortage of public health providers, a vulnerable human population, and a confused governmental infrastructure.

If there was any doubt as to the vulnerability of this region to mosquito-borne disease, consider that a year after Hurricanes Katrina and Rita hit, the number of West Nile virus cases jumped by 24 percent in Alabama, Louisiana, Mississippi, and Texas.[32] A parish in Louisiana reported mosquito populations 800 percent above those prior to the hurricanes.[33] With more that 40 percent of the mosquitoes testing positive for West Nile virus,[34] there was a silver lining: Rift Valley fever is not yet here.

Having played out a scenario in which Rift Valley fever is introduced to the United States under conditions less opportune than those following the devastation of the Gulf coast, the National Research Council summarized its findings. They concluded that this insect-borne disease is a tempting biological weapon; that the infrastructure of surveillance, diagnosis, communication, and response is inadequate to respond effectively; that the initial outbreak would cause major damage to the nation's economy and human health; that once the disease reached the wildlife population, the virus would be permanently ensconced in this natural reservoir (meaning an unending cost of enhanced pest management and public health programs to prevent major outbreaks); and that—as any good scientific committee is compelled to mention—there should be more research to understand the disease if we hope to prevent or control its spread. In other words, nothing stands between the American public and an epidemic of Rift Valley fever other than a thin membrane of the country's intelligence agencies, the shifting motives and capacities of terrorist organizations, the fickleness of the weather, and the hunger of North American mosquitoes. Fortunately, other insect-borne diseases have a few more obstacles to their being converted into weapons by terrorists.

The most important limitation to a terrorist's launching an insect-borne disease is the mismatch between the exotic pathogen and the native vectors. In many instances, the appropriate six- (or eight-) legged carrier of the disease is not found in Europe or North America, so both the pathogen and its vector would need to be introduced. This would seem to be a rather serious logistical problem, although the hurdle may be lower than one might suppose. The United States and other industrial nations have proved quite susceptible to invasion by biting insects. A sort of blood-filled welcome mat seems to greet the visiting vector.

In August 1985, a mosquito never before seen in the United States was found breeding in Houston, Texas.[35] The Asian tiger mosquito (*Aedes albopictus*), named for the impressive pattern of black-and-white striping on its body, was flourishing in the warm stagnant waters found within discarded tires. In fact, the best guess is that the insect arrived as a stowaway on a shipment of used truck tires from Asia, destined for recapping in the United States. Given the 290 million scrap tires in the United States (one for nearly every man, woman, and child in the country), the invader's future was incredibly bright. Within two years, the mosquito had invaded 17 states. So much for one of the most scientifically and technologically advanced nations on earth being able to contain the spread of a blood-feeding insect. But did the lack of control merely reflect a lack of concern among medical experts? Not hardly. This species was a known vector of dengue and was suspected to be a carrier of Western, Eastern, and LaCrosse encephalitis, yellow fever, and dog heartworm.

Developing and deploying a pathogen-vector weapon system might be possible in some cases, but what are a bioterrorist's options when finding, breeding, infecting, and releasing an exotic insect is impractical? In entomological terms, the problem is vector competence. Without an evolutionary association, microbes are not adapted to the unfamiliar species in a target country, so the pathogen is unable to replicate in the insect's tissues and reach levels that allow efficient transmission. Of course, the incompetent vector might carry a few microbes on its mouthparts from the blood of an infected host and function as a flying "dirty syringe," but this is generally an inefficient means of spreading disease. What's needed is a change in the vector's physiology to meet the pathogen's needs. And what evolution might take millennia to develop, scientists can now create in a matter of months.

The dream of engineering insects to do our bidding extends back more than half a century. On the heels of the Cuban missile crisis, President Kennedy ordered an acceleration of biological warfare research, expecting the military to make use of the most advanced technologies available at the time. In response to his commander-in-chief's clarion call, Major General Marshall Stubbs, head of the U.S. Army Chemical Corps, told Congress that research was underway to develop insect strains more resistant to cold and insecticides—presumably to extend entomological weapon systems into the northern reaches of the Soviet Union.[36] In 1962, these new strains were almost certainly being developed with tried-and-true artificial selection methods in which those insects that survived exposure to increasingly cold conditions in each generation were mated to boost the frequency of the relevant genes and the cold hardiness of the insects.

A decade later, scientists made the initial breakthroughs that allowed them to imagine a world in which organisms would be genetically tailored to our specifications. In biological factories, stainless steel vats of microbes would produce medicines, nutrients—and toxins. By the 1980s, not only were military scientists talking about making pathogens more deadly and easier to produce, they were beginning to think in terms of creating new insect-microbe associations. SIPRI's experts predicted the development of both more efficient and totally new insect-vector systems for purposes of waging biological warfare.[37] And today, we know that these pioneering scientists were not simply dreaming—nor were the critics of biotechnology merely having nightmares.

A series of breakthroughs beginning in the late 1990s have led to the successful genetic engineering of mosquitoes.[38] With methods for inserting genetic material into vectors, the next step was to find and insert a gene that rendered the vector incompetent. Now that we have this creature, the research will culminate with the release of these insects into the wild—where they will, we hope, outcompete the natives. But, of course, the methods that allow genetic engineers to produce a strain of ineffective vectors could well be used by others seeking the opposite result.

According to Letchworth there are no conceptual obstacles to creating some terrifying weapons. He maintains that "there's no reason why a mosquito could not be genetically engineered to transmit, even perhaps produce, the HIV virus."[39] And if there were concerns that the insect would not effectively compete with its naturally occurring brethren, Letchworth has the answer: "To allow the new vector to take over, design in some insecticide resistance genes." But to genetically engineer a match between a pathogen and a potential vector, the scientist is not limited to tweaking the insect's DNA.

In terms of a biological weapon system, it would be more promising to alter the genetics of the microbe so that it "fits" into the physiological system of an insect carrier that is already abundant within the enemy's homeland. Although such technological advances are beyond the reach of most terrorists, they are certainly well within the capacity of many nations. And scientific expertise is for sale on the world market. Concerns of blasphemy notwithstanding, playing God may soon be within reach of well-funded terrorist organizations. And the costs are not as high as one might imagine. Letchworth estimates that a facility to tailor-make viruses could be set up for less than $1 million, and the laboratory could be undetectable.

In 1999, *New Scientist* published an article on biological warfare that portrayed a hellish alliance of entomology, virology, and genetic engineering.[40] Imagine a field of corn infested with whiteflies (family Aleyrodidae). These waxy relatives of aphids make the plants appear as if they have dandruff. The farmer's yield is less than stellar, but the insects have not seriously damaged the crop. So the harvest is sent to a brewery, where the corn is used to make beer. The diabolic plot begins to crystallize as hundreds of people become sick. Victims suffer abdominal cramps and vomiting, others have problems breathing, and some die of respiratory arrest.

Faced with public panic, the government tries desperately to track down the cause of the illness and finally identifies botulism toxin. This substance is capable of killing at phenomenally low doses: an amount equal to the weight of a single kernel of popcorn is sufficient to kill 2,000 people. But a critical question looms: What is the source of the toxin? Eventually, researchers trace the origin of the poisoning to the tiny insects that infested the cornfield.

Whiteflies feed by piercing the plant tissues with needlelike mouthparts and in the process transmit microbes in their saliva. A terrorist organization funded rogue scientists to genetically modify a normally harmless plant virus to synthesize botulinum toxin, and the insect dutifully infected fields of corn. In the weeks it took to crack the case, the insects have reproduced and spread at a phenomenal rate. An initial population of six females has the potential to produce 125 billion offspring in a year.

Wave after wave of food scares sweep the nation. As ever more corn is infected, the Department of Agriculture pours billions into an emergency control program. Midwestern skies fill with smoke as enormous tracts of farmland are torched. The credibility of the government is in shambles. The terrorists declare victory.

Such a disastrous chain of events presumes an enemy with the capacity to genetically engineer organisms, or at least to gain access to the requisite methods through a second party. While the necessary training and biotechnology are becoming increasingly available, western scientists are not forfeiting the entomological arms race. Instead, they are turning the tables, enlisting insects for homeland defense. The remarkable abilities of these creatures are being harnessed to protect humans from the dangers posed by modern enemies as well as the buried legacies of past conflicts.

24

SIX-LEGGED GUARDIAN ANGELS

Following the attacks on the World Trade Center and the Pentagon on September 11, 2001, the United States responded with a massive retaliation against Afghanistan, followed by an invasion of Iraq. Except for the names of the some of the weapon systems (such as the F/A-18 Hornet fighter aircraft, which were subject to counterattack by Stinger missiles), insects played no part in these offensive operations. Defense against terrorism, however, is another story.

To prevent future terrorist attacks, the United States developed a staggeringly complex system of technologies and bureaucracies, culminating in the creation of the Department of Homeland Security, with an annual operating budget of $40 billion. Systems for detecting explosives, chemical weapons, and biological agents became central to the government's effort. And scientists are coming to understand that no instrument is more finely attuned to the environment, and more keenly responsive to trace chemicals, than the sensory system of insects. It should not be surprising that the U.S. government is investing millions of dollars in exploiting the potential of insects as guardians of the country.[1] But this is not the first venture of this sort—nearly a half century ago, military scientists pursued a similar line of research that was also driven by a sense of urgency and fear.

In June 1977, a feeble, gray-haired former professor of medicine limped across the border into West Germany as East German secret police watched.[2] Adolf-Henning Frucht was a pawn in a complicated spy exchange, the political winds of the Cold War having finally blown in his favor. Ten years earlier, the professor had been on his way to attend a conference on scientific administration in Prague. Frucht had been surprised at being asked to represent his East German medical institute at the event, given that the devoted researcher had little interest in such matters.

Frucht's surprise transformed into clarity when he realized that he'd walked into a trap, its jaws formed by the two State Security officials who escorted the frail academic from the train. After months of being grilled by the East German secret police, he was tried for espionage. The life sentence that he received was arguably fair. Frucht was, in fact, a spy.

In the early 1960s, his country's Institute for Industrial Physiology had approached the professor with a tempting offer. He was asked to direct his research toward developing a new method of detecting airborne toxins. East Germany was under siege by the West, and it was considered every scientist's patriotic duty to contribute to the defense of his nation. They believed that capitalist armies were likely to use every imaginable method to crush their opponents, including nerve gases.

Frucht began to piece together a line of research with considerable promise. He knew that nerve gases had been spawned by the insecticide industry and that organophosphate weapons would poison insects as well as humans. What he needed, however, was an insect that could serve as a canary in a coal mine, that would overtly reveal its intoxication before levels of nerve gas became lethal to people. Of course, some insects sing, but cricket trills and cicada buzzes were not sufficiently reliable to serve as an early warning system. However, one insect was identified by Frucht as having a form of communication that was dependable and intimately linked to its neurophysiology: the firefly (family Lampyridae).

Frucht discovered that even minuscule doses of nerve gases impeded light production in fireflies. Indeed, the poison functioned like a dimmer switch: the insects' light diminished in proportion to the amount of organophosphate in the environment. Although he was delighted with his scientific breakthrough, his subsequent political discovery was devastating.

With his research moving decisively toward the development of a monitoring system, Frucht began to receive visits from high-ranking officials. His conversations soon made it clear that the firefly project was anything but a means of protecting his countrymen. Rather, the East German government sought out his device to protect its own troops from nerve gases that they would use during a massive offensive against western Europe. In Frucht's mind, he was being used in preparing to launch World War III.

Sickened by his government's duplicity, Frucht began passing secrets to the West in an effort to avoid an imbalance of power that would undermine the tenuous peace. Not only was Frucht's firefly project of interest to western intelligence agencies, but his research put him in the position of being acutely aware of top-secret developments in chemical warfare. His detector would be used to protect communist forces from a new V-agent, the formula of which he passed to his British and American contacts. But like Frucht, the East German secret police were masters at detection, and his treason was eventually discovered.

Rather than running a world-class research laboratory, the former professor of medicine spent his days in solitary confinement or in menial labor. The prison library provided the only refuge for the mind of the imprisoned scientist. After a decade of surviving through sheer dint of will, he was freed in a prisoner exchange and hobbled into West Germany. Although Adolf-Henning Frucht gained his freedom, the world still is not liberated from the weapons that dimmed the glow of his fireflies.

Modern militaries are fully cognizant that nerve gases may be part of enemy arsenals, and U.S. field training manuals continue to warn soldiers that when all's quiet on the front—when the crickets are silenced (and, although not mentioned, when the fireflies are darkened)—a chemical attack may be under way.[3] And given the ease of acquiring organophosphates, the battlefield may not be the only setting for their use in today's world.[4] Although nerve gases are near the top of the terrorists' wish list, there is a weapon of even greater concern in the post-9/11 world.

The anthrax attacks on the offices of U.S. senators and major media in the weeks after the fiery assaults on New York City and Washington, D.C., riveted the attention of the government. Developing methods to detect biological warfare agents skyrocketed to the top of national security priorities. While using fireflies to directly monitor poisonous chemicals was a clever application of insect neurology, discovering how these creatures could provide a phenomenally sensitive means of detecting microbial agents required a deep understanding of firefly physiology. Years earlier, the American space program laid the foundation for modern technologies that exploit the biochemical "fire" of these remarkable beetles (fireflies are not actually flies).

In the 1960s, scientists at the Goddard Space Flight Center became keenly interested in organisms surviving at the earth's margins as a means of gaining insight into the nature of life on other planets.[5] In the heady days of the space race with the Soviets, American researchers could already imagine sending unmanned missions to Mars and beyond. But if such explorations came to pass, how would we know if there was extraterrestrial life?

NASA needed an instrument to detect living organisms, whatever their form or function. Of course, nobody knew what sort of biochemistry aliens might possess, but a reasonable starting point would be a chemical that was universal to life on earth. While deoxyribonucleic acid (DNA) was a good candidate, this molecule was enormously complex and varied in its structure. There was, however, another chemical found in every living system: adenosine triphosphate (ATP). All organisms—from bacteria and fungi to plants and animals—use ATP to store and release metabolic energy. So what the researchers needed was a simple and reliable detector of ATP. Enter the firefly—or at least the chemical pathway found in its rear end.

The bioluminescence of fireflies is the result of a series of reactions within the light organ at the tip of the insect's abdomen. There are five ingredients that, when combined, yield the chartreuse glow. The key substances are two rather cleverly named chemicals. Luciferin is the substrate, the material that generates light. Luciferase is the enzyme, the chemical that transforms luciferin from its inactive state into the light-emitting form.[6] But for luciferase itself to become activated and capable of transforming luciferin, three other chemicals must be present: magnesium, oxygen, and ATP. So in a test tube containing just luciferin, luciferase, magnesium, and oxygen, nothing happens. Add even a minuscule trace of ATP and the liquid begins to glow, and the more ATP that is added, the brighter the light.

The scientists at Goddard Space Flight Center figured that if they could devise a mechanism to collect environmental samples, disrupt cellular membranes to release the contents, inject this gunk into a reactor vessel containing the four essential ingredients of beetle bioluminescence (excluding, of course, ATP), and then use a photovoltaic sensor to measure the production of light, they'd have a life detector (see Figure 24.1).

This became the basic design of the Firefly, a 1-pound device that could be launched into the upper reaches of Earth's atmosphere, or someday put aboard a rocket's payload to another planet. The automated processing of a sample took about two minutes and the reaction itself yielded a burst of light in less than a second. The early version of the Firefly required relatively large amounts of ATP—as much as would be found in a thousand microbes. By 1975, a tenfold increase in sensitivity was achieved, and today it is possible to detect the light emitted from a sample containing a single cell.[7]

Early on, the U.S. military saw the potential of the original Firefly as a biological warfare detector.[8] An aerosol of pathogens would yield elevated levels of ATP in atmospheric samples. Although the device would miss viruses (these wickedly simple agents have no metabolism of their own, so they lack ATP), all other biological weapons could be detected at very low

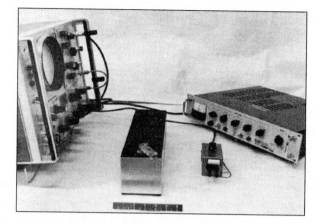

Figure 24.1. An early version of the Firefly from 1968, a device that utilized the bioluminescent chemistry of the insect for which it was named as the basis for detecting life—extraterrestrial organisms of interest to NASA or bacterial aerosols of interest to the military. The actual instrument with reaction chamber is in the center of the photograph, while an oscilloscope for signal readout is on the left and the power supply is on the right. (Courtesy of Spherix)

levels. Although the battlefield commander would not know exactly what pathogen had been released, the key to surviving an attack was more a matter of donning protective gear than of knowing the precise nature of the assailant. False alarms were also a possibility owing to natural pulses of airborne organisms, but these were a small price to pay compared to being surprised by wholesale germ warfare. It seems that the Soviet military concurred, as they independently developed a detector using the same principles as the Firefly.[9] But modern defense must take into account tactics much different from those of the Cold War era.

Although U.S. commanders worried that Saddam Hussein would use anthrax or other biological weapons during the Iraq War, the overall sense is that today's soldiers are unlikely to encounter a microbial mist in the course of battle. If pathogens are used by an enemy, the consensus seems to be that the target most likely will be the general public. What keeps defense planners awake at night is the realization that spreading germs within an unwary nation requires a single saboteur's access to the largely unprotected food processing and distribution system. But our nocturnal insect ally may hold the key to a good night's rest in the U.S. Department of Homeland Security.

Federal statutes that assign strict legal liability to every handler in the American food chain—from farm to table—have provided plenty of economic incentive for companies to devote considerable attention to safety issues. However, recent food poisoning incidents have revealed both the fallibility of the safeguards and the startling speed with which a contaminant can spread across the country. From a terrorist's perspective, the food distribution network is a nearly ideal target.

The traditional approach to monitoring for harmful organisms in the food industry has been to swab surfaces and then culture the samples in a nutrient medium. One problem with this method is that various microbes grow in different media, so no single assay detects every kind of pathogen. The other limitation is time. Growing the microbes takes days, and by the time a positive result is obtained the contaminated food may have been distributed throughout the nation. In the best of all worlds, a facility should be able to test its products instantaneously and continuously. And here's where firefly biochemistry has proved its mettle.

By adapting the instrument used by NASA to seek alien life, the Kikkoman Corporation has developed a device to rapidly and repeatedly monitor for the presence of ATP using the principles of firefly luminescence.[10] The concept is simple: there ought not to be any live organisms in food preparation areas, so life-free countertops, floors, and walls ensure consumer safety (of course, soy sauce is fermented, so there's plenty of ATP in the bottle). The problem is that we also eat the dead tissue of plants and animals, so background levels of ATP would normally swamp the presence of any living microbes in such foods. In an odd exchange of technological innovation, the solution to this problem emerged from NASA's further modification of Kikkoman's device.

The spacecrafts that were sent to explore Mars had to be built and launched under absolutely life-free conditions to avoid the possibility of introducing earthly microbes to another planet—a sort of reverse *Andromeda Strain*. And NASA's Jet Propulsion Laboratory found that the food industry's detector was nearly ideal for ensuring the sterility of the clean rooms where spacecraft were assembled.[11] The only problem was that if ATP were detected, it would make a big difference whether it came from dead or living organisms. The former were a matter of concern, but the latter were a potential disaster.

So microbiologists in the laboratory's Biotechnology and Planetary Protection Group developed a method for sorting the living from the dead. Prior to introducing a sample to the firefly cocktail and looking for the telltale glow, an enzyme is added to degrade ATP that has leaked from dead microbes. Only then are the cells broken apart to ensure that any ATP that makes it into the detector comes from a living organism.

In a further refinement serving the interests of both food safety and national security, researchers at New York's Rockefeller University are developing pathogen-specific enzymes that would rupture only targeted cells.[12] In this way, rapid and reliable tests for particular microbes, such as anthrax, are on the horizon. And with further engineering developments, some see a day in which firefly biochemistry becomes part of automated sensors attached to the water sources and air intakes of buildings.

Adolf-Henning Frucht's discovery that fireflies could warn people of nerve gases, just as canaries once warned coal miners of toxic vapors, has served as the foundation for a spectrum of technological devices to detect chemical and biological weapons. And the fundamental notion that evolution has produced phenomenally sensitive systems that can be adapted for military uses extends beyond the firefly. The capacity of dogs to locate odor sources, including explosives, is legendary. But another animal has an even more highly tuned sense of smell, a species whose capacity to detect infinitesimal traces of particular chemicals and whose potential for obedience training trumps the bloodhound. This other creature is man's best (six-legged) friend: the honey bee (*Apis mellifera*).

An effective guard has two essential attributes: vigilance and responsiveness. The sentry must remain alert throughout his watch and should decisively challenge an intruder. When the infiltrator is a chemical, rather than a human enemy, the same qualities are needed. However, having soldiers walking about while sniffing the air, and then shouting if they smell something amiss, would be absurd. Bees, on the other hand, make outstanding guards, as the U.S. Army found.

By the mid-1950s, Aberdeen Proving Ground in Maryland had become a deadly dump of military leftovers.[13] For the better part of two decades, the army had disposed of its unwanted chemical-warfare agents, unexploded munitions, and dregs from research and production facilities in the fields of this army garrison. Seepage had turned the pastures into wastelands, and the military became increasingly concerned with finding and remediating the most toxic areas. Sending out moon-suited soldiers to take environmental samples was one option, but this was dangerous, expensive, and inefficient. So the army recruited bees for the hazardous duty.

Bees are living dust mops. Their bodies are covered in hairs that, when magnified, look like split ends. This furry coat readily picks up an electrostatic charge, so the insects are like magnets for fine particles. And their capacity to collect contaminants from the environment does not end with their fuzz.

The workers are, well, busy as bees. Their search for pollen and nectar sources means that the insects crisscross an area of about ten square miles around the hive. The foraging bees maintain a remarkable metabolic rate, stoked by large amounts of oxygen. Per gram of body weight, a flying bee inhales air at about 50 times the rate of an exercising human. And when it's hot, bees drink copious amounts of water that they regurgitate into their nest and then fan with their wings to provide evaporative cooling. Put these activities together and a beehive becomes a veritable vacuum cleaner, sucking up and amplifying trace levels of contaminants in air, water, soil, and vegetation.

There's no doubt that bees are vigilant, but a good sentry is also responsive. And the military scientists were quick to discover a means by which these insects could reveal when they'd encountered toxins. In fact, bees perform their assigned duties with military precision. Their behavioral comings and goings are regimented as long as the insects are healthy. But when the workers are intoxicated—and bees are quite sensitive to a range of chemicals—the colony's behavior changes markedly. By moving hives to various areas of Aberdeen Proving Ground and placing infrared "bee counters" at the hive entrances to reveal changes in foraging activity, the military could ascertain whether a site contained toxins. Bees were such useful environmental samplers that researchers also developed automated methods to analyze pollen, wax, honey, and even the air within the hive for traces of toxic chemicals.

Currently, the Departments of Defense and Homeland Security are funding investigations of other insects as chemical samplers.[14] More than two dozen species of beetles (order Coleoptera), crickets (family Gryllidae) and moths (order Lepidoptera) are being studied for their ability to sweep various environments and collect contaminants. So far it appears that the honey bees are tough to match. Not only are they fanatical laborers, but they reliably return home after a day of work. Other insects are less cooperative and must be recovered using sticky papers, special lights, or baited traps. But using insects to wander through an area in the hope that they will accidentally encounter and inadvertently collect dangerous compounds is not terribly efficient—a bit like vacuuming your house at random rather than focusing on the high-traffic areas. At least for bees, such an approach fails to take advantage of one of these insect's most remarkable qualities. While a Hoover cannot be trained to find soiled areas of a carpet, bees can learn to seek out chemical "dirt."

The rather outlandish notion of turning bees into a full-fledged warning system has garnered the attention of America's most unconventional research organization, the Defense Advanced Research Projects Agency (DARPA). Boot camp for bees was just the sort of high-risk, high-return project for which the agency has become (in)famous. These are folks who came up with the idea of a mechanical elephant for the jungles of Vietnam, telepathy for psychic spying, and, most recently, the ignominious "terrorism futures market"—along with research behind the Internet, Global Positioning Systems, stealth technology, and the computer mouse. DARPA is pouring $60 million into 20 projects that attempt to exploit living systems for military applications. And if you're going to spend that kind of cash on developing a chemical detection system, you might as well pick a target that provides a lot of bang for the buck.

Few research and development projects seem more farfetched than using bees to detect land mines, but then not many ventures have greater potential to relieve human suffering than a low-cost, high-efficiency system for eliminating this bane of modern warfare. Not only are U.S. troops at risk from these devilish devices, but also vast swaths of valuable farmland have been rendered unusable. Officials estimate that there are 110 million unexploded land mines salted around the globe, or nearly one for every 50 people on earth. With 2,000 people killed and 20,000 maimed every month, the need is overwhelming to find these weapons for humanitarian, not to mention military, reasons. But land mines are solid objects, not vaporous substances, so how can a bee locate them beneath the soil?

In the early days of modern warfare, mines were encased in metal, so minesweepers were essentially metal detectors. But in recent times mines have become cheap mass-produced weapons, which means they're usually housed in plastic. Without metal to indicate their presence, the challenge is to detect faint traces of explosives that constantly leak from the mines. Such extraordinarily low-level vapor plumes have been exploited by scientists at Sandia National Laboratory in their development of handheld chemical "sniffers" to track down the buried booby traps. While these sophisticated instruments are effective, they are also expensive to manufacture, technically demanding to operate, and difficult to maintain. These are hardly the qualities that poor, undereducated, worn-torn countries seek when adopting new technologies. Machines are complicated and costly, but at least some animals are another story.

Dogs have proved to be highly sensitive and accurate mine detectors. The canines can be taught to associate the odor of explosives with a reward, so they become enthusiastic partners in the search for mines. Training and handling the dogs are not trivial demands, however. The animals are not cheap to maintain, and even with a very long leash (which seems advisable), the handler is in danger—not to mention man's best friend. Given these drawbacks, bees look pretty good. Moreover, the insects' antennae can beat the most sensitive doggy nose.

Honey bees can sniff out 2,3-DNT (the vaporous residue from military-grade TNT, which commonly serves as the explosive in mines) at unbelievably low levels.[15] An ounce of explosive buried in 40 pounds of sand emits a plume containing about 50 parts per trillion of 2,3-DNT in the air. This is the equivalent of a bathtub of chemical dumped into Lake Erie. And a honey bee can find this delicate aroma of explosive even when the simulated mine is 100 yards from the hive. What's more, teaching a worker bee to hunt down the leaky land mines is simpler than housebreaking a dog.

Evolution has shaped bees into quick studies. To survive, these insects must learn which plants are producing nectar amid a diverse and changing floral spectrum. Once a worker strikes pay dirt, she teaches her nestmates the location of the flowers through a remarkable "dance language." So conditioning these smart insects to associate a particular odor with food turned out to be remarkably simple.

The entomological training regimen consisted of moistening a sponge with a sugar solution to which just a hint of explosive had been added, then placing this chemical classroom where the workers were sure to find it.[16] The bees quickly learned that a whiff of TNT held the promise of a sweet snack. Moreover, the insect tutorials were phenomenally efficient—tens of thousands of bees could be trained in an hour. Once the insects made the connection, they became nearly infallible guides to hidden explosives.

The proving ground consisted of a simulated minefield, in which the scientists had planted and mapped the explosives. Given 60 minutes of searching, the trained bees detected 99 percent of the mines buried within 200 yards of their hive—a task that would take hours or days using sophisticated instruments. But even with the bees trained to selectively sweep a field for land mines, a significant hurdle remained before the detection system was operational.

The military had to know not only that there were mines somewhere in the area but they needed to pinpoint their location. After all, neither infantrymen nor villagers gain a whole lot from just knowing that there are land mines somewhere in the vicinity of a beehive. The first approach was to attach tiny, rice-size radio packs to the bees and follow their movements with an electronic tracking system. But this defeated the elegant simplicity and cost efficiency of using the bees. The breakthrough came when researchers stopped trying to outthink the bees.

Jerry Bromenshenk of the University of Montana, played an important role throughout the development of the mine-detection project.[17] As an entomologist who had studied bees for three decades, he understood how these insects could be transformed into entomological bloodhounds, leading humans directly to buried mines without the complexity of radio transmitters. The human overseer had merely to shed the role of a busybody scientist and take on the persona of a lazy foreman. Just watch the bees while they did all the work.

In the summer of 2003, Bromenshenk stood anxiously at the edge of a minefield where he'd placed ten colonies of his trained bees—without teensy backpacks. The movements of the bees would be tracked using human observers, video cameras, and LIDAR (a device that uses a laser in much the same way as radar or sonar use radio or sound waves). The latter two tracking systems were necessary for military researchers with a fondness for technology. But Bromenshenk was betting that a pair of binoculars would work just as well. His deep-seated hope was to field a system that could be used by both American troops and peasant farmers. And to Bromenshenk's delight, the three monitoring methods were equally capable of distinguishing a cluster of airborne bees. Not only did the insects locate the mines, but the number of bees hovering over a spot indicated the strength of the odor plume—more bees meant either a larger mine or a concentration of small mines (see Figure 24.2). However, Bromenshenk's celebration was dampened by a most troubling event.

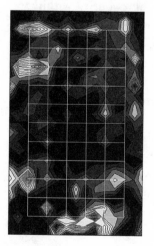

Figure 24.2. This map was derived from applying scanning LIDAR (similar to radar, except reflected light rather than radio waves is used to detect objects) to determine the densities of bees hovering over a 2½-acre minefield. The lighter areas indicate places where bees clustered, which matched areas with high visual counts (a labor-intensive method useful when the observer is not endangered by fused mines) and regions of chemical plumes from buried mines. (Courtesy of Joe Shaw)

In the world of mine detection, the most serious problem is a false negative, in which the system indicates that nothing is present when, in fact, there is a mine. But the reverse problem is also of concern. A false positive—the indication of an explosive when there is none—diminishes efficiency as disposal technicians focus attention on something that isn't really there. And so when the bees hovered insistently over a site within a test plot that had no mines, the researchers were concerned. However, analysis of a soil sample from the location where the bees raised a supposedly false alarm revealed a low level of contamination. Apparently, an earlier, improperly documented experiment had left behind a trace of 2,3-DNT, and the bees notified the military of their embarrassing oversight.

Bromenshenk saw that his bees could help pave the way for subsistence farmers—those people most likely to have suffered the ravages of war and the lingering effects of land mines—to rebuild their agriculture. They wouldn't need to import expensive equipment when the local bees could be put into service. With a sample of explosive-contaminated soil, a dribble of sweet syrup, a squirt bottle, and a bit of patience anyone could train the insects. Then, the farmers could simply watch where the bees gathered in the fields. Moreover, once local villagers caught on to using bees as mine detectors, the wholesale restoration of beekeeping might soon follow. And agricultural development experts had long known that returning pollinators to war-torn lands is essential to restoring a people's capacity to feed themselves.

Bromenshenk's enthusiasm became infective—at least within the military-entomological complex. Seeing the remarkable strides made with trained bees, scientists from the U.S. Department of Agriculture began to imagine how wasps could be converted into mine detectors. These insects might even be taught to respond like the beagles that sniff for contraband in airports. While the bees simply hover over an odor plume, wasps naturally perform distinctive behaviors when in the immediate proximity of a food source. Jim Tumlinson conceives of a platoon of wasps that, upon finding an explosive (or any other target chemical), settle down and rub their antennae on the precise origin of the smell—or attempt to sting the odor source, which they perceive to be prey. Joe Lewis, on the other hand, favors a more elaborate approach.[18]

Lewis invented a hand-held odor-detection device affectionately called the "Wasp Hound." The entomological linchpin consists of a set of six-legged detectors that are conditioned using the familiar sugar-and-spice approach (where the "spice" is whatever target smell the operator wants the insects to associate with the sweet reward). The contraption draws air into chambers that hold five trained wasps in Lilliputian squeeze chutes. Whenever the insects smell the odor, they duck their heads to receive a sweet reward—and, in so doing, trip an electric eye. Although one wasp might make the occasional mistake (after all, its brain is smaller than a typewritten period), the quintet is highly reliable.

But insects are not without their limitations. Unlike soldiers, bees and wasps do not perform their assigned duties at night, in the rain, or when it's cold. The insects' ability to locate mines also declines where there are very dry conditions or there is dense vegetation. And, at least so far, the only field tests have involved simulated minefields free of bomb fragments, spent munitions, and other chemical distractions. There are, however, ways of overcoming these limitations. Although some scientists have considered the possibility of genetically engineering a better honey bee, DARPA believes that an even more radical form of engineering may hold the key to the ultimate detector—an entomological cyborg.

25

INSECT CYBORGS AND ROBOFLIES

The U.S. military doesn't use the term "cyborgs," although this is precisely what their scientists and engineers are developing. Perhaps this sounds a bit too much like the stuff of science fiction. The Defense Advanced Research Projects Agency (DARPA) prefers to call their futuristic, insect-machine hybrids "vivi-systems."[1] The goal is to merge evolution and engineering, to take insects and "turn them into war-fighting technologies." DARPA's Controlled Biological and Biomimetic Systems Program is hoping to fuse the sensory, locomotory, energetic, and orientation capacities of insects with the very best of human ingenuity. As futuristic as such ventures might appear, the notion of using insects as critical components of machines saw its first military application during the Vietnam War.

In the jungles of Southeast Asia, finding the enemy before he found you made the difference between dying in an ambush and living to fight another day. Even the most experienced soldiers were often unable to sense the presence of Viet Cong guerrillas in the dense vegetation. Our senses are unable to penetrate more than a few meters into the forest, and we are easily fooled by camouflage. Not so for blood-feeding insects. These creatures can sniff out a host from long distances, and a bouquet of other odors does not distract them from the scent of a meal. One of the most sensitive detectors of a warm-blooded presence is the assassin bug (the insect that tormented victims of the Bug Pit in Uzbekistan). These insects know that they're on track when they detect a faint plume of carbon dioxide, and the bloodthirsty beasts further refine their search by honing in on a cocktail of chemicals found in mammalian breath and sweat.

Insects don't have any lips to smack in anticipation of a meal, but the bugs can make a soft buzzing sound by rubbing their sharp, elongated beaks over a series of ridges on their sternum in much the same way that we might strum the teeth of a comb. Although it seems that most assassin bugs use this sound to warn off predators or unwelcome suitors, the U.S. military found a species that sounded off when it detected a potential host. This would seem to be a peculiar behavior for a parasite, but the bug's murmur was nearly inaudible. Some entomologists speculate that the insect may live among an extended family and signal relatives of an impending feast. Whatever the creature was up to, the military had big plans for the bug that sang for its supper.

Wartime inventors developed a machine that turned assassin bugs into scouts.[2] The insects were placed in special capsules that were held within a device equipped with an audio amplifier. Air was drawn into the device and passed over the encapsulated assassin bug, a design feature that presumably provided an element of directionality and prevented the insect from getting excited by the presence of the human operator. When a host or enemy—depending on whether one takes the perspective of the insect or the soldier—came within five hundred feet of the device, the bug began to buzz, the amplifier made the sound audible to the operator, and the impending ambush was defused. At least that's the way it was supposed to work. The machine with the embedded insect was tested, but there's no evidence that it was ever used in the jungles of Vietnam. This early attempt at integrating insect and machine was rather crude—the entire insect was used as the detector. It was, in a sense, the conceptual predecessor of the "Wasp Hound" used for chemical detection. Today, however, the approach is more in line with that of Dr. Frankenstein, using only those body parts of greatest value in the creation of a benevolent monster.

In an Iowa State University laboratory, entomologist Tom Baker has built a device that is a true vivisystem—an insectan cyborg for locating land mines.[3] He found that the antennae of moths send neural impulses in response to volatile molecules associated with explosives. But rather than attempting to train moths, which aren't terribly clever insects to begin with, Baker simply lopped off the antennae and let technology do the thinking. The amputated antennae, which survive and function in their detached state for hours, are hooked up to microprocessors that convert the neural impulses into audible tones, such that the pitch drops as the cyborg encounters the scent of an explosive. The biggest problem is that the disembodied insect antennae respond to a wide range of stimuli, and a neurological signature unique to TNT has yet to be found. So the device works fine in the laboratory where there are few odors to compete with that of an explosive, but under field conditions the system operator is unable to distinguish the odor plume of a land mine from a plethora of other scents.

Integrating living tissues into machines is quite a challenge. After all, antennae and other body parts usually depend on being attached to the rest of the creature (see Figure 25.1). Mechanical devices aren't typically equipped with supplies of nutrients, blood, oxygen, and other conditions needed to sustain disembodied organs. So DARPA is taking the concept of entomological engineering to the next level. The ultimate goal is to dispense with the messiness of living systems entirely, transferring insect qualities into purely human creations.

Biomimetics is the most extreme version of the technological exploitation of the living world. In this field, scientists and engineers attempt to capture the essential functions of an organism in a machine. The coveted physical features are mechanically or electronically replicated without the disadvantages that come with the fickle, demanding, and unpredictable nature of living tissues.

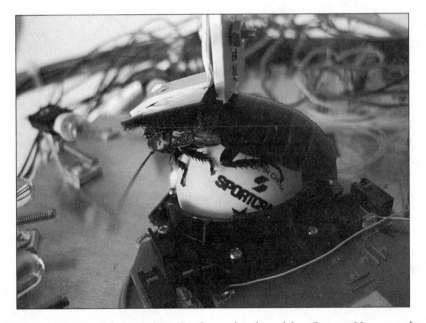

Figure 25.1. A cockroach-based cyborg developed by Garnet Hertz at the University of California, Irvine, demonstrating that exciting advances in insectan robotics do not depend on DARPA funding. The device uses a living Madagascar hissing cockroach atop a modified trackball to control a three-wheeled robot. Movements of the ball beneath the insect's feet are transferred into movements by the robot. Infrared sensors provide navigation feedback to the cockroach, creating a pseudo-intelligent system with the cockroach as the CPU. (Courtesy of Garnet Hertz)

For example, rather than trying to untangle and decipher the neural patterns generated by a moth's antenna, the goal is to build the essence of antennal neurophysiology into an electronic system. Such efforts began as early as 1968, when the Army Research Office paid the Philco Corporation to develop a "manmade nose"—a chemical sensor capable of detecting minute traces of particular odors.[4] Although the project never materialized into a working prototype, the olfactory model that the scientists used was not the snout of a bloodhound but the antenna of an aphid.

In recent years, DARPA funded a project to mimic the remarkable capabilities of an obscure beetle that is the envy of weapon designers.[5] Heat-seeking missiles and other "smart" bombs use infrared radiation to find their targets. The problem is that the detection systems can operate only at freezing temperatures, and sustaining these conditions requires expensive and heavy cooling systems that demand frequent maintenance. If a system could detect infrared signatures at ambient temperatures, the payoffs in terms of weight, cost, and durability would be tremendous. And that's exactly what the European jewel beetle (*Melanophila accuminata*) has to offer.

This beetle uses freshly burned trees as a food source for its young. The damaged trees are unable to mount a defense against the insects, so the larvae flourish beneath the charred bark. A smoldering forest triggers a mating frenzy. But the windfall of larval food occurs sporadically and the ideal conditions for egg laying don't last long, so evolution has provided the half-inch adults with an incredible capacity to locate forest fires.

The jewel beetle "feels" a distant fire by using an infrared sensor on its thorax—and the insect can sense flames from as far away as 40 miles. The beetles use a specialized organ that looks like a microscopic ear of corn within a tiny crater. Each "kernel" is a hardened dome that absorbs incoming infrared radiation. The radiation causes the dome to expand ever so slightly, with the change being sufficient to distort a sensory nerve attached to the structure. This nerve then sends a signal notifying the insect of a distant fire—and the impending orgy.

Helmut Schmitz, a zoologist at the University of Bonn in Germany, is attempting to recreate the beetle's detector using synthetic materials that absorb infrared radiation and expand enough to generate an electronic signal.[6] The prototype can detect a flame that is within 12 inches. A forest fire generates a much stronger signal, but the beetle's detector functions at a distance two hundred thousand times greater than the artificial device. So the challenge is to mimic the beetle's phenomenal sensitivity to microscopic changes in its sense organs, and that means finding a component that responds when the dome expands by a millionth of an inch. Schmitz and his colleagues believe that a capacitive sensor device might do the trick. And so work continues on reconstructing an infrared sensory system using manmade parts. However, when it comes to military applications of insect anatomy and physiology the real payoff may not be in mimicking neurology.

There's a very good reason that evolution produced legged, rather than wheeled or tracked, creatures. Wheels (even with advanced suspension systems) and tracks (such as those found on tanks) lack the dynamic properties necessary to traverse extremely uneven surfaces. Thus, insects have become the standard platform for the development of robotic locomotion across random rubble and pitted pathways—the conditions invariably encountered in the course of war and other disasters.

The original motivation for putting robots on tortuous terrain was to minimize the risk to humans, and this concern drove developments in two areas. First, military commanders (not to mention soldiers) much preferred allowing machines to take the risks of finding paths through minefields or locating mines during clearing operations. Second, search-and-rescue teams saw that using robots within damaged buildings and other unstable structures would shift the danger from man to machine. Moreover, if the robots could be miniaturized, they could navigate tight spaces inaccessible to lumbering humans.

The program director of DARPA's Controlled Biological and Biomimetic Systems, Alan Rudolph, put the situation succinctly: "Legged robotics will likely eventually dominate because they have a greater potential to deal with obstacles . . . if we can figure out how to build them."[7] And with tens of millions of dollars to entice the sharpest entomological and engineering minds, the military doesn't see failure as an option.

A menagerie of arthropods—including crabs, lobsters, and scorpions—has been used as models for locomotion. However, no creature has yielded better results to date than the lowly cockroach (order Blattodea). DARPA's golden child is a mechanical roach that owes its existence to the work of a scientist at the University of California, Berkeley, who painstakingly analyzed the dynamics of cockroach movement. Robert Full, a zoologist with a passion for biomechanics, discovered that the secret to a cockroach's ability to clamber over rough surfaces lies in an utter lack of grace.[8] The creature bumbles along by using its six legs in alternating sets of three. While we two-legged animals have a single point of contact, which makes tripping all too easy, the cockroach is supported by stable tripods (the front left, middle right, and hind left alternate with the front right, middle left, and hind right). Add to this that the insect uses a sprawled posture to maintain a low center of gravity, and it's virtually impossible to build an obstacle course that would topple a cockroach. Scaled up to human proportions, a cockroach dashing along at 50 steps per second is the equivalent of a human running the high hurdles at 200 miles per hour.

Using these insights, researchers at the University of Michigan and McGill University in Canada collaborated to capture the essence of cockroach movement in a robot.[9] In the late 1990s, entomological insight spawned a nightmarish machine known as RHex (short for "Robotic Hexapod"). Designed by engineering professors Martin Buehler and Dan Koditschek, the robot stumbled and thrashed its way over all sorts of debris. In the last few years, RHex metamorphosed into the Scout series, an evolving lineage of ever-improving, six-legged machines that scamper over obstacles and climb stairs. And further refinements in "roachbots" are on the horizon, thanks to work at other institutions.

As scientists come to understand how the cockroach leg performs as a limber, dynamic structure, new lines of engineering are developing. At Cleveland's Case Western University, a biologist has teamed up with an engineer to mimic a cockroach leg in excruciating detail.[10] Roy Ritzmann and Roger Quinn have developed a gigantic roach, with legs 17 times larger than the real insect. This upscaling allows them to design and program every aspect of movement. Using a complex pneumatic piston system to power the behemoth, they have constructed appendages with a sophisticated network of strain gauges linked to a computer that tells the system how to compensate for changing forces by adjusting the pressure applied to the leg joints as the robot walks. But this approach might be technological overkill (see Figures 25.2 and 25.3).

While some biologists began with the assumption that mimicking insect movement would depend on developing computational feedback systems to match the insect's nervous system, recent discoveries have shown that the key might lie in the insect's musculature, rather than in its brain.[11] Full has discovered that cockroach muscles don't merely move the legs but also adjust the stiffness of the individual segments and joints via a complicated but unconscious system of tugging and pulling. At least at high speeds, complex neural processing seems to fade in importance as pure, albeit staggeringly sophisticated, mechanics take over. Even without feedback from the nervous system, the 21 muscles of the cockroach leg function as elegantly reciprocating rubber bands, constantly tuning the femur, tibia, and knee to the flexibility needed to match the challenges of scurrying through kitchen cabinets, under sinks, and between walls.

Figure 25.2. BILL (Biologically-Inspired Legged Locomotion)-Ant mimics not only the movement of its namesake but also features actuated mandibles with force-sensing pincer plates. The actively compliant hexapod robot is capable of grasping and moving objects while reacting to external forces through sensors in its feet and pincers. BILL-Ant was developed in 2005 by William Lewinger at the Case Western Reserve University Center for Biologically-Inspired Robotics, directed by Roger Quinn. (Courtesy of Roger Quinn)

Full's insights concerning the inner workings of the cockroach leg have led Mark Cutkosky at Stanford University to pursue development of artificial, dynamic limbs using a process called "shape deposition manufacturing."[12] By incorporating solid-state structures embedded with electronics into plastic appendages, he's hoping to mimic the passive mechanical properties provided by the interacting muscles within the insect leg.

Unraveling the complexities and overturning the assumptions of the insect leg have proved critical to embedding the evolutionary brilliance of living organisms into the engineering of robots. But as unexpected as the process of insect walking has turned out to be, the real surprises came with ventures into insect flight—a phenomenon that DARPA has spent $50 million trying to capture and copy for military applications.

The U.S. Department of Defense has been salivating at the possibility of building tiny aircraft for reconnaissance and espionage.[13] Imagine a microspy the size of a house fly peeking into enemy strongholds, eavesdropping on

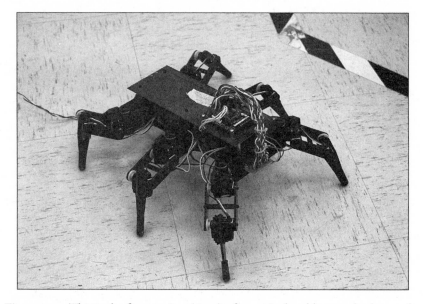

Figure 25.3. This is the first version (2005) of an articulated hexapod robot under development by Dr. Alejandro Ramirez-Serrano and his research team at the Autonomous Reconfigurable Robotics Systems Laboratory at the University of Calgary, Alberta, Canada. The hexapod and its control mechanisms are being developed to overcome obstacles five times the height of the robot. The robot features pan-tilt cameras for eyes, ultrasonic and infrared sensors, 20 servo motors, and four computers. (Courtesy of Alejandro Ramirez-Serrano)

conversations, or sniffing out stores of chemical weapons. Such possibilities led DARPA to lure aeronautical engineers (generous funding is effective bait for grant-starved academics) into directing their formidable intellects toward the development of "micro air vehicles," or MAVs—self-powered, aerodynamically stable flying machines no more than six inches long. The early ventures consisted of simply shrinking conventional aircraft, but it soon became evident that the principles that accounted for fixed-wing flight became irrelevant at this scale. For a Lilliputian plane to stay aloft it would need to fly at a phenomenal speed to attain the necessary lift, and generating such speeds—let alone controlling the thing as it screamed along—was deemed to be technically impossible.

Thinking in terms of biomimicry was the first crucial step in solving the problem. Robert Michelson of Georgia Institute of Technology's Aerospace Laboratory put the situation in simple terms: "Nothing in nature achieves sustained flight with fixed wings or with propellers. . . . All tiny creatures flap their wings."[14] Not unexpectedly, the engineers tried to mimic birds. But attempts to build diminutive ornithopters—bird-like flying machines—were quickly scrapped when scientists realized that the physics underlying flight in these animals is no different from that which accounts for the lift of airplane wings. What engineers needed was a whole new conceptualization of flight, a novel set of principles that would allow MAVs to generate lift. And then came the conceptual breakthrough: build an *entomopter*.

An insect is not just a small version of a bird when it comes to flight. Indeed, the old yarn about physicists being able to prove that a bumblebee can't fly—despite its evident ability to do so—is not entirely apocryphal. If the analysis is limited to large-scale aerodynamic forces, such as those that provide an airplane or bird wing with lift, no insect should be able to stay aloft. But as the scale decreases, entirely new properties emerge: small is different. The essence of flight is beautifully captured in a single mathematical term called the Reynolds number. The value of this parameter for any particular structure is a function of three components: speed, wing dimension, and density of the fluid.[15] When thinking of conventional flight, the air density plays a role at high altitudes where the atmosphere is thin. With insects, air density at the earth's surface is extremely relevant. When you're tiny, normal air becomes thick.

We don't usually think of air as a fluid, but a gas behaves like a very diffuse liquid. As an animal's body size decreases, the effective density of air increases. For a gnat (suborder Nematocera), air has the same resistance as oil does to us. This explains why you can drop an ant from a height of 10 feet—which would be like our falling from a mile above the earth—and it lands without injury. The little fella is dropping through the air the way we'd be sinking into a tank of honey. Insects don't so much fly through the air as swim through it.

At the scale of an insect, it makes more sense to call those flapping structures oars rather than wings. Everything that engineers had learned about design in terms of airfoil properties, wingspan factors, and surface smoothness had no bearing on insects. Thrips can "fly" using a structure composed of hairs sticking out along a shaft; try flying an airplane using palm fronds for wings. With a scientific understanding of insect aerodynamics, the challenge became building a machine that would fly like a fly flies.

Flies, rather than bees, became the model for MAVs (probably because the former are both safe and easy to raise). A creature that can take off backward, hover in place, dart sideways, and land upside down had to be the epitome of flight for an aeronautical engineer with a penchant for grand challenges. Studies showed that blow flies—disgusting insects all in all, but easily observed and mass-produced—created eddies by flapping and twisting their wings in a manner not unlike how one sculls while treading water. These microscopic whirlwinds provided the insect with lift, stability, and maneuverability. But constructing an entomopter the size of a fly was asking a lot, so the goal was phrased in slightly more realistic terms: develop a machine with flexible, 1-inch wings that could sustain autonomous flight.

With support from DARPA and the Office of Naval Research, Ron Fearing and his team at the University of California, Berkeley, have been working toward a "micromechanical flying insect."[16] Although there are no "roboflies" zipping down the corridors of Cory Hall, there has been substantial progress since they began in 1998. At least Fearing's group has shown that an entomopter with a tiny motor powered by lithium batteries, set into a carbon fiber "thorax," could flap a polyester wing fast enough (150 beats per second) to generate lift comparable to that of a blow fly. Other research teams at California Institute of Technology and Vanderbilt University have also made strides. Engineers at Harvard University recently launched a flylike robot that weighed a little more than the plastic head of a pushpin, but the entomopter was tethered to its power supply.[17]

In addition to a lightweight power supply, stability and maneuverability remain challenges, not to mention getting a flying robotic insect to navigate its way through a complex environment and carry a payload. But if everything continues to play out as hoped, the engineers could soon have a bitty device that zips along at a respectable seven miles per hour. Or maybe they've already succeeded.

A *Washington Post* article in October 2007 reported that people gathered at political rallies have been describing the appearance of insectlike flying devices since as early as 2004. One individual at the Republican National Convention in New York described "a jet-black dragonfly hovering about 10 feet off the ground, precisely in the middle of 7th Avenue."[18] Perhaps the person was paranoid (black helicopters giving way to black insects) or simply saw an actual dragonfly. However, several people at an antiwar rally in Washington, D.C., independently described large dragonflies trailing strings of small berrylike spheres and flying in formation. Not surprisingly, government agencies have declined to discuss the topic. If the CIA (the agency reportedly developed a gasoline-powered "insectothopter" in the 1970s, but scrapped the project because the dragonfly-like device was unstable) or other defense and security agencies have such a device, then mission creep has begun to change the tenor of the venture.

The military began its biomimetic project with the most altruistic of intentions, or so one is led to believe.[19] The goals included activities such as locating land mines and wounded soldiers. Finding injured people in bombed-out buildings is laudable, but it is a small step from search-and-rescue to a bit of harmless snooping and then to seek-and-destroy. Rather than mounting a tiny video camera on a robotic insect, one could arm it with a poisoned needle—a mechanical bee with a lethal sting. Just a milligram or two (about the weight of a grain of sand) of the right venom would be deadly, and this warhead would be much lighter than any other payload of military interest. The increasingly creepy game of "what if?" did not stop at pinpoint assassinations. As scientists have come to understand more about swarming behavior, another tactic mastered by insects becomes plausible—coordinated, collective attacks by MAVs. A swarm of entomopters sucked into a jet engine could quickly bring down enemy aircraft or disable a hijacked plane.

Such futuristic weapon systems would require enormous, but not at all inconceivable, engineering advances. And we can be sure that wealthy industrial nations will be the ones making the technological breakthroughs and deploying the entomechanical armaments. However, the most frightening and likely uses of insects as weapons in the modern world do not require sophisticated science. In the fast-changing, high-tech world of warfare, some of the most effective tactics are ironically the most primitive. Consider that in the Middle East, American smart bombs have given way to "improvised explosive devices," and Islamic suicide bombers function as cheap—if human life is perceived as having little value—guided missiles.

To engineer an MAV is well beyond the ability of today's terrorist organizations. Even carrying out germ warfare requires sophisticated technology such as autoclaves, incubators, sterile media, and other accoutrements. But insects could become the terrorists' six-legged box-cutters, their biological weapon of choice. These creatures are abundant, available, safe to handle, easy to transport, self-dispersing, self-perpetuating, and—if properly selected—phenomenally effective. The newest liquid explosives are surely cause for concern, but an insect net and a Ziploc bag could be sufficient to wreak environmental and economic havoc. One might wonder whether the U.S. government is able to protect the nation from the possible range of entomological attacks.

26

"VIGILANT AND READY"?

In today's world, entomological terrorism is not perceived as a clear and present danger. However, historical and recent events strongly suggest that western nations would be well advised to take seriously the possibility that insects could be used to attack people and agriculture. In this context, the United States has developed several lines of defense, but whether these are adequate is not at all clear.

The first—and arguably least effective—tactic is the law.[1] As early as the seventh century BCE, the rules of war prohibited destruction of forests, orchards, herds, and even beehives. The dishonorable nature of starving a populace became reified in the 20th century by the Geneva Protocol. In broader terms, the Biological and Toxin Weapons Convention (BTWC) forbids using living organisms, presumably including insects (although not explicitly named), as weapons. But there are serious problems with such accords.

To begin, these international agreements are legal contracts, meaning that the violation by one party frees the others from any obligation to the transgressor. The possibility of a tit-for-tat exchange with biological weapons is not pretty. Whether the release of crop-feeding insects would be grounds for firing back with yellow fever is not clear, but once a contract is broken, the only constraint is the moral integrity of the combatants.

Next, governments that have ratified international accords have a less than stellar record of compliance. Various countries used poison gas in World War I and biological agents in World War II. More recently, in a chilling but unwitting allusion to the entomological origins of nerve gas, Iraq's Major General Maher Abdul Rashid declared, "If you give me a pesticide to throw at these worms of insects [the Iranians] to make them breathe and become exterminated, I'd use it."[2] Despite having signed the Geneva Protocol, the Iraqis killed 20,000 Iranian soldiers with gas attacks from 1983 to 1988 and slaughtered hundreds of Kurds in March 1988 with a cocktail of sarin, tabun, VX, and mustard gas. According to U.S. intelligence sources, several countries that have ratified the BTWC are pursuing offensive biological weapons, as might be expected given that the convention lacks any substantive means for verifying compliance, let alone punishing violators.

Even ostensibly moral governments have been willing to use banned weapons when the nation's survival is at stake.[3] In 1944, Churchill told his service chiefs that he would seriously consider using gas if it would prove decisive should Britain face a life-or-death struggle or if it would substantially shorten the war. The same year, the American high command saw their forces slogging across blood-drenched islands and planned to hit Iwo Jima with poison gas. Only President Franklin D. Roosevelt's terse denial prevented the military from waging chemical warfare. And it must be noted that not all parties capable of waging entomological warfare are signatories of international accords that would prohibit biological weapons.

Allan Krass, writing for the Stockholm International Peace Research Institute, summarized the situation clearly:

> These comments [concerning prohibited tactics] are not intended to dismiss completely either the possibility of irrational or bizarre behaviour by insane or desperate leaders or the danger of covert operations by one state against the population or resources of another. But there is little point in considering insane or desperate acts in the context of a discussion of treaties, since such legal instruments would have little or no effect on the actions of states led by madness or driven to the point where their national survival is at stake.[4]

The point is, of course, that for many states, organizations, and individuals operating in the modern world, the ideological ends would justify the entomological means.

With the rule of law insufficient to protect nations from entomological weapons, governments are compelled to invest other resources into defending their people and assets. The general strategy consists of an initial phase that includes deterring, preventing, and detecting an attack, and if these steps fail, the next phase involves responding to and recovering from a strike. For the United States, at least 16 agencies have a stake in agricultural bioterrorism, and this hodgepodge of players approaches two dozen when the possibility of a direct assault on humans is considered.[5] In this alphabet soup of agencies, one acronym floats to the top of nearly every discussion of entomological warfare—the agency that serves as the first line of defense as well as a central player should an attacker slip past border guards: USDA.

The U.S. Department of Agriculture has three primary branches involved in entomological terrorism: the Animal and Plant Health Inspection Service (APHIS, which is the operational arm of the agency, although a significant portion of the service was subsumed under the Department of Homeland Security in 2003), the Agricultural Research Service (ARS, which is responsible for scientific developments), and the Cooperative State Research, Education, and Extension Service (CSREES, which controls federal funding to—and hence, the scientific priorities of—agricultural programs at the nation's universities). During the Cold War, USDA scientists collaborated with the Department of Defense in various largely clandestine projects.

The first public involvement of federal scientists in entomological warfare came in 1961, when the ARS published a report warning that foreign sabotage of crops and livestock was possible.[6] The potential agents included an arsenal of insects: Medfly, khapra beetle (*Trogoderma granarium*, a pest of stored grain), Asiatic rice borer (*Chilo suppressalis*), silver "Y" moth (*Autographa gamma*, a pest of tomato, bean, and potato), Sunn pest (*Eurygaster integriceps*, a bug that feeds on cereal crops), dura stem borer (*Sesamia cretica*, a moth with larvae that decimate sorghum and sugarcane), and five species of potato weevils (family Curculionidae). In addition to generating a list of likely entomological weapons, the analysts recommended that every county develop a defense board for detecting and combating an insect invasion. Although the ARS took the first high-profile position on agricultural terrorism, its current state of readiness is deplorable in the eyes of some experts.

Geoff Letchworth, who knows the workings of the ARS from the inside, asserts that the nation is not ready for a bioterrorist attack: "Not if I can write for you on a postcard a series of different ways to paralyze the agricultural industry of the United States, where we have no possibility of being able to respond; I'd say the resources are not adequate." He contends that while scientists understand the situation and have "appropriately evaluated costs and benefits to come up with the things that we ought to be spending time and energy on, USDA managers are hopelessly out of the loop."[7]

Letchworth finds the priorities of his former agency to be driven by agricultural special interests, arguing that "management puts effort into oriental gardens and horticulture and things that industry could do for itself, rather than focusing on an insurance function that private industry can not afford to pursue." But he acknowledges that USDA administrators have to work with what Congress allocates. Or, in rather more pointed terms, Letchworth describes the ARS as "the whore of Congress" while suggesting that this is not entirely bad. Politicians are responsive to the needs of agricultural producers, but industry—and hence the federal government and its agencies—is not primarily concerned with developing methods to ensure that the country is prepared for future risks, including bioterrorism. Letchworth's bottom line is that "the balance has gone way too far towards intervention and away from prevention." While the USDA's research branch is struggling to proportion its efforts in accordance with the possibility of an entomological attack, its sister agency faces similar challenges on the frontline of defending the country against insect incursions.

The most conspicuous activity of APHIS is at ports of entry, where inspection officers labor to prevent the accidental—and intentional—introduction of pests (see Figure 26.1). And insects are often at the top of their watch-and-worry list. If interdiction fails and a dangerous pest gains a foothold,

Figure 26.1. Customs and Border Protection specialists inspect shipments of grain, fruit, vegetables, lumber, meat, and flowers (yes, flowers) for harmful pests. While it might seem that protecting food, wood products, and livestock would be more important, floriculture in the United States is a $5.4 billion industry—more than Microsoft's quarterly profits. The value of an acre of flowers can exceed $300,000, or more than 1,500 times the value of an acre of corn. (Photo by Gerald L. Nino, U.S. Customs and Border Protection)

the agency launches a defensive juggernaut, usually designed to eradicate the organism before it spreads. Of the 17 emergency plant-protection programs enacted between 1995 and 2000, 11 targeted insects (and two of the targeted plant diseases were transmitted by insects), and of the 12 emergency animal protection programs, one-third involved insects.[8]

Even before the terrorist attacks on the American homeland, USDA officials and outside analysts were thinking in terms of national defense. However, assessments of the agency's state of readiness were remarkably discordant. While the federal government considered itself primed for action, external evaluators were skeptical, to say the least. In 1999, the head of the APHIS's Marketing and Regulatory Programs, Michael Dunn, described his agency as vigilant and ready when it came to the "intentional introduction of biological agents for terrorist purposes." The administrator assured the public that "in response to this threat, USDA is working closely with other federal agencies to monitor, identify, and safeguard areas vulnerable to bioterrorism."[9]

The top APHIS administrator, Ron Sequeira, was only a tad less confident. He listed a number of problems with the agency's capacity to conduct pest surveys, but nothing that couldn't be overcome with an upgrade of technology (and funding, of course). As for preventing sabotage, Sequeira proposed, in the tradition of all good bureaucrats, to study the possibility of coming up with a plan: "In order to respond to biological terrorism threatening animal and plant production, APHIS will consider development of a 'bioterrorism rapid response' strategy."[10]

Those outside the agency were not quite so convinced that the USDA was prepared for a terrorist attack. According to Jonathan Ban, writing for the Chemical and Biological Arms Institute in June 2000, U.S. agriculture was a ripe target:

> Given the tremendous economic, political, and strategic value of U.S. agricultural resources, the Washington policy community has been slow to realize their vulnerability to attack by an antagonistic state, economic or agricultural competitor, or terrorist, especially with biological weapons.[11]

The dueling viewpoints continued into 2001, when an article published by the American Institute of Biological Sciences asserted that "the poor level of biosecurity on the majority of farms today guarantees unchallenged and unhindered access to the determined, patient terrorist."[12] A paper issued by the Belfer Center for Science and International Affairs was more moderate, claiming that while "a determined group could conceivably carry out a devastating attack," an act of agricultural bioterrorism would be extremely difficult. The authors sought to dispel concern by concluding that "there is no evidence of terrorist groups with the motivation to carry out a catastrophic attack against U.S. agriculture."[13] September 11, 2001, proved otherwise for the nation's economic infrastructure.

The USDA was remarkably prescient in commissioning the National Research Council of the National Academy of Sciences to evaluate the country's preparedness for biological threats to agriculture in 2000. The terrorist attacks on the World Trade Center and the Pentagon took place while the committee of 12 scientists was in the midst of its study. Discussions of America's vulnerability were raised to a fever pitch, and the academy made a special exemption to their normal practice of maintaining strict confidentiality of studies until public release. In March 2002, the academy experts briefed White House, Homeland Security, and Department of Agriculture officials. The incentive for this early apprisal may not have been only a matter of political timeliness. The committee had determined that the nation was acutely vulnerable.

The sense of urgency that motivated the unprecedented briefing can be surmised from the contents of "Countering Agricultural Bioterrorism," the publicly available summary of the Council's investigation. Although some earlier analysts had claimed that U.S. agriculture was too dispersed to represent a viable target, the National Academy scientists clearly disagreed. Not only were fields and buildings soft targets—easily found, readily entered, and virtually unguarded—but American agricultural practices had painted a bull's-eye on farms, ranches, feedlots, and confinement facilities:

> In many ways, attacks on plants and animals may be easy to mount. Agricultural crops and animals are often grown, housed, or grazed in relatively high-density and uniform conditions, which make the spread of disease and infestations more rapid and effective. . . . Genetic homogeneity, often desirable in agriculture to optimize yields or nutritional content, adds to the vulnerability of crops and animals to epidemics.[14]

The authors studiously avoided alarmist language; the nation was not facing imminent doom or impending famine from an entomological attack. Although an enemy would not be able to defeat the United States by releasing insects, the report made clear that a successful invasion would do far more than inconvenience a few farmers:

> In the wake of the events of September 11, 2001, few would disagree that the United States, more than ever before, must be alert and prepared for the possibility of surreptitious attacks within its borders—attacks aimed not so much to achieve strategic military victories as to cause indiscriminate destruction, economic disruption, widespread injury, fear, uncertainty, and social breakdown.[15]

Even without the worrisome details of Appendix E (the section that was kept from the public and analyzed hypothetical attacks on U.S. agriculture), the council's overview bluntly and unambiguously identified serious weaknesses in the nation's defenses:

> The committee came to the following key conclusions: 1) the United States is vulnerable to bioterrorism directed against agriculture, 2) the nation has inadequate plans to deal with it, 3) the current U.S. system is designed for defense against unintentional biological threats to agricultural plants and animals, and 4) although strengthening the existing system is a resource-efficient and effective part of the response to bioterrorism, it is not sufficient. The committee recommends a concerted effort on the part of the U.S. government to develop a comprehensive plan to counter agricultural bioterrorism.[16]

These conclusions diametrically opposed the USDA's claims in 1999 that the agency was "vigilant and ready." Nor would it appear that the plans for a "bioterrorism rapid response" strategy ever materialized. In a striking indictment, the council concluded that

> coordination within and among key federal agencies, as well as coordination of federal agencies with state and local agencies and private industry, appears to be insufficient for effectively deterring, preventing, detecting, responding to, and recovering from agricultural threats.[17]

Either the USDA's earlier assurances that they were working with other federal agencies were based on wishful thinking or the cooperation had somehow unraveled. The agency's administrators may have been trying to conceal gaping holes in national security from would-be terrorists, but they didn't fool the National Research Council. Nor is it likely that an enemy with the capacity to do simple math would have believed that APHIS was providing a credible first line of defense.

According to publicly available figures, agricultural inspectors are overwhelmed. To get a feel for the scale of the problem, consider a recent study conducted by the agency.[18] APHIS had long considered the 83 ports of entry along the U.S.–Canada border to be low risk, with few serious infractions. To test this belief, the agency undertook an unusually intensive program of inspection over the Labor Day weekend in 2001. The careful search of 4,000 vehicles crossing into New York and Michigan resulted in the seizure of 6.5 tons of prohibited material and the interception of 200 pest organisms. In other words, at low-risk border crossings there was nearly four pounds of illegal material per vehicle and an invasive species in one of every 20 cars and trucks. Add to this the 39,000 trains, 141,000 aircraft, 200,000 ships, 463,000 buses, 584,000 commercial vehicles, and 4 million parcels and letters that enter the country every year, and a very worrisome picture takes shape.

Recognizing the size of the hole in the nation's defensive perimeter, Congress has rushed to plug the leak with money and bureaucracy. While there is no doubt that hundreds more inspectors at the borders will provide a greater level of protection, the impact of such increased staffing would be minimal. From a statistical perspective, carefully examining one out of every 8,300—rather than one in 10,000—incoming people and packages will make little difference to a prospective terrorist. And recent U.S. bureaucratic maneuvers may have made the nation more vulnerable.

In 2003, the federal government dismantled the USDA's program for protecting American agriculture in order to feed the resources into the Department of Homeland Security.[19] Twenty-five hundred inspectors were transferred from APHIS—the operational branch of the agency responsible for protecting agriculture from invasive pests—to the Department of Homeland Security (DHS). A Government Accountability Office (GAO) report in 2007 revealed that a majority of the APHIS inspectors who were assigned to the DHS's Customs and Border Protection (CBP) say that their ability to protect agriculture has been compromised by low morale, training deficiencies, equipment shortages, and manpower shortfalls.[20] The GAO report forced CBP to conduct an analysis of their staffing allocations, which revealed 33 percent fewer inspectors than were needed to protect to agriculture.[21]

According to testimony at a congressional hearing following the release of the report, nothing gets a lower priority than agriculture in the DHS hierarchy of concern. California's Representative Dennis Cardoza, chairman of the House Agriculture Subcommittee on Horticulture and Organic Agriculture, provided a scathing summary: "DHS is absolutely failing at its mission to prevent bug and pathogen infestations from coming into this country"; he went on to describe the federal agency's shortcomings as malfeasance, claiming that "the transfer has been a colossal mistake and a colossal waste of taxpayer money."[22]

Contrary to the recommendations of the National Research Council, the U.S. government has stacked its resources at the border, betting that inspectors can detect entomological weapons. This is an egregious strategic error, as is evident from the continued flow of pest insects into the country. Not unexpectedly, Customs and Border Protection has not conducted the studies needed to determine whether their program is working. There are, however, some compelling anecdotal reports from various states. According to Florida's commissioner of agriculture, there has been a 27 percent increase in new plant pests and diseases since 2003.[23] New York Senator Charles Schumer contends that at least seven new organisms—including the Swede midge (*Contarinia nasturtii*), a serious pest of vegetable crops—invaded his state between 2004 and 2006.[24] And California Senator Dianne Feinstein has noted that Fresno County suffered its first fruit fly outbreak while DHS was manning the borders.[25]

If the agricultural industry is unprepared, the situation is no better with respect to the medical community's readiness to defend the populace against entomological weapons. While the National Research Council's report focused on risks to crops and livestock, their analysis of Rift Valley fever—a disease of both animals and humans—makes clear that the country is vulnerable to insect-borne diseases. Indeed, the Centers for Disease Control and Prevention has identified 25 diseases with bioterrorism potential, seven of which are transmitted by insects.[26] However, there is scant evidence that vector-management programs are any more prepared than when West Nile virus arrived in 1999.

The reason for this state of affairs may be the nation's head-in-the-sand response to biological weapons. According to Lieutenant Colonel Terry Mayer, "The United States is ill-prepared to defend against or counter [biological warfare]—why? One view is that 'the United States has a tendency to wish the problem would go away because it seems too unsavory and too difficult to handle.' "[27] In other words, many Americans prefer psychological denial to national defense.

To be fair, some progress has been made. In 2003, government officials tested the health-care system by simulating a plague outbreak in Denver.

Joseph M. Henderson of the Centers for Disease Control and Prevention determined that the agency was better prepared than before the September 11 attacks, but "the health care system is the weakest link in the chain."[28] This conclusion exemplifies the odd competition among agencies to cast themselves as the least prepared public sector in order to obtain funding. And it appears that medicine is winning the race.

Biodefense is the hottest ticket in federal funding.[29] The Department of Health and Human Services was spending about $250 million on this program in 2001, an expenditure that rose twelvefold in 2002 and has been at or above $4 billion since 2003. And bioterrorism is the goose laying the golden egg for medical science. In 2005, the National Institutes of Health planned to devote $1.8 billion to biodefense research.

Former U.S. Senate Majority Leader Bill Frist and other political leaders have proposed a "Manhattan Project" to combat bioterrorism. They are dismayed at the relatively underfunded and absurdly fragmented effort to protect the American people from biological weapons. And they have a point. While the Missile Defense Agency has a $7.7 billion budget overseen by a single director, biodefense has a $5.5 billion budget managed by nearly 30 administrators in a dozen agencies.

So how should the nation prepare for the possibility of entomological terrorism? The experts are deeply divided, and their positions depend on how they characterize the enemy. From the perspective of the DHS, the focus is on spectacular events that would cost thousands of lives and billions of dollars. In considering whether insects might be used as weapons, Michael Oraze, director of agricultural and biological terror countermeasures for DHS, maintains that

> insects could be introduced and they would harm us, but I also believe that the terrorists don't spend their shots lightly. They take their time and do the big thing right, once—the catastrophic sort of attack that we will not suspect or be prepared to prevent. . . . We are spending our efforts on those who would harm us the most in the near term.[30]

Oraze believes that terrorists would use "their most potent weapon" in an attack, and insects don't have the same cachet as smallpox or anthrax. However, Oraze fails to consider that microbial weapons, particularly in aerosol form, are extremely difficult to develop and deliver. And given the demonstrable potency of insects, it seems that the government may once again be making the mistake of failing to expect the unexpected. But DHS is making an even higher stakes wager.

According to Oraze, winning the war on terror depends on countering our enemy's knockout punch. He admits that some analysts argue that terrorists are in it for the long haul, but that "looking ahead to a 10- or 20-year strategy that terrorists might use against us, given the all of the intelligence and the pressures of the day, it is not something that is yet on our radar." Thus patience might well be an enemy's most effective strategy. While DHS sees itself as guarding the nation from a roundhouse punch in the course of a winnable boxing match, others view the situation very differently. And from this other perspective, insects are a much greater cause for concern.

Lieutenant Colonel Robert Kadlec doesn't dismiss the possibility that Al-Qaeda would relish the opportunity to "kill a lot of people," but he argues that the better metaphor for the conflict between the western world and terrorists is a 100-year wrestling match, rather than a ten-round title fight:

> It's about fatigue and long-term struggle. A war of attrition is more likely to bring victory than the one-punch knockout, which is very difficult against the United States. . . . In that light you don't kill a million people. In fact, you don't want to because it creates another element of war, which is passion and retribution. By pursuing victory in a strategic, long-term fashion you win. The big hit is a nuclear detonation, but one needs to consider how to fight a war of economic and public health attrition.[31]

Americans have confidence that the government can protect their health and wealth, but centers of political stability and social value are vulnerable. Kadlec cautiously draws parallels to the American experience in Vietnam. He points out that despite the U.S. military's technological superiority, "the guys with sandals made of tires won" because they eroded our ability and willingness to fight. Kadlec refers to the "punji stick tactics" of the Viet Cong. The constant threat of stepping on these sharpened sticks tipped with feces or poison was exhausting; the real payoff was the psychological stress, rather than the damage that the booby trap inflicted. And insects make fine punji sticks.

Given Kadlec's view that America's focus on spectacular attacks and heroic interventions is misguided, it is not surprising that he sees the nation as being poorly prepared for acts of entomological terrorism:

> I would have to think that before 9/11 we were at a D- and we're now at a D+. How you define the problem is how you will find solutions. . . . If you

believe that others will inflict chronic harm on our nation, then you develop a strategy for a war of attrition—and we haven't defined the problem or our responses in that way. We've thought about car bombs and nuclear materials, but we haven't thought about weapons that are in the terrorists' domain and endemic to where they are living. Quite frankly, vectors are underappreciated.[32]

Kadlec argues that in the face of uncertainty—we simply don't know what diseases will be chosen by terrorists—the best defense would be to build a strong public health infrastructure. We cannot stop every traveler who is sick from entering the country, but we can stockpile vaccines, train health professionals, and educate the public. With a viable public health system, the nation would be poised to respond to whatever may come. And the same may well apply to agriculture. We can revitalize an anemic pest-management infrastructure—with adequately funded mosquito-abatement districts, for example—to respond to organisms that slip past our border guards.

Agriculture mirrors medicine's predilection for favoring the spectacle of the surgeon implanting an artificial heart over the dutiful work of a public health nurse monitoring blood pressure, providing dietary recommendations, and encouraging exercise. The latter approach is not high-tech or Hollywood heroic—the qualities that garner political support and social interest. But such a mindset is the best defense against entomological terrorism when one cannot know what agent will be used to sicken people or to decimate crops and livestock.

In terms of pest management, we need a system of educational programs and trained observers capable of recognizing new pests. But the federal budget for agricultural extension—the USDA's education and applied research function—has been losing ground for years. When a novel species is found, we need the expertise to make a rapid and definitive identification, but the nation's taxonomic expertise is appallingly limited. There must be either stockpiles of chemicals or the means for industry to rapidly respond to demand (as we saw during the West Nile virus outbreak, when a single city cornered the national market on insect repellent within weeks). Likewise, we need surge capacity in terms of aerial applicators, who increasingly struggle to stay in business. Finally, we must have the research and regulatory ability to move rapidly and effectively from chemical control to more sustainable practices, including biological control with carefully selected natural enemies.

Even if terrorists never attack with disease vectors or agricultural pests, the country can reap continuing benefits. New insect pests continue to infiltrate U.S. borders even without the assistance of terrorists. In many cases, we are deplorably slow to respond for lack of a strong pest-management infrastructure. Given the staggering losses from invasive species, an effective pest-management infrastructure would pay for itself whether or not our enemies resorted to smuggling particularly nasty insects into the country.

And so the highest-stake gamble in modern history may be whether the American government bets that terrorists will exploit their position with the occasional attempt to deliver a knock-out blow or with an incessant effort to sap our will to fight. Of course, the United States and other western nations need not devote their defensive resources to only one or the other of these strategies. But if history has any lessons to offer, putting no money on the insects is a very dangerous wager.

EPILOGUE

*Dusk descends on a sweltering New Orleans. A naked man lies in a fetal posi-
tion, sweating and moaning in an apartment a few blocks from Canal Street. His
jaundiced body is mottled with bruises where vessels have hemorrhaged. The pillow
and bedside are caked with what looks like coffee grounds but are drying gobs of
blackened, coagulated blood that he has vomited. The man's breathing is raspy and
labored as he slowly drowns in his own fluids.*

*The filthy window of the room is shut tightly, letting in no breath of air—and
letting out none of the tens of thousands of insects that cover the walls and the man's
body. The mosquitoes are* Aedes aegypti, *not the most common species along the Gulf
Coast, but easy to collect in huge numbers if one knows where to look. Anyone with a
course in medical entomology could build a simple trap and conscript a bloodthirsty
army.*

*Across the hall, another man cracks his door and peers out. Seeing nobody in
the hallway, he draws back into his room. A moment later he emerges with his
head covered in netting and wearing beekeepers' garb, then slips into the sickroom.
Brushing the whining mosquitoes away from the veil, he watches his suffering
compatriot. As a convulsion wracks the martyr's body, the feeding insects rise in a
ravenous cloud, droning their annoyance at having their meal disturbed.*

*Taking advantage of the moment, the garbed man crosses the room and opens
the window. Sensing the air currents and drawn toward the light, a cloud of
mosquitoes pours through the third-floor window, carrying a payload of yellow
fever into the sultry streets. The city's tropical heat, stagnant waters, crumbling
infrastructure, decrepit health-care system, and haggard people—nearly a quarter
million resolute souls after Katrina—will provide an ideal setting for an epidemic.
The man pulls a cell phone from his pocket and reads the coded text messages from
his associates in Houston and Miami. He smiles, brushes a mosquito from the key
pad, and dials the news desk at CNN.*

In 1981, William H. Rose, of the U.S. Army Test and Evaluation Command
at Dugway Proving Ground, wrote a report entitled *An Evaluation of
Entomological Warfare as a Potential Danger to the United States and European
NATO Nations*.[1] The document provided a chillingly prescient view of the
changing role of insects from weapons of war to tools of terrorism. The *Aedes*

aegypti/yellow fever "attack system" was seen as an ideal covert weapon for use against urban populations.

Twenty years later, the Biological Weapons Convention conference issued a report accusing rogue states of operating clandestine biological warfare programs. The villains included Iran, Iraq, Libya, Syria, and North Korea.[2] And the North Koreans, according to the Monterey Institute of International Studies, have been conducting research on entomological weapon systems, with yellow fever being the favored insect-borne disease.[3]

All of this might be taken as political paranoia, except that a recent authoritative analysis of biological threats to the United States put yellow fever at the top of list. Jack Woodall has the credentials to be taken very seriously in the field of bioterrorism. He is a virologist and epidemiologist currently serving as the director of the Nucleus for the Investigation of Emerging Infections Diseases in the Department of Medical Biochemistry at the Federal University of Rio de Janeiro. Woodall was previously the director of the New York State Department of Health's Arbovirus (short for arthropod-borne virus) Laboratory and has worked for the World Health Organization, the U.S. Centers for Disease Control and Prevention, and the East African Virus Research Institute in Uganda. With respect to biological warfare, Woodall cofounded the Swiss Disaster Relief Unit to respond to collateral damage to civilians in case biological, chemical, or nuclear weapons were used in the first Gulf War; he ran the World Health Organization's Iraq desk after that war; and he debriefed the leaders of the first chemical and biological inspection teams of UNSCOM.

In a 2006 article published in the *Scientist*, Woodall notes that, despite the near eradication of the yellow fever vector from the Americas, *Aedes aegypti* has reappeared in Florida, Louisiana, and Texas.[4] The six-legged home invader is stymied to some degree by the closed windows of air-conditioned buildings, but this line of defense is imperfect. Houston considers itself the most air-conditioned city in the world, but 6 percent of homes still lack this amenity, leaving nearly 120,000 people as prime targets.

There is a yellow fever vaccine, but travelers to endemic areas often don't bother to protect themselves—and they sometimes wish they had. In 2002, a 47-year-old man returned from an Amazonian fishing trip with a raging fever that progressed to the classic hemorrhaging of yellow fever. This fellow lived—and died—in Texas, which is now the home of *A. aegypti*.

Woodall warns that years of complacency and disease darlings of the media (bird flu being the current infatuation) have set up the American health-care system to miss cases of yellow fever, one of which will eventually seed an outbreak: "After all, what U.S. clinician is going to suspect yellow fever rather than malaria in a traveler returning home from the tropics with fever and vomiting? So how many times will the United States dodge the bullet?" And what if the bullet is not an accident?

To those who would dismiss this scenario as entomophobic hyperbole, consider West Nile virus and our impotent efforts to contain an insect-borne pathogen that arose from a single location to afflict people in 47 states. Consider the 654 dead and the 6,997 people suffering from debilitating neurological damage. Consider whether our ability to medicate humans and control insects makes entomological warfare and terrorism impossible in today's world.

Consider yourself lucky. So far.

SUGGESTED READINGS

Chapter 1

Biological and chemical warfare in early times is explored by Adrienne Mayor in *Greek Fire, Poison Arrows, and Scorpion Bombs: Biological and Chemical Warfare in the Ancient World* (New York: Overlook Duckworth, 2003). Edward Neufeld's paper, "Insects as Warfare Agents in the Ancient Near East," *Orientalia*, 49 (1980): 30–57, reconstructs entomological warfare via Jewish and Christian scripture. Overviews of bee-based weapons throughout history include John Free's *Bees and Mankind* (London: George Allen and Unwin Ltd., 1982) and Robert Sutherland's two-part "The Importance of Bees in War Time," *Southeastern Michigan Beekeepers' Association*, 13 (2003). A broader coverage of stinging insects can be found in a paper by John Ambrose, "Insects in Warfare," in *Army* (December 1974): 33–38.

Chapter 2

Adrienne Mayor's *Greek Fire, Poison Arrows, and Scorpion Bombs: Biological and Chemical Warfare in the Ancient World* (New York: Overlook Duckworth, 2003) describes how plant and insect toxins have been weaponized. The tale of Mithridates and his poisoned honey can be pieced together from a variety of sources, including Robin Seager's *Pompey the Great: A Political Biography* (Malden, Mass.: Blackwell, 2002), Peter Greenhalgh's *Pompey the Roman Alexander* (Columbia, Mo.: University of Missouri Press, 1981), and Frank Marsh's *A History of the Roman World from 146 to 30 BC* (London: Methuen, 1963). And Xenophon's misadventures with deli bal are nicely recounted in Mayor's book as well as Robert Root-Bernstein's article "Infectious Terrorism," *The Atlantic* (May 1991).

Chapter 3

Descriptions of how ants have been used as instruments of torture include *Once They Moved Like the Wind: Cochise, Geronimo, and the Apache Wars* by David

Roberts (New York: Simon & Schuster, 1994) and *Death in the Desert: The Fifty Years' War for the Great Southwest* by Paul Wellman (New York: Macmillan, 1935). The story of Nasrullah, Stoddart, Connolly, and the Bug Pit can be gleaned from Sir Alexander Burnes's *Travels into Bokhara* (London: Oxford University Press, 1973), Peter Hopkirk's *The Great Game: The Struggle for Empire in Central Asia* (New York: Kodansha, 1992), and Karl Meyer and Shareen Brysac's *Tournament of Shadows: The Great Game and the Race for Empire in Central Asia* (Washington, D.C.: Counterpoint, 1999). For more about the pit's denizens, consider Dunston Ambrose's *Assassin Bugs* (Fairfield, N.H.: Science Publishers, 1999).

Chapter 4

In *Greek Fire, Poison Arrows, and Scorpion Bombs: Biological and Chemical Warfare in the Ancient World*, Adrienne Mayor links fleas to the Ark of the Covenant. The bubonic plague outbreak that started in Kaffa has been variously interpreted by Andrew Robertson in "From Asps to Allegations: Biological Warfare in History," *Military Medicine*, 160 (1995): 369–373; and Eric Croddy in *Chemical and Biological Warfare: A Comprehensive Survey for the Concerned Citizen* (New York: Springer-Verlag, 2002). This infamous incident, along with similar ventures throughout history, is recounted by Erhard Geissler and John Ellis van Courtland Moon in *Biological and Toxin Weapons: Research, Development and the Use from the Middle Ages to 1945* (New York: Oxford University Press, 1999).

Chapter 5

Perhaps the best overview of Napoleon's entomological trials and tribulations is Robert Peterson's article "Insects, Disease, and Military History" in *American Entomologist*, 41 (1995): 147–160. David Chandler's *The Campaigns of Napoleon* (New York: Scribners, 1973) is a monumental account of the French leader's military ventures, including those in which insect-borne disease had the upper hand. The disastrous French occupation of Haiti—thanks in large part to the ravages of yellow fever—is recounted by Philippe Girard in *Paradise Lost: Haiti's Tumultuous Journey from Pearl of the Caribbean to Third World Hotspot* (New York: Palgrave Macmillan, 2005). The biology and ecology of the insect vectors of bubonic plague, yellow fever, and typhus can be found in *Entomology in Human and Animal Health* by Robert Harwood and Maurice James (New York: Macmillan, 1979).

Chapter 6

The role of yellow fever during the Spanish-American and Civil Wars is described by Harvey Schultz, in the chapter "100 Years of Entomology in the Department of Defense," in *Insect Potpourri: Adventures in Entomology* (Gainesville, Fla.: Sandhill Crane Press, 1992), and by Molly Crosby in *The American Plague* (New York:

Berkley, 2007). The importance of illness during the U.S. Civil War is powerfully portrayed by Paul Steiner in *Disease in the Civil War: Natural Biological Warfare in 1861–1865* (Springfield, Ill.: Charles C. Thomas, 1968). Gary Miller's 1997 paper, "Historical Natural History: Insects and the Civil War," *American Entomologist*, 43 (1997): 227–245, is an authoritative account of the importance of insect vectors. For information on malaria and its vectors, see *Medical and Veterinary Entomology*, edited by Gary Mullen and Lance Durden (New York: Elsevier, 2002).

Chapter 7

Biological warfare efforts in World War I are described by Mark Wheelis, in the chapter "Biological Sabotage in World War I," in *Biological and Toxin Weapons: Research, Development and the Use from the Middle Ages to 1945* (New York: Oxford University Press, 1999). Although the World Wide Web is notoriously suspect as a source of information, the Web site "Insects, Disease and History," entomology.montana.edu/historybug, edited by Drs. Gary Miller and Robert Peterson, is a first-rate source of historically and entomologically credible information, including the entry by David Tschanz on "Typhus Fever on the Eastern Front in World War I."

Chapters 8–11

The biological warfare program of the Japanese in World War II is extensively documented, but the entomological weapons development and delivery is most vividly recounted in *Unit 731 Testimony* by Hal Gold (Singapore: Yen Books, 1996), *Factories of Death: Japanese Biological Warfare, 1932–1945, and the American Cover-Up* by Sheldon Harris (New York: Routledge, 2002), and *A Plague Upon Humanity: The Secret Genocide of Axis Japan's Germ Warfare Operation* by Daniel Barenblatt (New York: HarperCollins, 2004). The chilling transcripts of the Khabarovsk Trial are available as *Materials on the Trial of Former Servicemen of the Japanese Army Charged with Manufacturing and Employing Bacteriological Weapons* (Moscow: Foreign Languages Publishing House, 1950). A broader historical perspective is provided by Robert Harris and Jeremy Paxman in *A Higher Form of Killing: The Secret History of Chemical and Biological Warfare* (New York: Random House, 2002).

Chapter 12

The weaponization of the Colorado potato beetle is described by Erhard Geissler and John Ellis van Courtland Moon in *Biological and Toxin Weapons: Research, Development and the Use from the Middle Ages to 1945* (New York: Oxford University Press, 1999) and by Benjamin Garrett in his 1996 paper, "The Colorado Potato Beetle Goes to War," *Chemical Weapons Convention Bulletin*, 33 (1996): 2–3.

The link between entomological warfare and the development of insecticides that led to the discovery of nerve gas is recounted by Eric Croddy in *Chemical and Biological Warfare: A Comprehensive Survey for the Concerned Citizen* (New York: Springer-Verlag, 2002).

Chapter 13

The Allied efforts to develop entomological weapons can be gleaned from various sources, including Ed Regis's *The Biology of Doom: The History of America's Secret Germ Warfare Project* (New York: Henry Holt, 2000); B. J. Bernstein's article "The Birth of the U.S. Biological-Warfare Program," *Scientific American,* 256 (1987): 116–121; Norman Covert's *Cutting Edge: The History of Fort Detrick,* available at detrick.army.mil/detrick/cutting_edge; and Milton Leitenberg's paper "Biological Weapons in the Twentieth Century: A Review and Analysis" for the 7th International Symposium on Protection against Chemical and Biological Warfare in Stockholm, Sweden, 2001. The massive delousing program was described by C. M. Wheeler in "Control of Typhus in Italy 1943–1944 by Use of DDT," *American Journal of Public Health,* 36 (1946): 119–129.

Chapters 14–16

Excellent descriptions of entomological warfare in Korea can be found in Robin Clarke's *The Silent Weapons* (New York: David McKay, 1968), John Cookson and Judith Nottingham's *A Survey of Chemical and Biological Warfare* (New York: Monthly Review, 1969), and Stephen Endicott and Edward Hagerman's *The United States and Biological Warfare: Secrets from the Early Cold War and Korea* (Bloomington: Indiana University Press, 1998). The original *Report of the International Scientific Commission for the Investigation of the Facts Concerning Bacterial Warfare in Korea and China* (Peking: International Scientific Commission, 1952) is hard to find but fascinating, and the most strident refutation of this document is Milton Leitenberg's 1998 paper, "The Korean War Biological Warfare Allegations Resolved," Center for Pacific Asia Studies at Stockholm University, Occasional Paper 36.

Chapter 17

The weaponization of insects during the Cold War is recounted in Alistair Hay's paper, "A Magic Sword or a Big Itch: An Historical Look at the United States Biological Weapons Programme," *Medicine, Conflict and Survival,* 15 (1999): 215–234. Portions of *An Evaluation of Entomological Warfare as a Potential Danger to the United States and European NATO Nations* by William H. Rose (Dugway, Utah: U.S. Army Test and Evaluation Command, 1981) can be viewed at thesmokinggun.com/archive/mosquito1.html. Other good sources include Ed Regis's *The Biology of Doom: The History of America's Secret Germ Warfare Project* (New York: Henry Holt, 2000) and Robert Harris and Jeremy Paxman's *A Higher Form of Killing: The Secret History of Chemical and Biological Warfare* (New York: Random House, 2002).

Chapter 18

The conscription of insects during the Vietnam War is recounted in the paper by John Ambrose, "Insects in Warfare," *Army* (December 1974): 33–38, and in my paper "Entomological Warfare: History of the Use of Insects as Weapons of War," *Bulletin of the Entomological Society of America*, 33 (1987): 76–82. The passive use of plague as a weapon is described by John Cookson and Judith Nottingham in *A Survey of Chemical and Biological Warfare* (New York: Monthly Review, 1969). The yellow rain controversy can be found in the articles by Thomas Seeley et al., "Yellow Rain," *Scientific American* (September 1985): 128–137, and by Jonathan Tucker, "The 'Yellow Rain' Controversy: Lessons for Arms Control Compliance," *The Nonproliferation Review* (Spring 2001): 25–41.

Chapters 19–20

The dengue outbreak is described in "Hemorrhagic Dengue in Cuba: History of an Epidemic" by Gustavo Kourí, María G. Guzmán, and José Bravo, *PAHO Bulletin,* 20 (1986): 24–30, and "U.S. Biological Warfare: The 1981 Cuba Dengue Epidemic" by Bill Schaap, *Covert Action* (Summer 1982): 28–31. The thrips case is recounted by Eric Croddy in *Chemical and Biological Warfare: A Comprehensive Survey for the Concerned Citizen* (New York: Springer-Verlag, 2002) and on the Web site afrocubaweb.com/biowar.htm. Litanies of Cuban accusations can be found at greenleft.org.au/back/2003/561/561p12.htm and cuba.cu/gobierno/documentos/2002/ing/m240502i.html. An American refutation is provided by Milton Leitenberg in "Biological Weapons in the Twentieth Century: A Review and Analysis," 7th International Symposium on Protection against Chemical and Biological Warfare, Stockholm, Sweden, 2001.

Chapter 21

The use of Medflies as weapons is addressed in *Countering Agricultural Bioterrorism* (Washington, D.C.: National Research Council of the National Academies, 2003). Newspaper coverage of the Breeders' case includes "Female Medfly Found in Sun Valley Close to Area Targeted Earlier," *Los Angeles Times* (January 4, 1990): B3; "Officials Advertise to Contact Mystery Group Claiming Medfly Releases," *Los Angeles Times* (February 10, 1990): A13; and "Mystery Letter Puts a Strange Twist on Latest Medfly Crisis," *Los Angeles Times* (December 3, 1988): B1. The Breeders' letter and associated government communications were acquired through the Freedom of Information Act. Further insights were provided through interviews with Pat Minyard, acting director of Plant Health and Pest Prevention Services, California Department of Agriculture, and Jim Reynolds, western regional director for USDA-APHIS.

Chapter 22

How insects might be used to attack agriculture is discussed in SIPRI's *Environmental Warfare: A Technical, Legal, and Policy Appraisal* (London: Taylor & Francis, 1984), ed. A. H. Westing, and "Food and Agricultural Security," *Annals of the New York Academy of Sciences*, 894 (1999), ed. T. W. Frazier and D. C. Richardson. The damage done by invasive species is described in "New York's Battle with the Asian Long-Horned Beetle" by R. A. Haack et al., *Journal of Forestry*, 95 (1997): 11–15; S. B. Vinson, "Invasion of the Red Imported Fire Ant: Spread, Biology, and Impact," *American Entomologist* (Spring 1997): 23–38; and *California Agricultural Research Priorities: Pierce's Disease* (Washington, D.C.: National Academies Press, 2004).

Chapter 23

The story of West Nile virus in the United States is summarized by Judith Miller, Stephen Engelberg, and William Broad in *Germs: Biological Weapons and America's Secret War* (New York: Simon & Schuster, 2002), and the possibility that WNV was bioterrorism is explored by Richard Preston in "West Nile Mystery," *The New Yorker* (October 18–25, 1999): 90–107. The potential of Rift Valley fever was explored in interviews with Drs. Geoff Letchworth (former director of the USDA-ARS Arthropod-Borne Animal Disease Research Laboratory), Charles Bailey (director of research at the National Center for Biodefense), and Robert Kadlec (adviser to the Senate Subcommittee on Bioterrorism and Public Health within Health, Education, Labor).

Chapter 24

The use of assassin bugs to detect the enemy in Vietnam is described in the paper by John Ambrose, "Insects in Warfare," *Army* (December 1974): 33–38. How firefly biochemistry and behavior have been used to develop detectors of poison gas and biological agents is recounted by Robin Clarke in *The Silent Weapons* (New York: David McKay, 1968), John Cookson and Judith Nottingham in *A Survey of Chemical and Biological Warfare* (New York: Monthly Review, 1969), and Robert Harris and Jeremy Paxman in *A Higher Form of Killing: The Secret History of*

Chemical and Biological Warfare (New York: Random House, 2002). The use of bees as mine detectors is explained by Jerry Bromenshenk et al. in "Can Honey Bees Assist in Area Reduction and Landmine Detection?" *Journal of Mine Action*, 7.3 (2003).

Chapter 25

Accounts of entomopters can be found in "Microspies" by P. Garrison, *Air and Space* (April/May 2000): 54–61; and "Micro Warfare," *Popular Mechanics* (February 2001): 62–65. The potential of genetic engineering of insects is considered in *Biotechnology and the Future of the Biological and Toxin Weapons Convention,* SIPRI Fact Sheet, November 2001; and Raymond Zilinskas's paper "Possible Terrorist Use of Modern Biotechnology Techniques," 2000 Conference on Biosecurity and Bioterrorism, Rome (mi.infn.it/~landnet/Biosec/zilinskas1. pdf). Various Web sites developed by university researchers provide insights on insect-based robotic systems, including enme.ucalgary.ca/~aramirez/AR2S-Lab-Projects.html (University of Calgary's hexapod device), conceptlab.com/roachbot/ (University of California Irvine's cockroach cyborg), and neuromechanics.cwru. edu/news/igertnews3.htm and biorobots.cwru.edu/projects/billant/ (Case Western Reserve University's cockroach and ant projects).

Chapter 26

Insect-based attacks on agricultural targets and U.S. preparedness are explored in a number of sources, including "A Plague of Locusts" by Benjamin Garrett *Nonproliferation, Demilitarization, and Arms Control*, 5–6 (2000): 11–12; "Agricultural Biological Warfare: An Overview" by Jonathan Ban, *Chemical and Biological Arms Institute*, 9 (June 2000); "Biological Terrorism Targeted at Agriculture: The Threat to U.S. National Security" by Rocco Casagrande, *Nonproliferation Review* (Fall/Winter 2000): 92–105; *Covert Biological Weapons Attacks Against Agricultural Targets: Assessing the Impact Against U.S. Agriculture,* by Jason Pate and Gavin Cameron, Discussion Paper 2001-9 of the Belfer Center for Science and International Affairs; *Countering Agricultural Bioterrorism* (Washington, D.C.: National Research Council of the National Academies, 2003); and "Assessing the Agroterror Threat" by Steve Nash, *The Scientist* (May 10, 2004): 50–51.

NOTES

Introduction

1. Anne Mitchell, "Africanized Killer Bees: A Case Study," *Critical Care Nurse,* 26 (2006): 23–31.

Chapter 1

1. John Ambrose, "Insects in Warfare," *Army* (December 1974): 33–38.
2. Greek mythology includes the story of the Myrmidons, created by Zeus. The god populated the island of Aegina by transforming the ants into a race of people. The Myrmidons were as fierce and loyal as the ants for which they were named (*myrmi-* means "ants"). According to legend, the ant-people fought alongside Achilles in the Trojan War (from the account provided by *Encyclopedia Mythica* at pantheon.org/articles/m/myrmidons.html; accessed January 10, 2008).
3. Edward Neufeld, "Insects as Warfare Agents in the Ancient Near East," *Orientalia,* 49 (1980): 30–57.
4. Bees have an unusual genetic condition called haplodiploidy, which results in sisters sharing three-quarters of their genes with one another. Human siblings, on the other hand, have only a quarter of their genes in common.
5. Neufeld, "Insects as Warfare Agents."
6. Bernard W. Anderson, *Understanding the Old Testament* (New York: Prentice-Hall, 1960).
7. Roger S. Wotton's paper "The Ten Plagues of Egypt," *in Opticon 1826,* 11 (August 2007), provides a scientific explanation of the events; his publication can be accessed at ucl.ac.uk/opticon1826/currentissue/article/RfP_Art_LIFE_Wotton_Plagues.pdf (accessed January 10, 2008). In addition, a reasonably concise and ecologically plausible account of the plagues of Egypt can be found in an excellent article at the University of Saskatchewan's Web site, geochemistry.usask.ca/bill/Courses/Climate/Disturbance%20&%20Declin

e%20prt.pdf (accessed January 10, 2008). A similar description is provided via the Xyroth Enterprises Web site, xyroth-enterprises.co.uk/10plague.htm (accessed January 10, 2008).

8. The precise translation is actually "vermin," although some texts refer to the creatures as "maggots." In either case, the most likely insect accounting for the passage would seem to be biting midges, based on what we know of the biology and ecology of the region and the aspects of preceding and subsequent events.

9. See note 7.

10. Technically speaking, insects are not vectors of disease; they are vectors of the pathogens that cause disease. Hence, biologists would insist that instead of saying, "Flies are vectors of African horse sickness," one should say, "Flies are the vectors of the virus that causes African horse sickness." Such precision requires an awkward wordiness, so in this book I'll refer to insects and their kin as vectors of disease. The reader is asked to understand that by this I mean that the creatures are transmitting the pathogen that is the cause of the disease.

11. See note 7. In addition, these explanations can be found in the National Geographic program *The Bible Uncovered: Exodus* (that aired in 2007 and 2008), in which I provided a segment on the plagues of Egypt that were directly attributable to insects.

12. Neufeld, "Insects as Warfare Agents."

13. Adrienne Mayor, *Greek Fire, Poison Arrows and Scorpion Bombs: Biological and Chemical Warfare in the Ancient World* (New York: Overlook Duckworth, 2003), chap. 6.

14. The African people provide perhaps the richest history of using bees as weapons. As recounted by Leonard Mosley, in *Duel for Kilimanjaro: The East African Campaign 1914–18* (London: Weidenfeld and Nicholson, 1963), during the First World War, Tanzanian natives, who were allied with their German colonists, set entomological booby traps for the British infantry. The Africans laced the bush with trip wires connected to the lids of concealed beehives. Adrienne Mayor (*Greek Fire*) describes an incident in the Second World War when, as Italian tanks rolled through the Ethiopian highlands, the Africans bombarded the invaders with beehives. Taking umbrage at this mistreatment, the insects assailed the drivers, and in the ensuing chaos, several tanks careened down the mountainside and were destroyed.

15. Neufeld, "Insects as Warfare Agents," and Ambrose, "Insects in Warfare."

16. Neufeld, "Insects as Warfare Agents."

17. David Whitehead, *Aineias the Tactician: How to Survive under Siege*, translation and commentary (Oxford, U.K.: Clarendon Press, 1990).

18. Mayor, *Greek Fire*, chap. 6.

19. A taxonomic clarification is in order. The phylum Arthropoda includes a plethora of organisms, with the most familiar groups being Crustacea (the crusta-

ceans are represented by crabs, lobsters, shrimp, and barnacles), Diplopoda (the millipedes), Chilopoda (the centipedes), Arachnida (including the Acari or mites and ticks, Scorpiones or scorpions, Araneae or spiders, and several more obscure orders), and Hexapoda (the insects). The reader will hopefully forgive my expansion of entomological warfare to include those organisms that are not taxonomically speaking insects but bear biologically relevant and militarily important similarities such as a propensity to inject venom (scorpions and spiders) and to transmit microbial pathogens (mites and ticks).

20. Mayor, *Greek Fire,* chap. 6.
21. Ibid. She notes that Aelianus's *On Animals* is available in translation in the Loeb Classical Library, published by Harvard University Press.
22. John B. Free, *Bees and Mankind* (London: George Allen & Unwin, 1982).
23. Ambrose, "Insects in Warfare."
24. Ibid.
25. Mayor, *Greek Fire,* chap. 6.
26. Ambrose, "Insects in Warfare."
27. The possible role of bee boles in defense is addressed in Michael Burgett's course on "Plagues, Pests and Politics," with the relevant information at ent. orst.edu/burgettm/ent300_lecture14.htm (accessed January 10, 2008). Gene Kritsky of the College of Mount St. Joseph also provided insights concerning the placement and function of bee boles.
28. The account can be found in an article by Conrad Bérubé at the Apiservices (apicultural) Web site, apicultura.com/articles/us/war_bees.htm (accessed January 10, 2008).
29. Ibid.
30. Tracey Rihll, *Catapult: A History* (Yardley, Pa.: Westholme Publishing, 2007), and Richard Holmes, ed., *The Oxford Companion to Military History* (New York: Oxford University Press, 2001), entry on "siege engines."
31. Ambrose, "Insects in Warfare"; Mayor, *Greek Fire,* chap. 6.
32. See note 27 and Andrew G. Robertson, "From Asps to Allegations: Biological Warfare in History, *Military Medicine,* 160 (1995): 369–373.
33. Ambrose, "Insects in Warfare."

Chapter 2

1. The account can be found in an article by Conrad Bérubé at the Apiservices (apicultural) Web site, apicultura.com/articles/us/war_bees.htm (accessed January 10, 2008).
2. So as not to confuse various terms, a *poison* is any chemical substance causing injury or death to a living organism (e.g., plutonium, mustard gas, ethanol, curare, and bee venom). A *toxin* is a biologically produced poison (e.g.,

ethanol, curare, and bee venom). A *venom* is a toxin that is injected by an organism's bite or sting (e.g., bee venom).

3. Adrienne Mayor, *Greek Fire, Poison Arrows and Scorpion Bombs: Biological and Chemical Warfare in the Ancient World* (New York: Overlook Duckworth, 2003), chap. 2. Hamish Robertson has a fine article on the subject of the San people's use of poisonous beetles at the Biodiversity Explorer Web site: biodiversityexplorer. org/beetles/chrysomelidae/alticinae/arrows.htm (accessed January 10, 2008).

4. Mayor, *Greek Fire,* chap. 2.

5. Ibid.; J. H. Frank and K. Kanamitsu, "*Paederus,* sensu lato (Coleoptera: Staphylinidae): Natural History and Medical Importance," *Journal of Medical Entomology,* 24 (1987): 155–191.

6. R. K. Armstrong and J. L. Winfield, "*Paederus fuscipes* Dermatitis: An Epidemic on Okinawa," *American Journal of Tropical Medicine and Hygiene,* 18 (1969): 147–150.

7. J. Piel, I. Höfer, and D. Hui, "Evidence for a Symbiosis Island Involved in Horizontal Acquisition of Pederin Biosynthetic Capabilities by the Bacterial Symbiont of *Paederus fuscipes* Beetles," *Journal of Bacteriology,* 186 (2004): 1280–1286.

8. Mayor, *Greek Fire,* chap. 2.

9. Ibid., chap. 3.

10. Ibid., chap. 5.

11. Ibid.

12. Ibid.; Robert S. Root-Bernstein, "Infectious Terrorism," *The Atlantic* (May 1991): 44–50.

13. Using bees against Romans became something of an ancient refrain. Not only were these insects effective in the tunnels under Eupatoria and in the passes above Colchis, but the classical poet Virgil kept Caesar's soldiers from looting his valuables by storing them in beehives—a tactic mirrored two millennia later, when Otto Wiltschko, an East German spy, posed as a beekeeper and secreted a radio receiver in one of the hives. Unfortunately for Otto, his career as a spy was ended by Austrian authorities, who were less intimidated by stinging insects than were Roman soldiers (sustainable-gardening-tips.com/ Bees-in-Wartime.html; accessed January 10, 2008).

Chapter 3

1. Richard Sair, *The Book of Torture and Executions* (Toronto: Golden Books, 1944); and a Wikipedia entry, en.wikipedia.org/wiki/Scaphism (accessed January 10, 2008).

2. David Quammen, *Natural Acts* (New York: Nick Lyons Books/Schocken Books, 1985).

3. Sair, *Book of Torture and Executions.*

4. Falun Dafa Clearwisdom Web site, clearwisdom.net/emh/articles/2004/6/11/49032.html (accessed May 30, 2009).

5. In *The Gulag Archipelago: 1918–1956*, Solzhenitsyn describes how Soviet jailers used bed bugs to torture prisoners:

> In the dark closet made of wooden planks, there were hundreds, maybe even thousands, of bedbugs, which had been allowed to multiply. The guards removed the prisoner's jacket or field shirt, and immediately the hungry bedbugs assaulted him, crawling onto him from the walls or falling off the ceiling. At first he waged war with them strenuously, crushing them on his body and on the walls, suffocated by their stink. But after several hours he weakened and let them drink his blood without a murmur.

Along with the psychological trauma came physical suffering. After days of this treatment, a victim would experience chills, burning of the eyes, a painfully swollen tongue, and muscle spasms upon swallowing. These bizarre symptoms were probably due to a combination of blood loss, dehydration, and reactions to the allergens, anticoagulants, and other chemicals in the bugs' saliva.

6. Gordon C. Baldwin, *A Story of the Chiricahua and Western Apache* (Tucson, Ariz.: Dale Stuart King, 1965); Paul I. Wellman, *Death in the Desert: The Fifty Years' War for the Great Southwest* (New York: Macmillan, 1935); David Roberts, *Once They Moved Like the Wind: Cochise, Geronimo, and the Apache Wars* (New York: Simon & Schuster, 1993).

7. Jesse Green, ed., *Zuni: Selected Writings of Frank Hamilton Cushing* (Lincoln: University of Nebraska Press, 1979).

8. The story of Stoddart and Conolly is pieced together from Edward A. Allworth, *The Modern Uzbeks: From the Fourteenth Century to the Present, A Cultural History* (Stanford, Calif.: Hoover Institution Press, 1990); Alexander Burnes, *Travels into Bokhara* (London: Oxford University Press, 1973); Peter Hopkirk, *The Great Game: The Struggle for Empire in Central Asia* (New York: Kodansha, 1992); Fitzroy Maclean, *Back to Bokhara* (Oxford, U.K.: Alden, 1959); and Karl E. Meyer and Shareen Blair Brysac, *Tournament of Shadows: The Great Game and the Race for Empire in Central Asia* (Washington, D.C.: Counterpoint, 1999).

9. Josef Wolff, "Ameer of Bokhara Nasir Ullah," *Report of Josef Wolff 1843–1845*, available at geocities.com/Athens/5246/amir.html (accessed January 10, 2008).

10. Hopkirk, *Great Game*.

11. Penny J. Gullan and Peter S. Cranston, *The Insects: An Outline of Entomology* (Malden, Mass.: Blackwell, 2000).

12. Maclean, *Back to Bokhara*.

13. Ibid.

14. Ibid.

15. Michael Isikoff and Evan Thomas, "The Lawyer and the Caterpillar," *Newsweek,* newsweek.com/id/194595 (accessed May 30, 2009); Ewen MacAskill, "Bush Officials Defend Physical Abuse Described in Memos Released by Obama," *Guardian,* guardian.co.uk/world/2009/apr/17/bush-torture-memos-obama-mukasey (accessed May 30, 2009).

16. The memo from Jay Bybee, then chief of the Justice Department's Office of Legal Counsel, can be found at luxmedia.vo.llnwd.net/o10/clients/aclu/olc_08012002_ bybee.pdf (accessed May 30, 2009). In part, the memo states:

> In addition to using the containment boxes alone, you also would like to introduce an insect into one of the boxes with Zubaydah. As we understand it, you plan to inform Zubaydah that you are going to place a stinging insect into the box, but you will actually place a harmless insect in the box, such as a caterpillar. If you do so, to ensure that you are outside the predicate act requirement, you must inform him that the insects will not have a sting that would produce death or severe pain. If, however, you were to place the insect in the box without informing him that you are doing so, then, in order to not commit a predicate act, you should not affirmatively lead him to believe that any insect is present which has a sting which could produce severe pain or suffering or even cause his death [section of text redacted] so long as you take either of the approaches we have described, the insect's placement in the box would not constitute a threat of severe physical pain or suffering to a reasonable person in his position. An individual placed in a box, even an individual with a fear of insects, would not reasonably feel threatened with severe physical pain or suffering if a caterpillar was placed in the box. Further, you have informed us that you are not aware that Zubaydah has any allergies to insects, and you have not informed us of any other factors that would cause a reasonable person in that same situation to believe that an unknown insect would cause him severe physical pain or death. Thus, we conclude that the placement of the insects in the confinement box with Zubaydah would not constitute a predicate act.

Rather less reliable sources indicate that American interrogators may have used insects to extract information in other contexts as well. In the course of a convoluted hearsay testimony given at a military tribunal in 2007, the father of a Guantanamo detainee alleged that Pakistani guards had confessed that American interrogators used ants to frighten children. Two boys—aged seven and nine—were suspecting of knowing the location of their father, Khalid Sheikh Mohammed, the purported mastermind of the 9/11 attacks. According to the detainee's father, "The boys were kept in a separate area upstairs and were denied food and water. . . . They were also mentally tortured by having ants or other creatures put on their legs to scare them and get them to say where their father was hiding." John Byrne, "Bush Memos Parallel Claim 9/11 Mastermind's Children Were Tortured with Insects," available at rawstory.com/08/blog/2009/04/17/bush-torture-memos-align-with-account-that-911-suspects-children-were-tortured/ (accessed May 30, 2009).

Chapter 4

1. Adrienne Mayor, *Greek Fire, Poison Arrows and Scorpion Bombs: Biological and Chemical Warfare in the Ancient World* (New York: Overlook Duckworth, 2003), chap. 3.
2. Ibid.
3. Ibid.
4. Ibid.
5. Ibid., chap. 4.
6. William H. Robinson, *Handbook of Urban Insects and Arachnids* (New York: Cambridge University Press, 2005).
7. Erhard Geissler and John Ellis van Courtland Moon, eds., *Biological and Toxin Weapons: Research, Development and the Use from the Middle Ages to 1945*, SIPRI Biological and Chemical Warfare Studies (New York: Oxford University Press, 1999).
8. Ibid.
9. The story of the siege of Kaffa and the consequent plague is assembled from Eric Croddy, *Chemical and Biological Warfare: A Comprehensive Survey for the Concerned Citizen* (New York: Springer-Verlag, 2002); Geissler and Moon, *Biological and Toxin Weapons*; Erhard Geissler, ed., *Biological and Toxin Weapons Today*, SIPRI Biological and Chemical Warfare Studies (New York: Oxford University Press, 1986); Mayor, *Greek Fire,* chap. 4; Andrew G. Robertson, "From Asps to Allegations: Biological Warfare in History," *Military Medicine,* 160 (1995): 369–373.
10. Geissler, *Biological and Toxin Weapons Today*, p. 14.
11. See note 9.
12. Robert Harwood and Maurice James, *Entomology in Human and Animal Health* (New York: Macmillan, 1979); and Gary Mullen and Lance Durden, eds., *Medical and Veterinary Entomology* (New York: Elsevier, 2002).
13. Johannes Nohl, *The Black Death: A Chronicle of the Plague* (Yardley, Pa.: Westholme, 2006).
14. Geissler and Moon, *Biological and Toxin Weapons*.
15. Ibid.

Chapter 5

1. Robert K. D. Peterson, "Insects, Disease, and Military History," *American Entomologist*, 41 (1995): 147–160.
2. Adrienne Mayor, *Greek Fire, Poison Arrows and Scorpion Bombs: Biological and Chemical Warfare in the Ancient World* (New York: Overlook Duckworth, 2003), chap. 1.
3. Ibid., chap. 4.
4. Hans Zinsser, *Rats, Lice and History* (Edison, N.J.: Transaction Publishers, 2007), p. 112.

5. Peterson, "Insects, Disease, and Military History."

6. Ibid.

7. Robert Harwood and Maurice James, *Entomology in Human and Animal Health* (New York: Macmillan, 1979); and Gary Mullen and Lance Durden, eds., *Medical and Veterinary Entomology* (New York: Elsevier, 2002).

8. Although the arrival of *A. aegypti* in the New World was almost surely unintentional, the invasion of Hawaii by mosquitoes may have been a bizarre act of entomological warfare. These islands were free of mosquitoes until 1826, when the *Wellington* anchored at Maui. All accounts agree that the whaling ship's drinking-water casks were swimming with mosquito larvae. But here the stories diverge. According to one version, the sailors inadvertently released the insects into the local streams while freshening the ship's water supplies. A more intriguing account has it that the randy whalers were looking forward to fraternizing with the island's women. Knowing sailors, the local chief forbade them access to his village, and the men took their revenge by intentionally dumping the infested casks into nearby streams. This explanation was kindly provided by Dr. Dennis A. LaPointe of the USGS–Pacific Island Ecosystem Research Center in Hawaii.

9. Molly Caldwell Crosby, *The American Plague: The Untold Story of Yellow Fever, the Epidemic that Shaped Our History* (New York: Berkley, 2006).

10. The symptoms of yellow fever are described at the Centers for Disease Control and Prevention Web site: cdc.gov/ncidod/dvbid/yellowfever/index.htm (accessed January 10, 2008).

11. Peterson, "Insects, Disease, and Military History."

12. But alas, the French had not learned their lesson. When Ferdinand de Lesseps brought 500 young French engineers and 20,000 workmen to dig a canal across Panama in 1884, one-third of the laborers died and not a single engineer lived to draw his first month's pay; Peterson, "Insects, Disease and Military History."

13. Peterson, "Insects, Disease, and Military History."

14. Harwood and James, *Entomology in Human and Animal Health*; and Mullen and Durden, *Medical and Veterinary Entomology*.

15. Harwood and James, *Entomology in Human and Animal Health*; and Mullen and Durden, *Medical and Veterinary Entomology*.

Chapter 6

1. Gary L. Miller, "Historical Natural History: Insects and the Civil War," *American Entomologist*, 43 (1997): 227–245; online version available at http://entomology.montana.edu/historybug/civilwar2/flies.htm.

2. Ibid.

3. Paul E. Steiner, *Disease in the Civil War: Natural Biological Warfare in 1861–1865* (Springfield, Ill.: Charles C. Thomas, 1968), chap. 1.

4. Ibid.

5. Miller, "Historical Natural History: Insects and the Civil War."

6. Steiner, *Disease in the Civil War*, chap. 1.

7. The Centers for Disease Control and Prevention Web site provides information on the etiology, symptoms, treatment, and prevention of both malaria (cdc.gov/malaria/index.htm) and typhoid (cdc.gov/ncidod/dbmd/diseaseinfo/typhoidfever_g.htm; accessed January 14, 2008).

8. Robert Harwood and Maurice James, *Entomology in Human and Animal Health* (New York: Macmillan, 1979); and Gary Mullen and Lance Durden, eds., *Medical and Veterinary Entomology* (New York: Elsevier, 2002).

9. Miller, "Historical Natural History: Insects and the Civil War."

10. Steiner, *Disease in the Civil War*, p. 214.

11. Ibid., p. 219.

12. Ibid., chap. 6.

13. David W. Tschanz, "Yellow Fever and the Strategy of the Mexican-American War," Insects, Disease and History Web site, Entomology Group of Montana State University, entomology.montana.edu/historybug/mexwar/mexwar.htm (accessed January 14, 2008).

14. Steiner, *Disease in the Civil War*, p. 9.

15. Ibid., p. 195.

16. Ibid., chap. 5.

17. Ibid., p. 139.

18. Ibid.

19. Ibid., p. 130.

20. Miller, "Historical Natural History: Insects and the Civil War."

21. Ibid.

22. Jeffrey A. Lockwood, "Entomological Warfare: History of the Use of Insects as Weapons of War," *Bulletin of the Entomological Society of America*, 33 (1987): 76–82.

23. The conceptual seed of converting crop pests into warriors may have been planted during the Revolutionary War. After Hessian mercenaries (employed by the British) passed through New York, the farmers reported that the wheat crop was being severely damaged by the larvae of tiny midges. The Americans named the insect invader the "Hessian fly" and blamed the foreigners for having brought the pest in their straw bedding. Although nobody accused the Germans of intentionally introducing the fly, the potential for devastating an enemy's crops with a foreign insect could not have escaped military minds. The history of the Hessian fly in the United States is addressed in Michael Burgett's course on "Plagues, Pests and Politics," with the relevant information at ent.orst.edu/burgettm/ent300_lecture14.htm (accessed January 14, 2008).

24. Molly Caldwell Crosby, *The American Plague: The Untold Story of Yellow Fever, the Epidemic that Shaped Our History* (New York: Berkley, 2006); and Andrew G. Robertson, "From Asps to Allegations: Biological Warfare in History," *Military Medicine*, 160 (1995): 369–373.

Chapter 7

1. Richard Holmes, ed., *The Oxford Companion to Military History* (New York: Oxford University Press, 2001), entries on the "Eastern Front (1914–1918)" and "World War I."

2. David W. Tschanz, "Typhus Fever on the Eastern Front in World War I," Insects, Disease and History Web site, Entomology Group of Montana State University, entomology.montana.edu/historybug/WWI/TEF.htm (accessed January 18, 2008).

3. J. H. Frank and K. Kanamitus, "*Paederus*, sensu lato (Coleoptera: Staphylinidae): Natural History and Medical Importance," *Journal of Medical Entomology*, 24 (1987): 155–191.

4. Tschanz, "Typhus Fever on the Eastern Front in World War I"; and Robert K. D. Peterson, "Insects, Disease, and Military History," *American Entomologist*, 41 (1995): 147–160.

5. George B. Johnson, "The Battle Against Infectious Diseases," an article on the bioterrorism section of his *Backgrounders* Web site, txtwriter.com/Backgrounders/Bioterrorism/bioterror2.html (accessed January 18, 2008).

6. Lice were not the only insects to flourish in the trenches, nor were all these coinhabitants harmful to the soldiers. The dark, wet conditions were suitable for cave-dwelling glow worms. And these insects—along with dogs, monkeys, and bears—were honored in Britain's "Animals in War" memorial. Glow worms (family Lampyridae) were credited with providing light necessary for the troops to read maps in nighttime preparation for decisive assaults during the war; "UK Honors Glow Worm Heroes," CNN Web site, edition.cnn.com/2004/WORLD/europe/11/24/uk.newwaranimals/index.html (accessed January 18, 2008).

7. Harvey A. Schultz, "100 Years of Entomology in the Department of Defense," in J. Adams, ed., *Insect Potpourri: Adventures in Entomology* (Gainesville, Fla.: Sandhill Crane Press, 1992); and David Payne, "The Other British War on the Western Front in the Great War: The Hygiene War," on the Western Front Association's Web site, westernfrontassociation.com/thegreatwar/articles/research/theotherbritishwar.htm (accessed January 18, 2008).

8. Excerpt from the war memoir of James Brady can be found at the British Broadcast Corporation's Web site, bbc.co.uk/schools/worldwarone/survivor/memoir01.shtml (accessed January 18, 2008).

9. Tschanz, "Typhus Fever on the Eastern Front in World War I"; Rosalyn S. Carson-DeWitt, "Trench Fever," online *Encyclopedia of Medicine* at findarticles.com/p/articles/mi_g2601/is_0013/ai_2601001395 (accessed January 18, 2008).

10. Lisa A. Jackson and David H. Spach, "Emergence of *Bartonella quintana* Infection among Homeless Persons," *Emerging Infectious Diseases*, 2 (1996): 141–144.

Chapter 8

1. Hal Gold, *Unit 731 Testimony* (Singapore: Yen Books, 1996), chap. 1.
2. Ishii's early life is described in Daniel Barenblatt, *A Plague Upon Humanity: The Secret Genocide of Axis Japan's Germ Warfare Operation* (New York: HarperCollins, 2004), chap. 1; Gold, *Unit 731 Testimony*; and Sheldon H. Harris, *Factories of Death: Japanese Biological Warfare, 1932–1945, and the American Cover-Up* (New York: Routledge, 2002), chap. 2.
3. Harris, *Factories of Death*, chap. 2; Ed Regis, *The Biology of Doom: The History of America's Secret Germ Warfare Project* (New York: Henry Holt, 2000), chap. 1.
4. Barenblatt, *Plague Upon Humanity*, chap. 1; Harris, *Factories of Death*, chap. 2.
5. Barenblatt, *Plague Upon Humanity*, chap. 1; Harris, *Factories of Death*, chap. 2.
6. Harris, *Factories of Death*, chap. 4; Regis, *Biology of Doom*, chap. 1.
7. Most of the military scientists in Japan and other nations referred to biological weapons as "bacteriological weapons" at this time, but this term technically includes only those arms that are based on bacteria. Scientists clearly meant to include other life forms, so "biological weapons/warfare" is a more inclusive term that covers not only bacteria but also fungi, rickettsia, viruses, and—of course—insects.
8. Gold, *Unit 731 Testimony*, chap. 1.
9. Barenblatt, *Plague Upon Humanity*, chap. 1; Harris, *Factories of Death*, chap. 1; and Richard Holmes, ed., *The Oxford Companion to Military History* (New York: Oxford University Press, 2001), entry on "Manchurian Campaign."
10. Barenblatt, *Plague Upon Humanity*, chap. 2; Gold, *Unit 731 Testimony*, chap. 2; Harris, *Factories of Death*, chap. 3.
11. From an account by Dong Zhen Yu in Harris, *Factories of Death*, chap. 3.
12. Ibid.
13. Robert Harris and Jeremy Paxman, *A Higher Form of Killing* (New York: Random House, 1983), chap. 4.

Chapter 9

1. The facilities and work at Pingfan are described in Daniel Barenblatt, *A Plague Upon Humanity: The Secret Genocide of Axis Japan's Germ Warfare Operation* (New York: HarperCollins, 2004), chaps. 3 and 4; Hal Gold, *Unit 731 Testimony* (Singapore: Yen Books, 1996), chaps. 2 and 3; Sheldon H. Harris, *Factories of Death: Japanese Biological Warfare, 1932–1945, and the American Cover-Up* (New York: Routledge, 2002), chaps. 4–6; Robert Harris and Jeremy Paxman, *A Higher Form of Killing* (New York: Random House, 1983), chap. 4; Tom Mangold and Jeff Goldberg, *Plague Wars: The Terrifying Reality of Biological Warfare* (New York: St. Martin's Griffin, 1999), chap. 3; and Ed Regis, *The Biology of Doom: The History of America's Secret Germ Warfare Project* (New York: Henry Holt, 1999), chap. 3.

2. Barenblatt, *Plague Upon Humanity*, chap. 8; Harris, *Factories of Death*, chap. 11.

3. Barenblatt, *Plague Upon Humanity*, chap. 4; Harris, *Factories of Death*, chap. 6; Harris and Paxman, *Higher Form of Killing*, chap. 4.

4. Barenblatt, *Plague Upon Humanity*, chap. 6; Harris, *Factories of Death*, chap. 6.

5. Barenblatt, *Plague Upon Humanity*, chap. 6.

6. Robin Clarke, *The Silent Weapons* (New York: David McKay, 1968), chap. 6.

7. Barenblatt, *Plague Upon Humanity*, chap. 4; Harris, *Factories of Death*, chap. 6.

8. Gold, *Unit 731 Testimony*, chap. 3; Harris, *Factories of Death*, chap. 2.

9. *Materials on the Trial of Former Servicemen of the Japanese Army Charged with Manufacturing and Employing Bacteriological Weapons.* Transcripts of the Khabarovsk Trial; testimony of Kawashima Kiyoshi (physician, major general, and chief of the Medical Service of the First Front Headquarters of the Japanese Kwantung Army) (Moscow: Foreign Languages Publishing House, 1950), p. 62.

10. Barenblatt, *Plague Upon Humanity*, chaps. 3 and 4; Gold, *Unit 731 Testimony*, chap. 2; Harris, *Factories of Death*, chap. 3; Mangold and Goldberg, *Plague Wars*, chap. 3; Regis, *Biology of Doom*, chap. 8.

11. Barenblatt, *Plague Upon Humanity*, chap. 3; Harris, *Factories of Death*, chap. 5.

12. Regis, *Biology of Doom*, p. 112.

13. Gold, *Unit 731 Testimony*, p. 44.

14. Barenblatt, *Plague Upon Humanity*, chap. 8; Harris, *Factories of Death*, chap. 9.

15. Barenblatt, *Plague Upon Humanity*, chaps. 3 and 6; Harris, *Factories of Death*, chaps 5 and 6.

16. Exactly what pathogen Ishii was studying is difficult to determine. He refers to an "epidemic hemorrhagic disease," which would have typically referred to one of the illnesses causes by a Hantavirus (e.g., hemorrhagic fever with renal syndrome, or HFRS). But these diseases are probably not transmitted by ticks (some Korean research suggested the possibility of rodents' mites as vectors, but current medical opinion does not support such a link). The more likely tick-borne disease would be encephalitis (which is not associated with hemorrhaging).

17. Barenblatt, *Plague Upon Humanity*, chap. 6; Harris, *Factories of Death*, chaps. 6 and 7.

18. *Materials on the Trial*, testimony of Kawashima Kiyoshi, p. 57.

19. Barenblatt, *Plague Upon Humanity*, chap. 6; Harris, *Factories of Death*, chap. 7.

20. Barenblatt, *Plague Upon Humanity*, chap. 6; Harris, *Factories of Death*, chap. 8.

21. Gold, *Unit 731 Testimony*, p. 44.

22. Barenblatt, *Plague Upon Humanity*, chap. 4; Harris, *Factories of Death*, chap. 8.

Chapter 10

1. Daniel Barenblatt, *A Plague Upon Humanity: The Secret Genocide of Axis Japan's Germ Warfare Operation* (New York: HarperCollins, 2004), chap. 5; Robin Clarke, *The Silent Weapons* (New York: David McKay, 1968), chap. 3; Stockholm International Peace Research Institute (SIPRI), *The Problem of*

Chemical and Biological Warfare: A Study of the Historical, Technical, Military, Legal and Political Aspects of CBW, and Possible Disarmament Measures. Vol. I: The Rise of CB Weapons (New York: Humanities Press, 1975).

2. Hal Gold, *Unit 731 Testimony* (Singapore: Yen Books, 1996), part 2.

3. Ed Regis, *The Biology of Doom: The History of America's Secret Germ Warfare Project* (New York: Henry Holt, 1999), chap. 1.

4. Barenblatt, *Plague Upon Humanity*, chaps. 6 and 8; Sheldon H. Harris, *Factories of Death: Japanese Biological Warfare, 1932–1945, and the American Cover-Up* (New York: Routledge, 2002), chap. 6.

5. Regis, *Biology of Doom*, chap. 1.

6. Gold, *Unit 731 Testimony*, part 2.

7. Ibid.; Andrew G. Robertson, "From Asps to Allegations: Biological Warfare in History," *Military Medicine,* 160 (1995): 369–373; SIPRI, *Problem of Chemical and Biological Warfare.*

8. Harris, *Factories of Death*, chap. 6.

9. Ibid., p. 101.

10. Barenblatt, *Plague Upon Humanity*, p. 135.

11. Harris, *Factories of Death*, p. 102.

12. Gold, *Unit 731 Testimony*, p. 215.

13. Barenblatt, *Plague Upon Humanity*, chap. 7.

14. Ibid.

15. Robert Harwood and Maurice James, *Entomology in Human and Animal Health* (New York: Macmillan, 1979); and Gary Mullen and Lance Durden, eds., *Medical and Veterinary Entomology* (New York: Elsevier, 2002).

16. Alan Macfarlane, *The Savage Wars of Peace: England, Japan and the Malthusian Trap* (Malden, Mass.: Blackwell, 1997).

17. Barenblatt, *Plague Upon Humanity*, chap. 7.

18. Seán Murphy, Alastair Hay, and Steven Rose, *No Fire, No Thunder: The Threat of Chemical and Biological Weapons* (New York: Monthly Review, 1984), chap. 3.

Chapter 11

1. Richard Holmes, ed., *The Oxford Companion to Military History* (New York: Oxford University Press, 2001), entries on the "Eastern Front (1941–45)" and "World War II."

2. Daniel Barenblatt, *A Plague Upon Humanity: The Secret Genocide of Axis Japan's Germ Warfare Operation* (New York: HarperCollins, 2004), chap. 8.

3. *Materials on the Trial of Former Servicemen of the Japanese Army Charged with Manufacturing and Employing Bacteriological Weapons,* Transcripts of the Khabarovsk Trial; testimony of Sato Shunji (physician, major general, and chief of the Medical Division of the 5th Army of the Japanese Kwantung Army) (Moscow: Foreign Languages Publishing House, 1950), pp. 339–340.

4. Hal Gold, *Unit 731 Testimony* (Singapore: Yen Books, 1996), part 2.

5. Barenblatt, *Plague Upon Humanity*, chap. 8.

6. Ibid.

7. Gold, *Unit 731 Testimony*, chap. 4.

8. Barenblatt, *Plague Upon Humanity*, chap. 8.

9. The possibility of launching an entomological attack via balloons made a lasting impression on the Americans. In 1960, studies were still being conducted by the U.S. Department of Agriculture and the Office of Naval Research on the survival of insects during high-altitude balloon flights. The official rationale for this research was to assess the biological effects of cosmic radiation in preparation for manned space flight. However, the veracity of this explanation was substantially undermined by the bizarre choice of insects as surrogates for humans. The government tested Oriental rat fleas and house flies—the insect vectors favored by the Japanese entomological warfare program; see W. N. Sullivan and C. N. Smith, "Exposure of House Flies and Oriental Rat Fleas on a High-Altitude Balloon Flight," *Journal of Economic Entomology*, 53 (1960): 247–248.

10. Barenblatt, *Plague Upon Humanity*, chap. 8.

11. Ibid., chap. 7.

12. Gold, *Unit 731 Testimony*, pp. 199–200.

13. Tien-wei Wu, "A Preliminary Review of Studies of Japanese Biological Warfare and Unit 731 in the United States," Century of China Web site, centurychina.com/wiihist/germwar/731rev.htm (accessed January 17, 2008).

14. Barenblatt, *Plague Upon Humanity*, chaps. 8 and 9; Sheldon H. Harris, *Factories of Death: Japanese Biological Warfare, 1932–1945, and the American Cover-Up* (New York: Routledge, 2002), chaps. 13–15.

15. Harris, *Factories of Death*, p. 294.

16. Ibid., p. 86. The claim that "scruples" precluded human testing of biological weapons in the United States was substantially, but not entirely, correct. San Quentin convicts were used as human guinea pigs by the Naval Research Unit at the University of California, Berkeley, to evaluate bubonic plague. None of the prisoners became seriously ill, although several developed mild symptoms. American scientists certainly did not engage in lethal studies of human subjects, but they did cross the line of human experimentation in this and several other instances; see Ed Regis, *The Biology of Doom: The History of America's Secret Germ Warfare Project* (New York: Henry Holt, 1999), chap. 7.

17. Barenblatt, *Plague Upon Humanity*, chap. 9; Harris, *Factories of Death*, chap. 16.

18. *Materials on the Trial*, pp. 397–398.

19. Barenblatt, *Plague Upon Humanity*, chap. 10.

20. *Materials on the Trial*, testimony of Kawashima Kiyoshi (physician, major general, and chief of the Medical Service of the First Front Headquarters of the Japanese Kwantung Army).

21. Author interview, March 14, 2008.

22. Harris, *Factories of Death*, chap. 16.

23. Ibid.
24. Gold, *Unit 731 Testimony*, chap. 5.
25. Barenblatt, *Plague Upon Humanity*, pp. 173–174.

Chapter 12

1. Erhard Geissler, "Biological Warfare Activities in Germany, 1923–45," in Erhard Geissler and John E. van Courtland Moon (eds.), *Biological and Toxin Weapons: Research, Development and Use from the Middle Ages to 1945*, SIPRI Biological and Chemical Warfare Studies (New York: Oxford University Press, 1999).
2. Geissler, "Biological Warfare Activities in Germany, 1923–1945," p. 111.
3. The biology and history of this insect can be found at the University of Vermont's Extension fact sheet, Vern Grubinger, *Colorado Potato Beetle*; uvm.edu/vtvegandberry/factsheets/potatobeetle.html (accessed January 17, 2008).
4. John Burdon Sanderson Haldane, "Science and Future Warfare," *Chemical Warfare Bulletin*, 24 (1938): 7–17.
5. Jonathan Ban, *Agricultural Biological Warfare: An Overview* (Washington, D.C.: Chemical and Biological Arms Control Institute, 2000); Benjamin C. Garrett, "The Colorado Potato Beetle Goes to War," *Chemical Weapons Convention Bulletin*, 33 (1996): 2–3; Geissler and Moon, *Biological and Toxin Weapons*, chap. 5.
6. Ban, "Agricultural Biological Warfare"; Garrett, "Colorado Potato Beetle Goes to War."
7. Garrett, "Colorado Potato Beetle Goes to War"; Geissler, "Biological Warfare Activities in Germany, 1923–45."
8. Stockholm International Peace Research Institute (SIPRI), *The Problem of Chemical and Biological Warfare: A Study of the Historical, Technical, Military, Legal and Political Aspects of CBW, and Possible Disarmament Measures. Vol. I: The Rise of CB Weapons* (New York: Humanities Press, 1975); Robin Clarke, *The Silent Weapons* (New York: David McKay, 1968), chap. 3.
9. Although nerve gas was not unleashed, the Nazis made despicable use of Zyklon B, a cyanide-based chemical that had been developed as an insecticide in the early 1930s. From 1942 to 1943, this poison was used to exterminate people in concentration camps. Auschwitz alone pumped nearly 22 tons of this horrifically misused insecticide into its gas chambers; see Seán Murphy, Alastair Hay, and Steven Rose, *No Fire, No Thunder: The Threat of Chemical and Biological Weapons* (New York: Monthly Review, 1984); SIPRI, *Problem of Chemical and Biological Warfare*. The only other use of an insecticide against humans in the Second World War came, as one might guess, at Pingfan. During a failed escape attempt, prisoners became trapped within the inner courtyard and the Japanese decided to subdue them using chloropicrin, a

fumigant for soil and wood pests. At low levels it functions as a tear gas, but the Japanese used enough to induce pulmonary edema. After the escapees drowned in their own fluids, the Japanese decided the insecticide was not a viable riot control agent; see Sheldon H. Harris, *Factories of Death: Japanese Biological Warfare, 1932–1945, and the American Cover-Up* (New York: Routledge, 2002), chap. 6.

10. Garrett, "Colorado Potato Beetle Goes to War"; Geissler, "Biological Warfare Activities in Germany, 1923–45."

11. Geissler, "Biological Warfare Activities in Germany, 1923–45," p. 98.

12. Garrett, "Colorado Potato Beetle Goes to War"; Geissler, "Biological Warfare Activities in Germany, 1923–45."

13. Geissler, "Biological Warfare Activities in Germany, 1923–45."

14. Ibid., p. 123.

15. Dale B. Gelman, Robert A. Bell, Lynda J. Liska, and Jing S. Hu, "Artificial Diets for Rearing the Colorado Potato Beetle, *Leptinotarsa decemlineata*," *Journal of Insect Science,* 7 (2001): 1–11.

16. Geissler, "Biological Warfare Activities in Germany, 1923–45," p. 124.

17. Garrett, "Colorado Potato Beetle Goes to War"; Geissler, "Biological Warfare Activities in Germany, 1923–45."

18. Geissler, "Biological Warfare Activities in Germany, 1923–45"; Jeffrey A. Lockwood, "Entomological Warfare: History of the Use of Insects as Weapons of War," *Bulletin of the Entomological Society of America*, 33 (1987): 76–82.

19. Garrett, "Colorado Potato Beetle Goes to War," p. 3.

20. "When the Nazis Tried to Starve Out Britain by Beetle-Bombing Crops," *International Herald Tribune* (February 25, 1970): 5.

21. Ibid.

22. Robert K. D. Peterson, "The Role of Insects as Biological Weapons," Insects, Disease and History Web site, Entomology Group of Montana State University, entomology.montana.edu/historybug/insects_as_bioweapons.htm (accessed January 17, 2008).

23. Garrett, "Colorado Potato Beetle Goes to War," p. 3.

24. Ibid.; Geissler, "Biological Warfare Activities in Germany, 1923–45."

25. Geissler, "Biological Warfare Activities in Germany, 1923–45."

26. Milton Leitenberg, *New Russian Evidence on the Korean War Biological Warfare Allegations: Background and Analysis*, a report from the Cold War International History Project, Woodrow Wilson International Center for Scholars Web site, kimsoft.com/2000/germberia.htm (accessed January 17, 2008).

27. Ibid.

28. Mark Wheelis, Lajos Rózsa, and Malcolm Dando, *Deadly Cultures: Biological Weapons Since 1945* (Cambridge, Mass.: Harvard University Press, 2006), p. 353.

29. United Nations, Report of the Secretary General, *Chemical and Bacteriological (Biological) Weapons and the Effects of their Possible Use*, Document A/7575/Rev. 1, S/9292/Rev. 1 (1969).

30. Ibid.

31. Dmitry Litovkin, "Valentin Yevstigneyev on Issues Relating to Russian Biological Weapons," *Yaderny Kontrol Digest*, 11 (Summer 1999), pircenter.org/board/article.php3?artid=77 (accessed January 17, 2008); Benjamin C. Garrett, "A Plague of Locusts," *Nonproliferation, Demilitarization, and Arms Control*, 6 (Fall 1999/Winter 2000): 11–12.

Chapter 13

1. Eric Croddy, *Chemical and Biological Warfare: A Comprehensive Survey for the Concerned Citizen* (New York: Springer-Verlag, 2002), chap. 8.

2. Ibid., chap. 4.

3. Valentin Bojtzov and Erhard Geissler, "Military Biology in the USSR, 1920–45," in Erhard Geissler and John E. van Courtland Moon (eds.), *Biological and Toxin Weapons: Research, Development and Use from the Middle Ages to 1945*, SIPRI Biological and Chemical Warfare Studies (New York: Oxford University Press, 1999).

4. Ibid.

5. Al Mauroni, *Chemical and Biological Warfare: A Reference Handbook* (Denver: ABC-CLIO, 2003), chap. 3.

6. Ibid., p. 60.

7. Croddy, *Chemical and Biological Warfare*, p. 255.

8. Ibid., p. 224.

9. Ed Regis, *The Biology of Doom: The History of America's Secret Germ Warfare Project* (New York: Henry Holt, 2000), p. 9.

10. Ibid., chap. 1.

11. Ibid.

12. Sheldon H. Harris, *Factories of Death: Japanese Biological Warfare, 1932–1945, and the American Cover-Up* (New York: Routledge, 2002), chap. 11.

13. Regis, *Biology of Doom*, chap. 1.

14. Ibid., p. 21.

15. Ibid.

16. Regis, *Biology of Doom*, chap. 1.

17. Harris, *Factories of Death*, chap. 11.

18. Ibid.

19. Barton J. Berenstein, "The Birth of the U.S. Biological-Warfare Program," *Scientific American*, 256 (1987): 116–121 (see 116).

20. J. B. S. Haldane, "Science and Future of Warfare," *Chemical Warfare Bulletin*, 24 (1938): 15.

21. Stockholm International Peace Research Institute (SIPRI), *The Problem of Chemical and Biological Warfare: A Study of the Historical, Technical, Military, Legal and Political Aspects of CBW, and Possible Disarmament Measures. Vol. I: The Rise of CB Weapons* (New York: Humanities Press, 1975).

22. Harris, *Factories of Death*, chap. 11.

23. Gradon B. Carter and Graham S. Pearson, "British Biological Warfare and Biological Defence, 1925–45," in Geissler and van Courtland Moon, *Biological and Toxin Weapons*.

24. Donald Avery, "Canadian Biological and Toxin Warfare Research, Development and Planning, 1925–45," in Geissler and van Courtland Moon, *Biological and Toxin Weapons*; Regis, *The Biology of Doom*, chap. 2.

25. Regis, *Biology of Doom*, chap. 2.

26. Tom Mangold and Jeff Goldberg, *Plague Wars: The Terrifying Reality of Biological Warfare* (New York: St. Martin's Griffin, 1999), chap. 4.

27. Regis, *Biology of Doom*, chap. 6.

28. William Patrick, Chief of Product Development Division at Fort Detrick, argued that dry powders, rather than insect vectors, were ultimately the ideal approach to dispersing pathogens because one didn't have "the problem of putting a living system within another living system" (author interview, March 14, 2008).

29. Joel Carpenter, "The First Intercontinental Weapon System: Japanese Fu-Go Balloons," at the *Project 1947* Web site, project1947.com/gfb/fugo.htm (accessed January 18, 2008).

30. Regis, *Biology of Doom*, chap. 5.

31. Ibid., chaps. 5 and 6.

32. Avery, "Canadian Biological and Toxin Warfare Research, Development and Planning, 1925–45."

33. The scientists were most likely mass producing *Drosophila melanogaster*, which is commonly called a fruit fly. However, this species is in the family Drosophilidae, which is technically the pomace flies, although sometimes called the small fruit flies. The fruit flies, properly speaking, are members of the family Tephritidae, which includes such notorious pests as the Mediterranean fruit fly, *Ceratitis capitata*.

34. Avery, "Canadian Biological and Toxin Warfare Research," p. 205.

35. Stanley P. Lovell, *Of Spies and Stratagems* (Englewood Cliffs, NJ: Prentice-Hall, 1963), chap. 6.

36. Ibid., chap. 13.

37. Ibid.

38. Ibid.

39. Harvey A. Schultz, "100 Years of Entomology in the Department of Defense," in J. Adams (ed.), *Insect Potpourri: Adventures in Entomology* (Gainesville, Fla.: Sandhill Crane Press, 1992).

40. F. L. Soper, W. A. Davis, F. S. Markham, and L. A. Riehl, "Typhus Fever in Italy, 1943–1945, and Its Control with Louse Powder," *American Journal of Hygiene*,

45 (1947): 305–334; C. M. Wheeler, "Control of Typhus in Italy 1943–1944 by Use of DDT," *American Journal of Public Health*, 36 (1946): 119–129.

41. Wheeler, "Control of Typhus in Italy," p. 122.

42. Schultz, "100 Years of Entomology in the Department of Defense."

43. Stockholm International Peace Research Institute, *The Problem of Chemical and Biological Warfare: A Study of the Historical, Technical, Military, Legal and Political Aspects of CBW, and Possible Disarmament Measures. Vol. VI: Technical Aspects of Early Warning and Verification* (New York: Humanities Press, 1975).

44. We think of mites as having eight legs, and the adults do. However, the culprit on the Pacific Islands was the larval stage of the chigger, which feeds on warm-blooded animals. This early stage of the mite has six legs, and only later in its development does the creature gain another pair of legs.

45. Schultz, "100 Years of Entomology in the Department of Defense," p. 64.

46. Ibid.

47. Ibid.

48. Croddy, *Chemical and Biological Warfare*, chap. 10.

49. Schultz, "100 Years of Entomology in the Department of Defense."

Chapter 14

1. Richard Holmes, ed., *The Oxford Companion to Military History* (New York: Oxford University Press, 2001), entry on the "Korean War."

2. Stephen Endicott and Edward Hagerman, *The United States and Biological Warfare: Secrets from the Early Cold War and Korea* (Bloomington: Indiana University Press, 1998), chap. 5.

3. Ibid., p. 146.

4. Ibid., p. 148.

5. Ibid., p. 78.

6. Ibid., chap. 5.

7. Ibid., p. 77.

8. Ibid., chap. 5.

9. Milton Leitenberg, "The Korean War Biological Warfare Allegations Resolved," Center for Pacific Asia Studies at Stockholm University, Occasional Paper 36 (May 1998), p. 10.

10. Endicott and Hagerman, *United States and Biological Warfare*, chaps. 4 and 5.

11. Ibid., chap. 4.

12. Holmes, *Oxford Companion*, "Korean War"; Ed Regis, *The Biology of Doom: The History of America's Secret Germ Warfare Project* (New York: Henry Holt, 1999), chap. 10.

13. Eric Croddy, *Chemical and Biological Warfare: A Comprehensive Survey for the Concerned Citizen* (New York: Springer-Verlag, 2002), chap. 8.

14. Ibid.

15. Endicott and Hagerman, *United States and Biological Warfare*, p. 200.
16. Ibid., p. 50.
17. Croddy, *Chemical and Biological Warfare*, p. 229.
18. Endicott and Hagerman, *United States and Biological Warfare*, chap. 1.
19. Ibid., p. 11.
20. Ibid., chap. 1.
21. Leitenberg, "Korean War Biological Warfare Allegations Resolved," p. 9.
22. Croddy, *Chemical and Biological Warfare*, chap. 8; Endicott and Hagerman, *United States and Biological Warfare*, chap. 12.
23. Croddy, *Chemical and Biological Warfare*, p. 230.
24. Leitenberg, "Korean War Biological Warfare Allegations Resolved," p. 11.

Chapter 15

1. Eric Croddy, *Chemical and Biological Warfare: A Comprehensive Survey for the Concerned Citizen* (New York: Springer-Verlag, 2002), chap. 8; Ed Regis, *The Biology of Doom: The History of America's Secret Germ Warfare Project* (New York: Henry Holt, 1999), chap. 11.
2. Daniel Barenblatt, *A Plague Upon Humanity: The Secret Genocide of Axis Japan's Germ Warfare Operation* (New York: HarperCollins, 2004), chap. 10.
3. *Report of the International Scientific Commission for the Investigation of the Facts Concerning Bacterial Warfare in Korea and China* (Peking: International Scientific Commission, 1952).
4. Regis, *Biology of Doom*, chap. 10.
5. *Report of the International Scientific Commission*, Appendix H.
6. Ibid., Appendix I.
7. Ibid., p. 126.
8. Ibid.
9. Ibid., p. 127.
10. Ibid.
11. Ibid.
12. Ibid., Appendix AA.
13. Ibid., p. 33.
14. Ibid., p. 24.
15. Ibid., p. 27.
16. Ibid., p. 290.
17. Ibid., pp. 48–50.
18. John Cookson and Judith Nottingham, *A Survey of Chemical and Biological Warfare* (New York: Monthly Review, 1969), p. 61.
19. *Report of the International Scientific Commission*, p. 48.
20. Ibid., p. 57.
21. Ibid.

22. Harvey A. Schultz, "100 Years of Entomology in the Department of Defense," in J. Adams (ed.), *Insect Potpourri: Adventures in Entomology* (Gainesville, Fla.: Sandhill Crane Press, 1992).

23. Ibid., p. 19.

24. Ibid., pp. 11–12.

25. Ibid., pp. 19–20.

26. Ibid., p. 59.

Chapter 16

1. Ed Regis, *The Biology of Doom: The History of America's Secret Germ Warfare Project* (New York: Henry Holt, 1999), chap. 11.

2. Chen Wen-Kuei, "My Experiences in Anti-Bacteriological Warfare," *Chinese Medical Journal*, 70 (1952, Suppl.): 38–46.

3. Milton Leitenberg, "The Korean War Biological Warfare Allegations Resolved," *Center for Pacific Asia Studies at Stockholm University*, Occasional Paper 36 (May 1998), p. 4.

4. Ibid.

5. Stephen Endicott and Edward Hagerman, *The United States and Biological Warfare: Secrets from the Early Cold War and Korea* (Bloomington: Indiana University Press, 1998), p. 155.

6. *Report of the International Scientific Commission for the Investigation of the Facts Concerning Bacterial Warfare in Korea and China* (Peking: International Scientific Commission, 1952), p. 503.

7. Endicott and Hagerman, *United States and Biological Warfare*, p. 167.

8. Ibid.

9. Ibid., chap. 11.

10. Ibid., p. 174.

11. Ibid., p. 191.

12. Robin Clarke, *The Silent Weapons* (New York: David McKay, 1968), p. 24.

13. Endicott and Hagerman, *United States and Biological Warfare*, chap. 5; Stockholm International Peace Research Institute (SIPRI), *The Problem of Chemical and Biological Warfare: A Study of the Historical, Technical, Military, Legal and Political Aspects of CBW, and Possible Disarmament Measures, Vol. V: The Prevention of CBW* (New York: Humanities Press, 1991).

14. Regis, *Biology of Doom*, chap. 11.

15. Ibid., p. 154.

16. SIPRI, *Problem of Chemical and Biological Warfare*.

17. John Cookson and Judith Nottingham, *A Survey of Chemical and Biological Warfare* (New York: Monthly Review, 1969), pp. 62–63.

18. Ibid., chaps. 1 and 7; Endicott and Hagerman, *United States and Biological Warfare*, chap. 12; Regis, *Biology of Doom*, chap. 11.

19. Sheldon H. Harris, *Factories of Death: Japanese Biological Warfare, 1932–1945, and the American Cover-Up* (New York: Routledge, 2002), chap. 16; Leitenberg, "Korean War Biological Warfare."

20. Leitenberg, "Korean War Biological Warfare," p. 17.

21. Endicott and Hagerman, *United States and Biological Warfare*, p. 25.

22. Robert K. D. Peterson, "The Role of Insects as Biological Weapons," Insects, Disease and History Web site, Entomology Group of Montana State University, entomology.montana.edu/historybug/insects_as_bioweapons.htm (accessed January 21, 2008).

23. Endicott and Hagerman, *United States and Biological Warfare*, p. 193.

24. SIPRI. *Problem of Chemical and Biological Warfare.*

25. Seán Murphy, Alastair Hay, and Steven Rose, *No Fire, No Thunder: The Threat of Chemical and Biological Weapons* (New York: Monthly Review, 1984), p. 34.

26. Leitenberg, "Korean War Biological Warfare."

27. Interview with author, January 17, 2006.

28. Interview with author, January 18, 2006.

29. Leitenberg, "Korean War Biological Warfare"; Tom Mangold and Jeff Goldberg, *Plague Wars: The Terrifying Reality of Biological Warfare* (New York: St. Martin's Griffin, 1999), chap. 31.

30. Leitenberg, "Korean War Biological Warfare," p. 21.

31. Ibid., pp. 21–22.

32. The text of the 1951 memorandum from Lieutenant General Matthew Ridgeway is available at the Military History Network Web site, milhist.net/global/whywearehere.html (accessed January 21, 2008).

Chapter 17

1. Eric Croddy, *Chemical and Biological Warfare: A Comprehensive Survey for the Concerned Citizen* (New York: Springer-Verlag, 2002), chap. 8; Ed Regis, *The Biology of Doom: The History of America's Secret Germ Warfare Project* (New York: Henry Holt, 1999), chap. 9.

2. Robin Clarke, *The Silent Weapons* (New York: David McKay, 1968), chap. 5; Robert Harris and Jeremy Paxman, *A Higher Form of Killing: The Secret History of Chemical and Biological Warfare* (New York: Random House, 2002), chap. 7.

3. Alastair Hay, "A Magic Sword or a Big Itch: An Historical Look at the United States Biological Weapons Programme," *Medicine, Conflict and Survival,* 15 (1999): 215–234; William H. Rose, *An Evaluation of Entomological Warfare as a Potential Danger to the United States and European NATO Nations* (U.S. Army Dugway Proving Ground: U.S. Army Test & Evaluation Command, 1981), available at Smoking Gun Archive, thesmokinggun.com/archive/mosquito1.html (accessed January 22, 2008).

4. Hay, "Magic Sword or a Big Itch."

5. Harris and Paxman, *Higher Form of Killing*, chap. 7.

6. Hay, "Magic Sword or a Big Itch"; Jeffrey A. Lockwood, "Entomological Warfare: A History of the Use of Insects as Weapons of War," *Bulletin of the Entomological Society of America*, 33 (1987): 76–82; Rose, *Evaluation of Entomological Warfare.*

7. Hay, "Magic Sword or a Big Itch."

8. "A History of Biological Warfare," available at the Gulf War Veterans Web site, gulfwarvets.com/biowar.htm (accessed January 22, 2008).

9. Regis, *Biology of Doom*, p. 177.

10. Clarke, *Silent Weapons*, p. 10.

11. Ibid., pp. 117–118.

12. Hay, "Magic Sword or a Big Itch."

13. Rose, *Evaluation of Entomological Warfare.*

14. Tom Mangold and Jeff Goldberg, *Plague Wars: The Terrifying Reality of Biological Warfare* (New York: St. Martin's Griffin, 1999), chap. 4.

15. Harris and Paxman, *Higher Form of Killing*, p. 169.

16. Hay, "Magic Sword or a Big Itch."

17. Ibid.; Seán Murphy, Alastair Hay, and Steven Rose, *No Fire, No Thunder: The Threat of Chemical and Biological Weapons* (New York: Monthly Review, 1984), chap. 3.

18. Murphy, Hay, and Rose, *No Fire, No Thunder*, p. 39.

19. Harris and Paxman, *Higher Form of Killing*, chap. 7.

20. Regis, *Biology of Doom*, chap. 11.

21. During Soviet inspections of U.S. facilities in 1991, the Russians requested access to the "mosquito room" at Pine Bluff Arsenal. Finding a massive water-filled vat with newly refitted plumbing, the Soviets were certain they'd found evidence of an active, offensive program in entomological warfare. The Americans admitted that the pool had been updated, but they were able to show that the modified tank was not being used to produce mosquitoes for military operations but to raise catfish for civilian research; Mangold and Goldberg, *Plague Wars*, chap. 15.

22. "History of Biological Warfare."

Chapter 18

1. Alastair Hay, "A Magic Sword or a Big Itch: An Historical Look at the United States Biological Weapons Programme," *Medicine, Conflict and Survival*, 15 (1999): 215–234.

2. The first time that herbicides were used in warfare, the British sprayed 2,4-D and 2,4,5,-T in Tanganyika and Kenya. The ultimate objective was to protect their troops by depriving tsetse flies (*Glossina*)—the vectors of sleeping sickness—of shady vegetation that protected them from the blistering

midday heat; Simon M. Whitby, *Biological Warfare Against Crops* (New York: Palgrave, 2002), chap. 8.

3. John Cookson and Judith Nottingham, *A Survey of Chemical and Biological Warfare* (New York: Monthly Review, 1969), p. 67.

4. United Nations, *Chemical and Bacteriological (Biological) Weapons and the Effects of Their Possible Use,* United Nations Report to the Secretary General, A/7575/Rev. 1, S/9292/Rev. 1 (1969).

5. While rats have long been considered a reservoir for the pathogen, recent studies suggest that all mammals are dead-end hosts. However, the rickettsia responsible for scrub typhus can persist even without rats' serving as warm-blooded incubators. It seems that chiggers can function as both a reservoir and vector. So, even without rats, the pathogen can be passed between generations of mites.

6. Harvey A. Schultz, "100 Years of Entomology in the Department of Defense," in J. Adams (ed.), *Insect Potpourri: Adventures in Entomology* (Gainesville, Fla.: Sandhill Crane Press, 1992).

7. Robin Clarke, *The Silent Weapons* (New York: David McKay, 1968), chap. 4; Jeffrey A. Lockwood, "Entomological Warfare: A History of the Use of Insects as Weapons of War," *Bulletin of the Entomological Society of America*, 33 (1987): 76–82.

8. Roger Sutherland, "The Importance of Bees in War Time," available at honeyflowfarm.com/newsletters/2003/november/novhoney.htm (accessed January 22, 2008).

9. Tom Mangold and John Penycate, *The Tunnels of Cu Chi* (New York: Berkeley, 1986), chap. 11.

10. FAO Agriculture and Consumer Protection, "The Giant Honeybee, *Apis dorsata,*" FAO Corporate Document Repository of the Food and Agriculture Organization of the United Nations, available at fao.org/docrep/X0083E/X0083E02.htm (accessed January 22, 2008).

11. Sutherland, "Importance of Bees in War Time."

12. John T. Ambrose, "Insects in Warfare," *Army* (December 1974): 33–38.

13. "When Killing Just Won't Do," an excerpted glossary from *Nonlethal Weapons: Terms and References*, a report published by the USAF Institute for National Security Studies, available at the *Harper's* magazine Web site, harpers.org/archive/2003/02/0079475 (accessed January 22, 2008).

14. John Mann, *Murder, Magic and Medicine* (New York: Oxford University Press, 1994), p. 48.

15. Thomas D. Seeley, Joan W. Nowicke, Matthew Meselson, Jeanne Guillemin, and Pongthep Akratanakul, "Yellow Rain," *Scientific American* (September 1985): 128–137; Sterling Seagrave, *Yellow Rain: A Journey Through the Terror of Chemical Warfare* (New York: M. Evans, 1981).

16. What caused the Hmong's maladies remains a mystery. But lest we be too smug about experts mistaking insect artifacts for serious threats, just a few years ago the Secret Service thought they'd found evidence that somebody

had fired a .45 caliber bullet into the wooden window frame of a government building—only to learn that the hole was the work of the Eastern carpenter bee, *Xylocopa virginica*; Al Greene, "The .45 caliber bee," *Pest Control Technology* online magazine, available at pctonline.com/articles/article .asp?ID=2627&IssueID=108 (accessed January 22, 2008).

17. Tom Mangold and Jeff Goldberg, *Plague Wars: The Terrifying Reality of Biological Warfare* (New York: St. Martin's Griffin, 1999), chap. 7.

18. Ken Alibek and Steve Handelman, *BIOHAZARD: The Chilling True Story of the Largest Covert Biological Weapons Program in the World—Told from Inside by the Man Who Ran It* (New York: Random House, 1999).

Chapter 19

1. Robert K. D. Peterson, "The Role of Insects as Biological Weapons," included on the Insects, Disease and History Web site, Entomology Group of Montana State University, entomology.montana.edu/historybug/insects_as_bioweapons .htm (accessed January 23, 2008).

2. Jeffrey A. Lockwood, "Entomological Warfare: A History of the Use of Insects as Weapons of War," *Bulletin of the Entomological Society of America*, 33 (1987): 76–82.

3. Simon M. Whitby, *Biological Warfare Against Crops* (New York: Palgrave, 2002), chap. 11.

4. Author interview, March 14, 2008.

5. Gustavo Kourí, María G. Guzmán, and José Bravo, "Hemorrhagic Dengue in Cuba: History of an Epidemic," *Pan American Health Organization Bulletin*, 20 (1986): 24–30.

6. Ibid.

7. William Schaap, "U.S. Biological Warfare: The 1981 Cuba Dengue Epidemic," *Covert Action Information Bulletin*, 17 (Summer 1982): 28–31.

8. Ellen Ray and William H. Schaap, *Bioterror: Manufacturing Wars the American Way* (New York: Ocean Press, 2003), p. 37.

9. Schaap, "U.S. Biological Warfare."

10. William H. Rose, *An Evaluation of Entomological Warfare as a Potential Danger to the United States and European NATO Nations* (U.S. Army Dugway Proving Ground: U.S. Army Test and Evaluation Command, 1981), p. 5, available at the Smoking Gun Archive, thesmokinggun.com/archive/ mosquito1.html (accessed January 23, 2008).

11. Ibid., p. 5.

12. Ibid.

13. Ibid., p. 59.

14. Soviet secrecy and American openness create the appearance of western nations' having been more heavily involved in biological warfare. However, quite the opposite is most surely the case, as revealed by Ken Alibek, a Soviet colonel

and the First Deputy Chief of the Soviet Union's offensive biological weapons program, who defected to the United States in 1992. Indeed, so similar were the U.S. and Soviet programs that he has concluded that there had to have been a spy within Fort Detrick; interview with Col. Charles Bailey, January 18, 2006. As for entomological warfare, the Soviets considered covert uses of mosquitoes and ticks as vectors of human and livestock diseases and various insects as carriers of plant pathogens. Other ventures included the development of automated mass-rearing facilities, the use of attractants to influence the movement patterns of introduced insects, and the application of radar for tracking migrations of mass releases. By the early 1980s, the Soviets abandoned their insect-vector program in favor of direct delivery systems for pathogens at a scale never contemplated by the Americans; Jonathan Ban, "Agricultural Biological Warfare: An Overview," in *The Arena*, 9 (Washington, D.C.: Chemical and Biological Arms Control Institute, 2000).

15. Schaap, "U.S. Biological Warfare."
16. Raisa Pages, "Cuba: Washington's Objective Is to 'Cause Hunger, Desperation, and Overthrow the Government,'" *Green Left Weekly*, November 12, 2003, available at greenleft.org.au/2003/561/29229 (accessed January 23, 2008).
17. "Biological Warfare Waged by the U.S. Against Cuba," available at the *Cuba Solidarity Campaign* Web site, poptel.org.uk/cuba-solidarity/CubaSi-January/Bio.html (accessed January 23, 2008).
18. "Working paper," submitted by Cuba to the Ad Hoc Group of the States Parties to the Convention of the Prohibition of the Development, Production and Stockpiling of Bacteriological (Biological) and Toxin Weapons and on Their Destruction, Twentieth Session, *Action Brought by the People of Cuba Against the Government of the United States of America for Economic Damages Caused to Cuba*, Geneva, June 21, 2000, bradford.ac.uk/acad/sbtwc/ahg52wp/wp417.pdf (accessed January 23, 2008).
19. Ibid., p. 9.
20. A transcript of the interview is available from the Cuban Web site cuba.cu/gobierno/documentos/2002/ing/m240502i.html (accessed January 23, 2008).
21. Ibid.
22. "Biological Warfare Waged by the U.S. Against Cuba."
23. "Working paper."
24. Raisa Pages, "Biological Warfare Against Cuba: Bee-Eating Insect Causes Losses of Two Million Dollars," *Granma International*, August 3, 2001, blythe .org/nytransfer-subs/2001-Caribbean-Vol-3/BIOLOGICAL_WARFARE_AGAINST_CUBA (accessed January 23, 2008).

Chapter 20

1. Eric Croddy, *Chemical and Biological Warfare: A Comprehensive Survey for the Concerned Citizen* (New York: Springer-Verlag, 2002), chap. 9; John Dinger,

"Cuba: No Use of Biological Weapons," U.S. Department of State press statement, May 6, 1997; *Note Verbale from the Permanent Mission of Cuba to the United Nations*, Addressed to the Secretary-General, April 29, 1997, as Item 80 of the preliminary list of the 52nd session of UN General Assembly, available at afrocubaweb.com/biowar.htm (accessed January 23, 2008).

2. *Note Verbale from the Permanent Mission of Cuba.*

3. Simon M. Whitby, *Biological Warfare Against Crops* (New York: Palgrave, 2002), chap. 4.

4. *Note Verbale from the Permanent Mission of Cuba.*

5. Dinger, "Cuba: No Use of Biological Weapons."

6. Milton Leitenberg, "Biological Weapons in the 20th Century: A Review and Analysis," available from the Federation of American Scientists Web site, fas .org/bwc/papers/bw20th.htm (accessed January 23, 2008).

7. Croddy, *Chemical and Biological Warfare*; Dinger, "Cuba: No Use of Biological Weapons"; *Note Verbale from the Permanent Mission of Cuba to the United Nations.*

8. *Note Verbale from the Permanent Mission of Cuba.*

9. Ibid.

10. Ibid; Whitby, *Biological Warfare Against Crops*, chap. 4.

11. Whitby, *Biological Warfare Against Crops*, chap. 4.

12. Leitenberg, "Biological Weapons in the 20th Century."

13. Croddy, *Chemical and Biological Warfare*, p. 245.

14. Ibid.

15. Whitby, *Biological Warfare Against Crops*, p. 65.

16. Milton Leitenberg, "Biological Weapons in the 20th Century," Section 9, "Undermining the International Regime: False Allegations of BW Use," available at fas.org/bwc/papers/review/under.htm (accessed January 23, 2008).

17. "Interview: Valentin Yevstigneyev on Issues Relating to Russian Biological Weapons," *Yaderny Kontrol [Nuclear Control]) Digest*, 11 (Summer 1999): 50.

18. Leitenberg, "Biological Weapons in the 20th Century," Section 9; Colum Lynch, "U.N. Employee Planned Locust Plague, Iraq Says," *Washington Post*, July 7, 1999; Judith Miller, "U.N. Backs Mine Expert Expelled by Iraq," *New York Times*, July 9, 1999.

19. Juan O. Tamayo and Meg Laughlin, "Cuba Files $181 Billion Claim Against U.S.," *Miami Herald*, June 2, 1999, available at nocastro.com/archives/cuba181m.htm.

20. "Cuba Profile: Biological," available at the Nuclear Threat Initiative Web site, nti.org/e_research/profiles/Cuba/Biological/3481.html (accessed January 23, 2008).

21. Tamayo and Laughlin, "Cuba Files $181 Billion Claim."

22. Thomas W. Frazier, "Natural and Bioterrorist/Biocriminal Threats to Food and Agriculture," in Food and Agricultural Security Special Issue, *Annals of the New York Academy of Sciences*, 894 (1999): 5.

Chapter 21

1. Robert S. Root-Bernstein, "Infectious Terrorism," *Atlantic Monthly* (May 1991): 44–50.
2. Malathion is much more readily detoxified by humans than by insects, and this differential susceptibility allows the insecticide to be used at rates that kill pests without being proportionately dangerous to mammals. The lethal dose for an adult human would be about 1 ½ cups of malathion. Lower doses may suppress the immune system and cause developmental and reproductive abnormalities. By comparison, a closely related compound, parathion, is deadly enough to have become the weapon of choice for assassins operating within South Africa's apartheid government. After breaking into a residence, a killer would smear the insecticide onto the victim's underwear. The chemical could then enter the body through large hair follicles under the arms and in the crotch; Tom Mangold and Jeff Goldberg, *Plague Wars: The Terrifying Reality of Biological Warfare* (New York: St. Martin's Griffin, 1999), chap. 23.
3. Stephanie Chavez and Richard Simon, "Mystery Letter Puts a Strange Twist on Latest Medfly Crisis," *Los Angeles Times* (Orange County Edition; December 3, 1988): B1.
4. Copy of the letter obtained by the author via a FOIA request to the U.S. Department of Agriculture Office of Inspector General.
5. Interview with author, October 13, 2004.
6. Copy of the report obtained by the author via a FOIA request to the U.S. Department of Agriculture Office of Inspector General.
7. Chavez and Simon, "Mystery Letter."
8. John Johnson, "Female Medfly Found in Sun Valley Close to Area Targeted Earlier," *Los Angeles Times* (January 4, 1990): B3.
9. Ashley Dunn, "Officials Advertise to Contact Group Claiming Medfly Releases," *Los Angeles Times* (February 10, 1990): A13.
10. From the report obtained by the author via a FOIA request to the U.S. Department of Agriculture Office of Inspector General.
11. Ibid.
12. Dunn, "Officials Advertise to Contact Group."
13. Jason Pate and Gavin Cameron, "Covert Biological Weapons Attacks Against Agricultural Targets: Assessing the Impact Against U.S. Agriculture," Discussion Paper 2001–9, Belfer Center for Science and International Affairs (2001), p. 11.
14. Author interview with James Reynolds, October 21, 2004.
15. Author interview with Pat Minyard, October 13, 2004.
16. Not everyone is as certain that the U.S. agricultural sector has avoided foreign terrorists. The sweet potato whitefly (*Bemisia tabaci*) arrived in the Imperial Valley in the summer of 1991 and destroyed $300 million worth

of crops. Most experts believe that the whitefly was accidentally introduced. However, Lieutenant Colonel Robert Kadlec (a former member of the U.S. delegation to the Biological Weapons Convention and inspector with the United Nations Special Commission to Iraq) has called attention to this particular strain's geographic origin (Asia or Africa), unusually broad host range, remarkable resistance to insecticides, and extraordinarily voracious feeding (combined with the capacity to transmit plant diseases)—all being consistent with a clandestine attack; Robert P. Kadlec, "Biological Weapons for Waging Economic Warfare," in *Battlefield of the Future*, ed. Barry R. Schneider and Lawrence E. Grinter, Air War College Studies in National Security, vol. 3 (Maxwell Air Force Base, Alabama, 1995), chap. 10, available at airpower.maxwell.af.mil/airchronicles/battle/chp10.html.

17. Faith Lapidus, "Could Insects Be Used as Instruments of Biological Warfare?" *Voice of America*, January 29, 2003, available from World News Web site, worldnewssite.com/News/2003/January/2003-01-29-38-Could.html.

18. Nicholas J. Neger, "The Need for a Coordinated Response to Food Terrorism," in Food and Agricultural Security Special Issue, *Annals of the New York Academy of Sciences*, 894 (1999).

Chapter 22

1. Floyd P. Horn and Roger G. Breeze, "Agriculture and Food Security," in Food and Agricultural Security Special Issue, *Annals of the New York Academy of Sciences*, 894 (1999): 11.

2. Robert P. Kadlec, "Biological Weapons for Waging Economic Warfare," in *Battlefield of the Future*, ed. Barry R. Schneider and Lawrence E. Grinter, Air War College Studies in National Security, vol. 3 (Maxwell Air Force Base, Alabama, 1995), chap. 10, available at airpower.maxwell.af.mil/airchronicles/battle/chp10.html.

3. "State: No Fine in Ant-Bite Death," *St. Petersburg Times*, July 13, 2001, available at sptimes.com/News/071301/news_pf/State/State__No_fine_in_ant .shtml.

4. The biology, history, and damage of the red imported fire ant are synthesized from C. R. Allen, D. M. Epperson, and A. S. Garmestani, "Red Imported Fire Ant Impacts on Wildlife: A Decade of Research," *American Midland Naturalist*, 152 (2004): 88–103; A. Flores and J. Core, "Putting Out the Fire," *Journal of Agricultural Research*, 52 (2004): 12–14; M. T. Henshaw, N. Kunzmann, C. Vanderwoude, M. Sanetra, and R. H. Crozier, "Population Genetics and History of the Introduced Fire Ant, *Solenopsis invicta* (Buren) in Australia," *Australian Journal of Entomology*, 44 (2005): 37–44; S. W. Taber, *Fire Ants* (College Station, Tex.: Texas A&M University Press, 2000); S. B. Vinson, "Invasion of the Red Imported Fire Ant: Spread, Biology, and Impact," *American Entomologist*, 43 (1997): 23–38.

5. U.S. Department of Agriculture Forest Service, "Effects of Urban Forests and Their Management on Human Health and Environmental Quality: Asian Longhorned Beetle," available at the USDA Forest Service Web site, fs.fed.us/ne/syracuse/Data/Nation/data_list_alb.htm (accessed January 24, 2008).

6. The biology, history, and damage of the Asian longhorned beetle are synthesized from R. A. Haack, K. R. Law, V. C. Mastro, H. S. Ossenburgen, and B. J. Raimo, "New York's Battle with the Asian Long-Horned Beetle," *Journal of Forestry*, 95 (1997): 11–15; D. J. Nowak, J. E. Pasek, R. A. Sequeira, D. E. Crane, and V. C. Mastro, "Potential Effect of *Anoplophora glabripennis* (Coleoptera: Cermabycidae) on Urban Trees in the United States," *Journal of Economic Entomology*, 94 (2001): 116–122; S. W. Ludwig, L. Lazarus, D. G. McCullough, K. Hoover, S. Montero, and J. C. Sellmer, "Methods to Evaluate Host Tree Suitability to the Asian Longhorned Beetle, *Anoplophora glabripennis*," *Journal of Environmental Horticulture*, 20 (2002): 175–180; National Agricultural Pest Information System, "Asian Longhorned Beetle," available at the Purdue University Web site ceris.purdue.edu/napis/pests/alb/; M. T. Smith, J. Bancroft, G. Li, R. Gao, and S. Teale, "Dispersal of *Anoplophora glabripennis* (Cerambycidae), *Environmental Entomology*, 30 (2001): 1036–1040.

7. D. Cappaert, D. G. McCullough, T. M. Poland, and N. W. Siegert, "Emerald Ash Borer in North America: A Research and Regulatory Challenge," *American Entomologist*, 51 (2005): 152–163.

8. Kadlec, "Biological Weapons for Waging Economic Warfare."

9. Kadlec's scenarios were not limited to the United States. As the world's third-largest producer of cotton, Pakistan derives nearly 60 percent of its export income from this one commodity. If India, for example, wanted to cripple its neighbor without the international condemnations associated with military action, an insect invasion would be ideal. By Kadlec's estimates, even a 15 percent loss in production would seriously undermine Pakistan's economy. Such dependence on a single export crop is not unusual in nonindustrial countries, and Kadlec suggests that the United States should be very concerned with entomological attacks on the agriculture—and hence the socioeconomic stability—of our allies in the developing world; Kadlec, "Biological Weapons for Waging Economic Warfare."

10. The biology, history, and damage of the glassy-winged sharpshooter and Pierce's disease are synthesized from P. C. Andersen, B. V. Brodbeck, and R. F. Mizell, III, "Plant and Insect Characteristics of *Homalodisca coagulata* on Three Host Species: A Quantification of Assimilate Extraction," *Netherlands Entomological Society*, 107 (2003): 57–68; Committee on California Agriculture and Natural Resources, *California Agricultural Research Priorities: Pierce's Disease* (Washington, D.C.: National Academies Press, 2004); R. A. Redak, A. H. Purcell, R. S. Lopes, M J. Blua, R. F. Mizell III, and P. C. Andersen,

"The Biology of Xylem Fluid-Feeding Insect Vectors of *Xylella fastidiosa* and Their Relation to Disease Epidemiology," *Annual Review of Entomology*, 49 (2004): 243–262; University of California, *Grape Pest Management* (Berkeley, Calif.: U.C. Ag Sciences Publications, 1981); University of California–Riverside, "Glassy-Winged Sharpshooter Resources," Insect Information available from the University of California-Riverside Web site, entomology.ucr.edu/information/gwss/ (accessed January 24, 2008).

11. Ranajit Bandyopadhyay and Richard A. Frederiksen, "Contemporary Global Movement of Emerging Plant Diseases," in Food and Agricultural Security Special Issue, *Annals of the New York Academy of Sciences*, 894 (1999).

12. Barry H. Thompson, "Where Have All My Pumpkins Gone? The Vulnerability of Insect Pollinators," in Food and Agricultural Security Special Issue, *Annals of the New York Academy of Sciences*, 894 (1999).

13. United Nations, *Chemical and Bacteriological (Biological) Weapons and the Effects of Their Possible Use,* United Nations Report to the Secretary General, A/7575/Rev.1, S/9292/Rev. 1 (1969).

14. Toyin Ajayi, "Smallpox and Bioterrorism," *Stanford Journal of International Relations*, 3 (2002), available at a Stanford Web site, stanford.edu/group/sjir/3.2.02_ajayi.html.

15. The biology, history, damage, and control of the screwworm are synthesized from W. G. Bruce and W. J. Sheely, "Screwworms in Florida," *University of Florida Agricultural Extension Service Bulletin,* 86 (1936); R. C. Bushland, E. F. Knipling, and A. W. Lindquist, "Eradication of the Screw-Worm Fly by Releasing Gamma Ray-Sterilized Males Among the Natural Population," *Proceedings of the International Conference on the Peaceful Uses of Atomic Energy,* 12 (1956): 216–220; W. G. Eden and C. Lincoln, "The Southwestern Screwworm Eradication Program," *A Review Conducted at the Direction of the Congress of the United States* (Fayetteville, Ark.: University of Arkansas, Agricultural Experiment Station, 1974.); O. H. Graham and J. L. Hourrigan, "Eradication Programs for the Arthropod Parasites of Livestock," *Journal of Medical Entomology* 13 (1977): 643–647.

16. Veterinary Services, "Screwworm," available at the U.S. Department of Agriculture's Animal and Plant Health Inspection Service Web site, aphis.usda.gov/lpa/pubs/fsscworm.html (accessed January 24, 2008).

17. National Research Council of the National Academies, *Countering Agricultural Bioterrorism* (Washington, D.C.: The National Academies Press, 2003).

18. Actually, the government's stern legal warning regarding dissemination of the document cites the wrong part of the federal code, referring to a provision exempting material that is "related solely to the internal personnel rules and practices of an agency." The federal law that the agency presumably meant to use exempts information that is "to be kept secret in the interest of national defense or foreign policy." One can only hope that the nation's commitment to security is more keenly supervised than its attention to legality. The error

aside, I was granted access to the material with the understanding that I would recount the contents only through paraphrasing.

Chapter 23

1. Erhard Geissler, *A New Generation of Biological Weapons in Biological and Toxin Weapons Today* (New York: Oxford University Press, 1986).
2. Arthur W. Rovine, "Contemporary Practice of the United States Relating to International Law," *American Journal of International Law*, 69 (1975): 382–405, esp. p. 404.
3. Alastair Hay, "A Magic Sword or a Big Itch: An Historical Look at the United States Biological Weapons Programme," *Medicine, Conflict and Survival*, 15 (1999): 215–234.
4. Eric Croddy, *Chemical and Biological Warfare: A Comprehensive Survey for the Concerned Citizen* (New York: Springer-Verlag, 2002), chap. 7.
5. Judith Miller, Stephen Engelberg, and William Broad, *Germs: Biological Weapons and America's Secret War* (New York: Simon & Schuster, 2002), chap. 11.
6. New York City's pest-management infrastructure was not particularly primed for response to a vector-borne disease outbreak. A September 8, 1999, story in the *New York Times* asserted that "New York City has one of the least aggressive mosquito-control programs in the region, with virtually no preventive maintenance and a budget that is less than 6 percent of that in nearby Suffolk County, whose human population is only one-fifth the size of New York's." While the city's mosquito-abatement program was rather limited, the fact remains that New York's capacity to mobilize resources in response to a health emergency was perhaps unmatched in the country.
7. Ibid.
8. Division of Vector Borne Infectious Diseases, "West Nile Virus," Centers for Disease Control and Prevention Web site, cdc.gov/ncidod/dvbid/westnile (accessed January 25, 2008).
9. "CIA concludes that Virus Outbreak Not Bio-Terrorism," Institute for Counter-Terrorism Web site, 212.150.54.123/spotlight/det.cfm?id=339 (accessed January 25, 2008); Richard Preston, "West Nile Mystery," *The New Yorker* (October 18–25, 1999): 90–107.
10. Preston, "West Nile Mystery."
11. Ibid.
12. Michael Jordan, "Experts Debunk Iraqi Role in West Nile Virus Spread," *Jewish Daily Forward*, February 8, 2002 (no longer posted online, the author can provide a hard copy).
13. Sharon L. Spradling, *The Practicality of Using West Nile Virus as a Biological Weapon*, Research report submitted to the Air Command and Staff College, Air University, Maxwell Air Force Base, April 2001.
14. Ibid.

15. Ibid.; Jordan, "Experts Debunk Iraqi Role in West Nile Virus Spread."

16. Preston, "West Nile Mystery."

17. Ibid., p. 100.

18. "CIA Concludes That Virus Outbreak Not Bio-Terrorism," Institute for Counter-Terrorism.

19. Jason Pate and Gavin Cameron. "Covert Biological Weapons Attacks Against Agricultural Targets: Assessing the Impact Against U.S. Agriculture," Discussion Paper 2001–9, Belfer Center for Science and International Affairs, 2001, p. 17.

20. Geissler, *A New Generation of Biological Weapons.*

21. FAO Agriculture and Consumer Protection, "Preparation of Rift Valley Fever Contingency Plans," FAO Corporate Document Repository of the Food and Agriculture Organization of the United Nations, available at www.fao.org/DOCREP/005/Y4140E/y4140e04.htm (accessed January 25, 2008); Tara K. Harper, "TKH Virology Notes: Rift Valley Fever," Tara K. Harper's Web site, tarakharper.com/v_rift.htm#out (accessed January 25, 2008); Karen Miller, "Rift Valley Fever," Science@NASA Web site, science.nasa.gov/headlines/y2002/17apr_rvf.htm (accessed January 25, 2008); Special Pathogens Branch, "Rift Valley Fever," Centers for Disease Control and Prevention Web site, cdc.gov/ncidod/dvrd/spb/mnpages/dispages/rvf.htm (accessed January 25, 2008).

22. Miller, "Rift Valley Fever."

23. Ibid.

24. FAO Agriculture and Consumer Protection, "Preparation of Rift Valley Fever Contingency Plans."

25. Although Bailey considers Rift Valley fever as the arthropod-borne disease of greatest concern to national defense, he also has raised the specter of Crimean-Congo hemorrhagic fever. The disease was named for the disjunctive regions where it was found in the 1940s and 1950s. The viral pathogen is transmitted by various ticks—some with close cousins in the United States. About five days after being bitten, the victim experiences fever, chills, headache, vomiting, and severe muscular pains. After another few days, internal bleeding culminates in massive hemorrhaging in the stomach and intestines, leading to death in one-third of the patients. Even if the disease is eradicated from the local human population, cattle, sheep, and small mammals provide a reservoir for the virus; interview with author, January 18, 2006.

26. Interview with author, January 18, 2006.

27. A similar scenario was described by William Patrick, former chief of the Product Development Division at Ft. Detrick, who imagined infection of 30,000 mosquito eggs with Venezuelan equine encephalitis, which "nobody at our security stations ever looks for" (author interview, March 14, 2008).

28. Interview with author, October 23, 2007.

29. Doug McGinnis, "Looking for Loopholes," *Beef Magazine*, July 1, 2004, available at beefmagazine.com/mag/beef_looking_loopholes.

30. Jocelyn Selim, "Virus Code Red," *Discover Magazine*, 26 (2005): 14.

31. Amy L. Becker, "Scientists Worry That Rift Valley Fever Could Reach the U.S.," Center for Infectious Disease Research and Policy Web site, cidrap-summit.org/cidrap/content/biosecurity/ag-biosec/news/july2104riftvalley.html (accessed January 25, 2008).

32. Mike Stobbe, "CDC Reports West Nile Cases Up," July 17, 2006, Associated Press, available at RedOrbit Breaking News Web site, redorbit.com/news/health/331507/cdc_reports_west_nile_cases_up.

33. Melissa Lee Phillips, "After the Flood: West Nile?" *Mosquito Views* (Spring 2006): 1–3.

34. Amanda Gardner, "West Nile Laying Low, So Far," *Health Day News*, July 14, 2006, available at the Healing Well Web site, news.healingwell.com/index.php?p=news1&id=533805.

35. Christopher William Ratigan, *The Asian Tiger Mosquito (Aedes albopictus): Spatial, Ecological, and Human Implications in Southeast Virginia*, Master's thesis, Virginia Polytechnic Institute and State University, Blacksburg, Virginia. To be more precise, the Asian tiger mosquito had been seen before in the United States. As early as 1946 the insect was reported in isolated incidents involving very few larvae, which mosquito control programs managed to effectively suppress; J. J. Pratt, R. H. Heterick, J. B. Harrison, and L. Haber, "Tires as a Factor in the Transportation of Mosquitoes by Ships," *Military Surgeon*, 99 (1946): 785–788.

36. Robert K. D. Peterson, "The Role of Insects as Biological Weapons," Insects, Disease and History Web site, Entomology Group of Montana State University, entomology.montana.edu/historybug/insects_as_bioweapons.htm.

37. Geissler, *New Generation of Biological Weapons*.

38. A. A. James, "Engineering Mosquito Resistance to Malaria Parasites: The Avian Malaria Model," *Insect Biochemistry and Molecular Biology*, 32 (2002): 1317–1323.

39. Interview with author, October 23, 2007.

40. Debora MacKenzie, "Run, Radish, Run," *New Scientist* (December 1999): 36–39.

Chapter 24

1. Jane Black, "Enlisting Insects in the Military," *Business Week*, November 5, 2001, available at businessweek.com/bwdaily/dnflash/nov2001/nf20011115_8187.htm; Mimi Hall, "Bugs, Weeds, Houseplants Could Join War on Terror," *USA Today*, May 27, 2003, available at usatoday.com/news/nation/2003-05-27-bugs-cover_x.htm; Faith Lapidus, "Could Insects Be Used as Instruments of Biological Warfare?" January 29, 2003, World News Web site, worldnews-site.com/News/2003/January/2003-01-29-38-Could.html.

2. Robert Harris and Jeremy Paxman, *A Higher Form of Killing: The Secret History of Chemical and Biological Warfare* (New York: Random House, 2002), chap. 10.

3. U.S. Army Soldier and Biological Chemical Command, *Guidelines for Use of Personal Protective Equipment by Law Enforcement Personnel During a Terrorist Chemical Agent Incident* (2003), Appendix G: "Overview of Chemical Warfare Agents."

4. For example, on March 20, 1995, just before the height of rush hour, an apocalyptic cult called Aleph released sarin into the Tokyo subway system. The attack was poorly conceived and executed, but it still managed to kill a dozen people and injure 5,000; Kyle B. Olson, "Aum Shinrikyo: Once and Future Threat?" Centers for Disease Control and Prevention Web site, cdc.gov/nci-dod/EID/vol5no4/olson.htm (accessed January 25, 2008).

5. Robin Clarke, *The Silent Weapons* (New York: David McKay, 1968), chap. 9.

6. This enzyme cannot be artificially synthesized but must be extracted from the insect—and there's not much there. Luciferase can be purchased from chemical supply houses at a cost of $50 per milligram (about the weight of a poppy seed), almost 4,000 times the price of gold.

7. Carl-Göran Hedén, ed., *The Problem of Chemical and Biological Warfare: A Study of the Historical, Technical, Military, Legal and Political Aspects of CBW, and Possible Disarmament Measures, Vol. VI: Technical Aspects of Early Warning and Verification* (Stockholm: Almqvist & Wiksell, 1975).

8. Clarke, *Silent Weapons*, chap. 9.

9. Tom Paulson, "Bioweapon Fears Spur Peaceful Retraining Effort," *Seattle Post-Intelligencer*, December 21, 1999, available at seattlepi.nwsource.com/local/biow21.shtml.

10. Trudy E. Bell, "Technologies Developed by NASA's Office of Biological and Physical Research to Keep Air, Water, and Food Safe for Astronauts in Space Can Also Help Protect People on Earth from Bioterrorism," NASA Space Research Web site available at spaceresearch.nasa.gov/general_info/homeplanet_lite.html (accessed January 25, 2008).

11. Ibid.

12. Kevin Coughlin, "Researchers Leading Attack on Anthrax," *Seattle Times*, October 14, 2002, available archives.seattletimes.nwsource.com/cgi-bin/texis.cgi/web/vortex/display?slug=anthsearch14&date=20021014.

13. Black, "Enlisting Insects in the Military."

14. Hall, "Bugs, Weeds, Houseplants Could Join War on Terror."

15. Jerry J. Bromenshenk, Colin B. Henderson, and Garon C. Smith, "Biological Systems," Appendix S in Jacqueline MacDonald and J. R. Lockwood (eds.), *Alternatives for Landmine Detection* (Santa Monica: Rand, 2003); J. J. Bromenshenk, C. B. Henderson, R. A. Seccomb, S. D. Rice, R. T. Etter, S. F. A. Bender, P. J. Rodacy, J. A. Shaw, N. L. Seldomridge, L. H. Spangler,

and J. J. Wilson, "Can Honey Bees Assist in Area Reduction and Landmine Detection?" *Journal of Mine Action*, 7.3 (2003), available at maic.jmu.edu/journal/7.3/focus/bromenshenk/bromenshenk.htm.

16. Black, "Enlisting Insects in the Military."

17. Bromenshenk et al., "Biological Systems"; Bromenshenk et al., "Can Honey Bees Assist"; Sandia National Laboratories, "University of Montana Researchers Try Training Bees to Find Buried Landmines," sandia.gov/media/minebees.htm (accessed January 25, 2008).

18. Guy Gugliotta, "The Robot with the Mind of an Eel: Scientists Start to Fuse Tissue and Technology in Machines," *Washington Post*, April 17, 2001.

Chapter 25

1. Adrienne Mayor, *Greek Fire, Poison Arrows and Scorpion Bombs: Biological and Chemical Warfare in the Ancient World* (New York: Overlook Duckworth, 2003), chap. 6.

2. Ibid.

3. Guy Gugliotta, "The Robot with the Mind of an Eel: Scientists Start to Fuse Tissue and Technology in Machines," *Washington Post*, April 17, 2001.

4. Robin Clarke, *The Silent Weapons* (New York: David McKay, 1968), chap. 9.

5. Jane Black, "Enlisting Insects in the Military," *Business Week*, November 5, 2001, available at businessweek.com/bwdaily/dnflash/nov2001/nf20011115_8187.htm; Paul Stone, "Creatures Feature Possible Defense Applications," *American Forces Press Service*, July 28, 1999, available at defenselink.mil/specials/bees/natures.html.

6. John Roach, "U.S. Military Looks to Beetles for New Sensors," *National Geographic News*, March 14, 2003, available at news.nationalgeographic.com/news/2003/03/0314_030314_secretweapons3.html.

7. Stone, "Creatures Feature Possible Defense Applications."

8. Daniel McCabe, "Building a Better Robot," *McGill Reporter*, June 26, 2000, available at http://www.mcgill.ca/reporter/32/18/buehler.

9. Ibid; Peter Menzel and Faith D'Aluisio, *Robo Sapiens: Evolution of a New Species* (Cambridge, Mass.: MIT Press, 2001), chap. 3.

10. Menzel and D'Aluisio, *Robo Sapiens*, chap. 3.

11. Ibid.

12. Ibid.

13. Peter Garrison, "Microspies," *Air and Space* (April/May 2000): 54–61; Stone, "Creatures Feature Possible Defense Applications"; Jim Wilson, "Micro Warfare," *Popular Mechanics* (February 2001): 62–65.

14. Garrison, "Microspies," p. 60.

15. Insects can move their wings at phenomenal speeds, as exemplified by a midge that beats its wings 1,046 times per second. But the key to insect flight, as we shall see, is not wing-beat frequency but fluid density.

16. Menzel and D'Aluisio, *Robo Sapiens*, chap. 3.
17. Rachel Ross, "Robotic Insect Takes Off," *Technology Review*, July 19, 2007, available at technologyreview.com/Infotech/19068.
18. Rick Weiss, "Dragonfly or Insect Spy? Scientists at Work on Robobugs," *Washington Post* (October 9, 2007): A03.
19. Black, "Enlisting Insects in the Military"; Menzel and D'Aluisio, *Robo Sapiens*, chap. 3.

Chapter 26

1. Arthur H. Westing, ed., *Cultural Norms, War and the Environment* (New York: Oxford University Press, 1988).
2. Robert Harris and Jeremy Paxman, *A Higher Form of Killing: The Secret History of Chemical and Biological Warfare* (New York: Random House, 2002), p. 241.
3. Harris and Paxman, *Higher Form of Killing*, chap. 5.
4. Allan S. Krass, "The Environmental Modification Convention of 1977: The Question of Verification," in Arthur H. Westing (ed.), *Environmental Warfare: A Technical, Legal and Policy Appraisal* (London: Taylor & Francis, 1984), pp. 65–76.
5. National Research Council of the National Academies (NRC), *Countering Agricultural Bioterrorism* (Washington, D.C.: National Academies Press, 2003).
6. Robin Clarke, *The Silent Weapons* (New York: David McKay, 1968), chap. 7.
7. Author interview, October 12, 2007.
8. Anne Kohnen, "Responding to the Threat of Agroterrorism: Specific Recommendations for the United States Department of Agriculture," Discussion Paper 2000-29, John F. Kennedy School of Government, Harvard University, Cambridge, Mass.
9. Michael V. Dunn, "The Threat of Bioterrorism to U.S. Agriculture," in Food and Agricultural Security Special Issue, *Annals of the New York Academy of Sciences*, 894 (1999): 186.
10. Ron Sequeira, "Safeguarding Production Agriculture and Natural Ecosystems Against Biological Terrorism: A U.S. Department of Agriculture Emergency Response Framework," in Food and Agricultural Security Special Issue, *Annals of the New York Academy of Sciences*, 894 (1999): 65.
11. Jonathan Ban, "Agricultural Biological Warfare: An Overview," *Chemical and Biological Arms Control Institute* (2000): 1.
12. Radford G. Davis, "Agricultural Bioterrorism," AIBS ActionBioscience Web site, actionbioscience.org/newfrontiers/davis.html (accessed January 28, 2008).
13. Jason Pate and Gavin Cameron, "Covert Biological Weapons Attacks Against Agricultural Targets: Assessing the Impact Against U.S. Agriculture," Discussion Paper 2001–9, John F. Kennedy School of Government, Harvard University, Cambridge, Mass., p. 3.

14. NRC, *Countering Agricultural Bioterrorism*, p. 22.

15. Ibid., p. 17.

16. Ibid., p. 94.

17. Ibid., p. 5.

18. Ibid., p. 47.

19. The possibility that the Department of Homeland Security is making the country less secure, at least in some particular realms, also applies to the ARS. According to Letchworth, "Regulations have narrowed the base of laboratories, people, and samples as scientists leave the field. There were serious discussions of emptying the freezers [of the Arthropod Borne Animal Diseases Research Laboratory] into the autoclave to destroy the specimens rather than trying to meet the data requirements of Homeland Security to be done with it. This is like burning books in the name of security. We are becoming less and less secure because we're losing people, facilities, and reference isolates." Interview with author, October 12, 2007.

20. Chris Clayton, "Reports Cite Flaws in Ag Inspections," Farms.com Web site, farms.com/readstory.asp?dtnnewsid=1711660 (accessed January 28, 2008); Roger Johnson (North Dakota Agriculture Commissioner), written statement before the House Agriculture Committee on Transferring USDA's Animal and Plant Health Inspection Service (APHIS) to the proposed Department of Homeland Security, June 26, 2002, agdepartment.com/Testimony/Testimony%202002/APHIS-Homeland%20Security%20Testimony.pdf (accessed January 28, 2008).

21. In response to the reports of the GAO and the USDA's Office of Inspector General, the agency has formed a "joint task force" that has developed action plans with recommendations for the USDA and DHS. At the same time, my communications with Melissa O'Dell (December 10, 2007), an APHIS public affairs specialist, made it clear the official position of APHIS is that the agency is even better prepared to protect the country in partnership with DHS than it was prior to the reorganization. In short, it seems that the USDA takes the position that there really is no problem, but if there was a problem, then they're fixing it.

22. Kate Campbell, "Congressional Hearing Triggers Alarm over Lax Pest Inspections," October 24, 2007, California Farm Bureau Federation Web site, cfbf.com/agalert/AgAlertStory.cfm?ID=920&ck=6D0F846348A856321729A2F36734D1A7 (accessed January 28, 2008).

23. Clayton, "Reports Cite Flaws in Ag Inspections."

24. WNBZ, "Schumer: More Funding Needed for Invasives," May 26, 2006, http://www.wnbz.com/May%202006/052606.htm (accessed January 28, 2008).

25. Dianne Feinstein, "Commentary: Agricultural Inspections Are Crucial to Our Nation's Crops," April 18, 2007, California Farm Bureau Federation Web site, cfbf.com/agalert/AgAlertStory.cfm?ID=804&ck=DC5689792E08EB2E219DCE49E64C885B (accessed January 28, 2008).

26. Emergency Preparedness and Response, "Bioterrorism Agents/Diseases," Centers for Disease Control and Prevention Web site, bt.cdc.gov/agent/agentlist.asp (accessed January 28, 2008).

27. Terry N. Mayer, "The Biological Weapon: A Poor Nation's Weapon of Mass Destruction," in *Battlefield of the Future*, ed. Barry R. Schneider and Lawrence E. Grinter, Air War College Studies in National Security, vol. 3 (Maxwell Air Force Base, Alabama, 1995), chap. 8.

28. "Bioterrorism: Are We Prepared?" AIBS ActionBioscience Web site, action-bioscience.org/newfrontiers/henderson.html (accessed January 28, 2008).

29. Dana Wilkie, "Biodefense Squeezes U.S. Science Budgets," *The Scientist* (March 2004): 52–53.

30. Interview with author, January 19, 2006.

31. Interview with author, January 17, 2006.

32. Ibid.

Epilogue

1. William H. Rose, *An Evaluation of Entomological Warfare as a Potential Danger to the United States and European NATO Nations*, U.S. Army Test and Evaluation Command, U.S. Army Dugway Proving Ground (1981), available at the Smoking Gun Archive, thesmokinggun.com/archive/mosquito1.html (accessed January 29, 2008).

2. Jenni Rissanen, "Acrimonious Opening for BWC Review Conference," *BWC Review Conference Bulletin*, November 19, 2001.

3. Alexia Treble, "Chemical and Biological Weapons: Possession and Programs Past and Present," Monterey Institute of International Studies' James Martin Center for Nonproliferation Studies Web site, cns.miis.edu/research/cbw/possess.htm (accessed January 29, 2008).

4. Jack Woodall, "Why Mosquitoes Trump Birds," *The Scientist* (January 2006): 61.

INDEX

Italicized page numbers indicate illustrations.